Advances in Geographical and Environmental Sciences

Series Editor

R. B. Singh, University of Delhi, Delhi, India

Advances in Geographical and Environmental Sciences synthesizes series diagnostigation and prognostication of earth environment, incorporating challenging interactive areas within ecological envelope of geosphere, biosphere, hydrosphere, atmosphere and cryosphere. It deals with land use land cover change (LUCC), urbanization, energy flux, land-ocean fluxes, climate, food security, ecohydrology, biodiversity, natural hazards and disasters, human health and their mutual interaction and feedback mechanism in order to contribute towards sustainable future. The geosciences methods range from traditional field techniques and conventional data collection, use of remote sensing and geographical information system, computer aided technique to advance geostatistical and dynamic modeling.

The series integrate past, present and future of geospheric attributes incorporating biophysical and human dimensions in spatio-temporal perspectives. The geosciences, encompassing land-ocean-atmosphere interaction is considered as a vital component in the context of environmental issues, especially in observation and prediction of air and water pollution, global warming and urban heat islands. It is important to communicate the advances in geosciences to increase resilience of society through capacity building for mitigating the impact of natural hazards and disasters. Sustainability of human society depends strongly on the earth environment, and thus the development of geosciences is critical for a better understanding of our living environment, and its sustainable development.

Geoscience also has the responsibility to not confine itself to addressing current problems but it is also developing a framework to address future issues. In order to build a 'Future Earth Model' for understanding and predicting the functioning of the whole climatic system, collaboration of experts in the traditional earth disciplines as well as in ecology, information technology, instrumentation and complex system is essential, through initiatives from human geoscientists. Thus human geosceince is emerging as key policy science for contributing towards sustainability/survivality science together with future earth initiative.

Advances in Geographical and Environmental Sciences series publishes books that contain novel approaches in tackling issues of human geoscience in its broadest sense — books in the series should focus on true progress in a particular area or region. The series includes monographs and edited volumes without any limitations in the page numbers.

More information about this series at http://www.springer.com/series/13113

Mukunda Mishra · R. B. Singh
Editors

COVID-19 Pandemic Trajectory in the Developing World

Exploring the Changing Environmental and Economic Milieus in India

IGU UGI

Springer

Editors
Mukunda Mishra
Department of Geography
Dr. Meghnad Saha College
Itahar, West Bengal, India

R. B. Singh
Department of Geography
University of Delhi
New Delhi, Delhi, India

ISSN 2198-3542 ISSN 2198-3550 (electronic)
Advances in Geographical and Environmental Sciences
ISBN 978-981-33-6439-4 ISBN 978-981-33-6440-0 (eBook)
https://doi.org/10.1007/978-981-33-6440-0

This Springer imprint is published by the registered company Springer Nature Singapore Pte Ltd.
The registered company address is: 152 Beach Road, #21-01/04 Gateway East, Singapore 189721, Singapore

Foreword

It is my great privilege to offer a brief perspective on this most timely and pertinent volume, edited by Dr. Mukunda Mishra and Professor R. B. Singh, entitled: *COVID-19 Pandemic Trajectory in the Developing World: Exploring the Changing Environmental and Economic Milieus in India* and published by Springer Nature Singapore Pte Ltd. as part of the series "Advances in Geographical and Environmental Sciences."

Human civilization is witnessing the unprecedented COVID-19 pandemic, which has been—and continues to be—devastating in its health, social, and economic consequences. Medical professionals globally are working at the frontline in the fight against this pandemic. However, beyond the immediacy of an acute health crisis, the pandemic has brought with it other threats to humankind associated with, for example, the global economy, society, culture, psychology, and politics.

Some benefits of the long period of lockdown have become apparent in that many countries have witnessed environmental reinvigoration and lowering of pollution levels. On the other hand, the restrictions on many aspects of society implemented in order to constrain the spread of the coronavirus have precipitated a global economic recession. The fallout of the latter is of grim significance to the entire population of the world. The diverse manifestations of the pandemic highlight the need for concerted, and coordinated global efforts to combat the coronavirus and its associated effects.

In these troublesome times that have touched all of us in one way or another, it is heartening to know that scientists have responded by conducting high-quality and highly relevant research into various elements of the pandemic. Science is a fundamentally collaborative enterprise, and the fact that, in such a short period of time, the editors have been able to bring together an impressive group of scholars to explore the range of issues associated with COVID-19 is quite remarkable.

The coronavirus pandemic continues to develop with devastating consequences and the contributions in this volume provide ample evidence of its diverse characteristics. This work therefore serves as an important scientific record of the attempts to understand and, ultimately, resolve the evolving impacts of a global crisis.

I extend my heartfelt greetings to the authors, publishing editors, and all those associated with the publication process. Moreover, I congratulate the team at Springer Nature Singapore Pte Ltd. for their endeavor in launching the book.

October 2020

Dr. Michael E. Meadows
President, International Geographical Union (IGU)
Professor, Department of Environmental & Geographical Science, University of Cape Town
Cape Town, South Africa

Preface

A pandemic carries manifold impacts on civilization, and obviously, the human health system faces the initial surge. The unfortunate spillover of the *Severe acute respiratory syndrome coronavirus 2* (SARS-CoV-2), making a cross-species transmission to jump from its initial host species and spreading within the human population, has emerged as an unprecedented health crisis in the recorded history of civilization that breaks all statistics of fatalities. The malicious virus has reached all the continents and discharged its hostile behavior against the human immune system against which the practitioners of medical sciences and the associated human resources are combating. The vaccine of the virus is at the *human trial* phase when this book is being written. Naturally, the physical health challenges and the emergency medical and clinical services form one facet of the challenge. The other facets of the pandemic are also broad and formidable too. Those serve the matter of the discussion contained in this book.

The initial spillover, causing the urban health challenge, gradually transformed into a global health challenge, getting the shape of the deadliest pandemic. The world witnessed the complete disruption of the world's most reliable health infrastructures of the European nations and then the USA triggered by the surge of fatalities and infection. The disease epicenter has been transferred to Latin America, especially Brazil, and then South-Asia, especially India. The last phases of the disease are significant because the clusters of the disease have started concentrating over the country groups that accommodate comparatively less public healthcare points and life-support infrastructures in comparison to the population they hold. It not only bottlenecked the public health system, instead, it posed many more challenges on the border horizons of society, economy, education, and policymaking. This volume brings this pandemic trajectory within its folds.

The long period of lockdown all over the globe has resulted in environmental reinvigoration and diminishing of the pollution level. Populous cities are becoming "greener" and "cleaner" than how they were before the lockdown. However, the debate emerges what the world would think ahead—the boon for the natural environment or the bane for the economy as the form of income insecurity, unemployment, and hunger. The question arises—whether the "green recovery" will be *long-standing*? Whether the indiscriminate use of alcohol-based hand sanitizers, soaps, hand washing liquids as part of newly imposed healthcare practices will bring forth new ecological challenges? Whether the mammoth waste generated from the use of plastic-based Personal Protective Equipment (PPE) and safety gear, used face-masks, test kits will pose challenges to safe solid-waste management? The reality is that, beyond the contemporary health challenges, the economic crisis will loom large throughout the globe, carrying new threats for the low-income countries and marginalized sectors in the post-pandemic world. Authors of different chapters have tried to explore the economic and environmental milieus emerging out of the pandemic in India, in Bangladesh, or in Brazil, who are merely not some countries; they are representing the developing world instead.

The global economic recession, with its unprecedented magnitude, is the most severe shock of the pandemic beyond human health challenges. This economic shock also impacted the world nations asymmetrically, depending on the capability of their economy to neutralize the shock. With an increasing number of COVID-19 infections, the Government of India has locked down transport services, closed all public and private offices, factories, and restricted mobilization. The consequence is very alarming. There is a huge job loss, and most of them have occurred in the unorganized sectors and informal sectors. Moreover, the possibility for a large volume of the population living just above the poverty line or recently recovered from the poverty situation has a greater likelihood to get included below the Poverty Line (BPL). It will cause undue pressure on the national economy due to the financial involvement of the government in the economic reform for the newly added population below the poverty line. The labor sector under the Mahatma Gandhi Rural Employment Guarantee Act (MGNREGA), 2005, is probably the worst impacted as they are not provided jobs due to lockdown. As most of the labor sectors are associated with the construction companies and daily wage earners, they are to face hardship. Travel restrictions and quarantines affecting hundreds of millions of people have left Indian factories short of labor, and it puts havoc in the production system. Besides, the tourism and associated hospitality industry face the severe shock of the lockdown and transport restriction. The tourism sectors are facing a severe economic crisis. As a whole, the developing economy of India faced hardship, which is reflected by a negative 23.9% downfall in the 1st Quarter of the Financial Year 2020.

There are ample shreds of evidence that make it clear that the closure of educational institutions is often considered as the first non-pharmaceutical intervention for the implementation in a pandemic, as students are effective *agents* in spreading the malicious virus. However, the long-term closure of the educational institutions and the *experiments* with the alternative *online* educational pedagogy have been transforming the education system to a *new normal*. This transformation exhibits itself to be a new challenge for the developing nations, making the system more vulnerable to witness the digital divides. Affecting all the dimensions of human development, the COVID-19 has posed a severe threat to all parts of the globe and questions the Sustainable Development Goals (SDG), particularly for the developing world.

However, every dark cloud has a silver lining. The warlike health emergency has been emerging as the catalyst, particularly for the developing world, toward a positive societal and economic transformation that carries long-term benefits. The drastic up-gradation of the public health infrastructures and life-support systems while combating the coronavirus would give a long-term benefit for the developing countries. The increasing government expenditure on public health would cater to Research and Development in the public health sector. There are opportunities, as well. Above all, the experience of combating the pandemic from different angles is worth mentioning that will help the world nations self-reliant to combat future crises.

Itahar, India Mukunda Mishra
New Delhi, India R. B. Singh

Acknowledgments

This book is the outcome of collective efforts. Our heartfelt recognition goes collectively to all of our friends, colleagues, students, and all others without whose active support in different aspects this book would not have seen the light of day. Yes, they all have made it possible with their respective roles.

We express our deepest sense of indebtedness to Dr. Michael E. Meadows, *Professor in the Department of Environmental & Geographical Science at the University of Cape Town & Honourable President of the International Geographical Union (IGU)*, for his kind bits of advice and for providing his valuable time to write the "Foreword" for this volume.

In a contributory volume, authors are the key stakeholders. We convey our sincere thanks to all contributors for offering us the opportunity to include their works in this volume. Here, we must mention the name of Dr. Sanjaya Bhatia, *Head of the United Nations Office for Disaster Risk Reduction (UNDRR) Global Education and Training Institute (GETI) & the Office for Northeast Asia (ONEA)* and Professor Wenwu Zhao, *Professor, Faculty of Geographical Science, Beijing Normal University* for their prompt response and active cooperation to our invitation in contributing to this volume.

We are grateful to the administrative authorities of our respective institutions for their administrative supports.

We also thank all the reviewers for their constructive comments and advice that led to substantial improvements to this volume's content and quality.

Constructive editorial advice and constant support from Ms. Yosuke Nishida, *the Editor, Earth Sciences, Geography & Environment, Springer Japan*, remains unparalleled. We acknowledge the support of the entire team of Springer Nature associated with the publishing process, disseminating their respective roles with utmost perfection.

Mukunda Mishra
R. B. Singh

Contents

Part I Pandemic Trapped World

1 **COVID-19 Pandemic Trajectory: Challenges and Opportunities for India** 3
R. B. Singh and Mukunda Mishra

2 **Lessons from the Success Story: The Systematic Approach to Fight Against COVID-19 in Beijing** 35
Yi Han, Jie Li, Kun Jia, Yanxu Liu, and Wenwu Zhao

3 **The Geography of COVID-19 in Rio de Janeiro, Brazil: Conflicts, Tensions, and Challenges** 51
Andrews José de Lucena, Leandro Dias de Oliveira, Pablo Ibanez, Gustavo Mota de Sousa, and André Santos da Rocha

Part II Emerging Debates

4 **Boon for the Environment and Bane for the Economy: Emerging Debate in Pandemic Stuck India** 73
Tanmoy Sarkar and Jayanta Mondal

5 **Blessing in Disguise in the Megacities: Environmental Co-benefits in Air Quality Amid Covid-19 Lockdown in Kolkata** 101
Kushagra Rajendra, Namita Yadav, and Dipanjana Chakraborty

6 **Questioning the Green Recovery: A Take on Post-COVID Scenario** 117
Anindya Basu and Lopamudra Bakshi Basu

7 **Restructuring State, Society, and Human Development: Projecting Post-COVID Pandemic Equations** 145
Siddhartha Sankar Manna

8 The *New Normal* of the Education System: Issues of Rights and Sustainability in Pandemic Trapped India 173
Srashta Chowdhury and Sushma Rohatgi

Part III Human, Development and Environment

9 Economic Lockdowns and Challenges of Rural Livelihood: Indian Scenario 209
Sujit Mandal

10 The Contagion Effects of COVID-19 and Public Transportation System: Conceptualizing the Shifting Paradigm in India 231
Biswajit Paul and Subir Sarkar

11 Inter-state Labor Migration in India: The Normal and Reverse Phase 257
Rupai Hembram and Uttam Garai

12 Human Discourses in Action: Community Health Workers' Contribution to Health Security and Pandemic Preparedness 275
Moumita Mondal

Part IV Preparedness and Policies

13 The Public Health System Resilience Addendum: A Tool to Help Governments Manage Biological Hazards Better and Prepare for an Uncertain Future 297
Sanjaya Bhatia

14 Instilling Self-Sustainment: The Key Survival Strategy Amid the Pandemic 307
Suresh Kumar and Raj Kumar

15 Community Resilience: A Potential Answer to the Emerging Pandemic 323
Somenath Halder

Editors and Contributors

About the Editors

Mukunda Mishra is an Assistant Professor and the designated Vice Principal of Dr. Meghnad Saha College in West Bengal, India. The college is affiliated to the University of Gour Banga. Dr. Mishra completed his postgraduate studies in Geography and Environmental Management at Vidyasagar University by securing the top rank in both the B.Sc. and M.Sc. panels of merit. He was awarded a Ph.D. in Geography from the same university. He was selected for the prestigious National Merit Scholarship by the Ministry of Human Resource Development, Government of India. His research chiefly focuses on analyzing unequal human development, and on creating multi-criteria predictive models. He has written a book entitled *Contouring Human Development: Methods and Applications Using an Indian District as Case Study*, published by Springer Nature on that particular aspect. He has more than ten years of hands-on experience in dealing with development issues at the ground level in various eastern Indian districts. He has published more than 20 research articles in journals of repute.

R. B. Singh is the Former Professor of Geography at the Delhi School of Economics, University of Delhi; the Secretary General and Treasurer of the International Geographical Union (IGU); the chair of Council for Scientific and Industrial Research (CSIR)–Central Food Technological Research Institute of the Government of India, and a member of the International Council for Science (ICSU) and the scientific committee Urban Health and Well-Being. He was awarded the prestigious Japan Society for the Promotion of Scientific Research Fellowship and has presented papers and chaired sessions in more than 40 countries. He has published 15 books, 40 edited research volumes, and more than 215 research papers. He has supervised 34 Ph.D. and 79 M.Phil. students. In 1988 the Unesco/the International Social Science Council awarded him the research and study grants in social and human sciences.

Contributors

Anindya Basu Department of Geography, Diamond Harbour Women's University, Sarisha, West Bengal, India

Lopamudra Bakshi Basu Department of Geography, Heramba Chandra College, Kolkata, West Bengal, India

Sanjaya Bhatia Global Education and Training Institute (GETI), Incheon, South Korea

Dipanjana Chakraborty Department of Geography, Sitalkuchi College, Sitalkuchi, West Bengal, India

Srashta Chowdhury Department of Geography, University of North Bengal, Darjiling, West Bengal, India

Uttam Garai Department of Geography, Sidho-Kanho-Birsha University, Purulia, West Bengal, India

Somenath Halder Department of Geography, Kaliachak College, Malda, West Bengal, India

Yi Han State Key Laboratory of Earth Surface Processes and Resource Ecology, Faculty of Geographical Science, Beijing Normal University, Beijing, China; Faculty of Geographical Science, Institute of Land Surface System and Sustainable Development, Beijing Normal University, Beijing, China

André Santos da Rocha Laboratory of Economic and Political Geography (LAGEP), Federal Rural University of Rio de Janeiro (UFRRJ), Seropédica, Brazil

Leandro Dias de Oliveira Laboratory of Economic and Political Geography (LAGEP), Federal Rural University of Rio de Janeiro (UFRRJ), Seropédica, Brazil

Gustavo Mota de Sousa Integrated Laboratory of Applied Physical Geography (LIGA), Federal Rural University of Rio de Janeiro (UFRRJ), Seropédica, Brazil

Rupai Hembram Department of Geography, Dr. Meghnad Saha College, Itahar, Uttar Dinajpur, West Bengal, India

Pablo Ibanez Laboratory of Economic and Political Geography (LAGEP), Federal Rural University of Rio de Janeiro (UFRRJ), Seropédica, Brazil

Kun Jia State Key Laboratory of Remote Sensing Science, Faculty of Geographical Science, Beijing Normal University, Beijing, China

Raj Kumar Department of Political Science, SPM College for Women, Delhi, India

Suresh Kumar Department of African Studies, University of Delhi, Delhi, India

Jie Li State Key Laboratory of Remote Sensing Science, Faculty of Geographical Science, Beijing Normal University, Beijing, China

Yanxu Liu State Key Laboratory of Earth Surface Processes and Resource Ecology, Faculty of Geographical Science, Beijing Normal University, Beijing, China;
Faculty of Geographical Science, Institute of Land Surface System and Sustainable Development, Beijing Normal University, Beijing, China

Andrews José de Lucena Integrated Laboratory of Applied Physical Geography (LIGA), Federal Rural University of Rio de Janeiro (UFRRJ), Seropédica, Brazil

Sujit Mandal Faculty of Science, Diamond Harbour Women's University, Diamond Harbour, West Bengal, India

Siddhartha Sankar Manna Department of Political Science, University of Gour Banga, Malda, West Bengal, India

Mukunda Mishra Dr. Meghnad Saha College, Itahar, Uttar Dinajpur, West Bengal, India

Jayanta Mondal State Aided College Teacher, Department of Geography, Dr. Meghnad Saha College, Ranipur, West Bengal, India

Moumita Mondal Department of Geography, Rammohan College, Kolkata, India

Biswajit Paul Department of Geography and Applied Geography, University of North Bengal, Darjeeling, West Bengal, India

Kushagra Rajendra School of Earth and Environmental Sciences, Amity University Haryana, Gurgaon, Haryana, India

Sushma Rohatgi Department of Geography, University of North Bengal, Darjiling, West Bengal, India

Subir Sarkar Department of Geography and Applied Geography, University of North Bengal, Darjeeling, West Bengal, India

Tanmoy Sarkar Department of Geography, Gazole Mahavidyalaya, Gazole, West Bengal, India

R. B. Singh Department of Geography, Delhi School of Economics, University of Delhi, New Delhi, India

Namita Yadav School of Earth and Environmental Sciences, Amity University Haryana, Gurgaon, Haryana, India

Wenwu Zhao State Key Laboratory of Earth Surface Processes and Resource Ecology, Faculty of Geographical Science, Beijing Normal University, Beijing, China;
Faculty of Geographical Science, Institute of Land Surface System and Sustainable Development, Beijing Normal University, Beijing, China

Abbreviations

ADB	Asian Development Bank
AFRO	African Region [WHO]
AISHE	All India Survey on Higher Education
AMRO	Region of the Americas [WHO]
AQI	Air Quality Index
ASHA	Accredited Social Health Activist
AUH	Ayurveda, Unani, and Homeopathy
BAT	Best Available Technology
BDBV	Bundibugyo Ebola Virus
BEP	Best Environmental Practices
CCHF	Crimean-Congo hemorrhagic fever
CDC	Centres for Disease Control and Prevention
CFTRI	Central Food Technological Research Institute
CHC	Community Health Centre
CHW	Community Health Workers
CIDRAP	Centre for Infectious Disease Research and Policy
CMC	Calcutta Medical College
COVID	Coronavirus Disease
CPCB	Central Pollution Control Board [India]
CR	Community Resilience
CRI	Community Resilience Index
CSA	Composite Score Assessment
DAISY	Digitally Accessible Information System
DALY	Disability-Adjusted Life-Years
DCCC	Dedicated COVID Care Centre
DCH	Dedicated COVID Hospital
DCHC	Dedicated COVID Health Centre
DGFT	Director General of Foreign Trade
DISD	Division for Inclusive Social Development
DRR	Disaster Risk Reduction
EBOV	Ebola Virus [Zaïre Type]
EDRM	Emergency and Disaster Risk Management

EHF	Ebola Hemorrhagic Fever
EMRO	Eastern Mediterranean Region [WHO]
EPI	Environmental Performance Index
ESA	European Space Agency
ESD	Education for Sustainable Development
EURO	European Region [WHO]
FEE	Foreign Exchange Earnings
FOSSEE	Free/Libre and Open Source Software for Education
FTA	Foreign Tourist Arrival
FWA	Family Welfare Assistant
GAHP	Global Alliance on Health and Pollution
GDP	Gross Domestic Product
GER	Gross Enrolment Ratio
GFC	Global Financial Crisis
GHG	Greenhouse Gas
GHG	Greenhouse gases
GNI	Gross National Income
GRAP	Graded Response Action Plan
GVA	Gross Value Added
HDI	Human Development Index
HEI	Higher Educational Institutes
HLPF	High Level Political Platform
HPAI	highly pathogenic avian influenza
IBC	Insolvency and Bankruptcy Code
ICCPR	International Covenant on Civil and Political Rights
ICESCR	International Covenant of Economic, Social and Cultural Rights
ICMR	Indian Council of Medical Research
ICRA	International Credit Rating Agency
ICT	Information and Communication Technology
ICU	Intensive Care Unit
IIP	Index for Industrial Production
ILO	International Labour Organization
IMD	Indian Meteorological Department
IMF	International Monetary Fund
INR	Indian National Rupee
IPT	Intermediate Public Transport
IRINS	Indian Research Information Network System
KMC	Kolkata Municipal Corporation
LFS	Labour Force Surveys
MCR	Making Cities Resilient [Campaign]
MERS	Middle East Respiratory Syndrome
MGNREGA	Mahatma Gandhi Rural Employment Guarantee Act
MIC	Medical Council of India
MoEFCC	Ministry of Environment, Forest and Climate Change [India]
MoHFW	Ministry of Health and Family Welfare [India]

MOOC	Massive Open Online Course
NAAQS	National Ambient Air Quality Standards [India]
NCAP	National Clean Air Programme
NCD	Non-Communicable Diseases
NCEUS	National Commission on Enterprises in Unorganised Sector
NDL	National Digital Library
NDMA	National Disaster Management Authority India]
NDRF	National Disaster Response Fund
NeGP	National e-Governance Plan
NEP	National Education Policy
NHAI	National Highway Authority India
NHM	National Health Mission [India]
NHP	National Health Policy [India]
NIOS	National Institute of Open Schooling
NiV	Nipah Virus
NRIM	National Rural Internet Mission
NYP	National Youth Policy
OECD	Organization for Economic Cooperation and Development
PHC	Primary Health Centre
PIAP	Public Internet Access Programme
PM	Particulate Matter
POSCO	Protection of Children from Sexual Offences Act
PPE	Personal Protection Equipment
PTR	Pupil–Teacher Ratio
PWF	Plastic Waste Footprint
RBI	Reserve Bank of India
RMSA	Rashtriya Madhyamik Shiksha Abhiyan [India]
RTE	Right to Education
SAICM	Strategic Approach to International Chemicals Management
SARS	Severe Acute Respiratory Syndrome
SARS-CoV-2	Severe Acute Respiratory Syndrome Coronavirus 2
SAT	Sustainability Assessment of Technologies
SDG	Sustainable Development Goal
SEARO	South-East Asia Region [WHO]
SFDRR	Sendai Framework for Disaster Risk Reduction
SIRU	Strategic Investment Research Unit
SOP	Standard Operation Procedures
SUDV	Sudan Ebola Virus
SUP	Single-Use Plastics
SWAYAM	Study Webs of Active-Learning for Young Aspiring Minds
TAFV	Tai Forest Ebola Virus
TERI	The Energy and Resources Institute
UDHR	Universal Declaration of Human Rights
UHS	Unified Health System
UNDESA	United Nations Department of Economics and Social Affairs

UNDP	United Nations Development Programme
UNEP	United Nations Environment Programme
UNESCO	United Nations Educational, Scientific, and Cultural Organization
WHO	World Health Organization
WPRO	Western Pacific Region [WHO]
ZIKV	Zika Virus

List of Figures

Fig. 1.1 The daily COVID-19 infection caseload sorted by WHO Regions for the period of 24 February to 15 August 2020 as (**a**) absolute number with the markers for total cases of infection and death in the world and (**b**) percentage share by regions 13
Fig. 1.2 *The European Epidemic*: showing the world distribution of confirmed cases of infection (equal numbers of countries in each class) and death toll due to COVID-19 as on 30 March 2020 (*Data Source* WHO 2020) 14
Fig. 1.3 *The devastation in the New World*: World distribution of confirmed cases of infection (equal numbers of countries in each class) and death toll due to COVID-19 as on 15 June 2020 (*Data Source* WHO 2020) 14
Fig. 1.4 Distribution of the million cities in the world with their population as pie and the level of urbanization as choropleth (*Data Source* United Nations 2018) 15
Fig. 1.5 *Shadows of risk over Brazil and India*: confirmed cases of infection and death toll due to COVID-19 as on 15 August 2020 (*Data Source* WHO 2020) 15
Fig. 1.6 Day-wise COVID-19 caseload in India along with the marker for total confirmed cases and total death toll up to 15 August 2020 (*Data Source* WHO 2020) 17
Fig. 1.7 Total confirmed caseloads in ten Indian states with worse condition for the period between 15 April and 15 August 2020 (*Data Source* Prepared by the authors, based on the data from official website of the Ministry of Health & Family Welfare, Govt. of India, i.e., https://www.mohfw.gov.in/) 18

Fig. 1.8 Countries by the import source for India and the percentage of the imported volume of different commodities from the leading import source, i.e., the People's Republic of China (*Data Source* Drawn by the authors, based on official portal of Export Import Data Bank, Department of Commerce, Govt. of India, i.e., https://tradestat.com merce.gov.in/eidb/ergncomq.asp) ... 19
Fig. 1.9 Quarterly GDP growth of India up to the 1st quarter of FY 2020–2021 (*Data Source* Govt. of India) ... 20
Fig. 1.10 Domestic general government health expenditure for its citizens for selected countries with their world rank within brackets, 2017 (*Data Source* WHO 2018) ... 23
Fig. 1.11 The graph of confirmed, recovered, and fatality relating to COVID-19 in India (*Data Source* Ministry of Health & Family Welfare, Govt of India 2020) ... 25
Fig. 1.12 Remarkable growth in PPE production: From 0 on 1st March, India produces nearly 4.5 lakh PPE kits every single day as on 18 May 2020, and in 60 days, the PPE industry in India has witnessed 56 times growth (*Data Source* Ministry of Health & Family Welfare, Govt. of India 2020) ... 26
Fig. 1.13 The key vision areas, centered by the Digital India Programme (*Source* Ministry of Electronics & Information Technology, Government of India) ... 30
Fig. 2.1 COVID-19 development trend and its countermeasures in Beijing ... 40
Fig. 2.2 Distribution of risk levels in key areas of Beijing (as of 17:00 on 26 June) ... 44
Fig. 2.3 Suggestions based on the systematic approach to combating epidemics of Beijing ... 46
Fig. 3.1 Deaths from COVID-19 per day in Brazil (from 15 March to 18 September 2020) and evolution of the moving average (*Source* Consortium of media outlets in Brazil) ... 55
Fig. 3.2 The moving average of deaths in Brazil States on September 18, 2020 ... 56
Fig. 3.3 COVID-19 deaths in Brazil by notification city on September 18, 2020 ... 57
Fig. 3.4 Brazil and Rio de Janeiro State Map: Brazil in grey color; Rio de Janeiro State in orange color; Metropolitan Area in red color ... 58
Fig. 3.5 The Metropolitan Area of Rio de Janeiro and its subdivisions: Rio de Janeiro capital (red color); Baixada Fluminense (green color); Seropédica city (dark green color); the cities of the Metropolitan East (yellow color) ... 59

Fig. 3.6 Deaths from COVID-19 per day in Rio de Janeiro State (from 15 March to 18 September 2020) and evolution of the moving average (*Source* Consortium of media outlets in Brazil) 60
Fig. 3.7 Confirmed cases in Rio de Janeiro State on March 15, 2020 61
Fig. 3.8 Confirmed cases, deaths numbers, and case fatality (the case fatality ratio is calculated by the number of deaths × 100 over the number of confirmed cases) ratio in Rio de Janeiro State on April 15, 2020 62
Fig. 3.9 Confirmed cases, deaths numbers, and case fatality ratio in Rio de Janeiro State on May 15, 2020 63
Fig. 3.10 Confirmed cases, deaths numbers, and case fatality ratio in Rio de Janeiro State on June 15, 2020 64
Fig. 3.11 Confirmed cases, deaths numbers, and case fatality ratio in Rio de Janeiro State on July 15, 2020 65
Fig. 3.12 Confirmed cases, deaths numbers and case fatality ratio in Rio de Janeiro State on August 15, 2020 66
Fig. 3.13 Confirmed cases, deaths numbers, and case fatality ratio in Rio de Janeiro State on September 15, 2020 67
Fig. 3.14 Area View of Main Building—Federal Rural University of Rio de Janeiro/UFRRJ 67
Fig. 4.1 Average temperature in the month of April in (**a**) 1990; (**b**) 1995; (**c**) 2000; (**d**) 2005; (**e**) 2010; (**f**) 2015; and (**g**) 2020 76
Fig. 4.2 Carbon dioxide concentration in the month of April in (**a**) 2005; (**b**) 2010; and (**c**) 2015 79
Fig. 4.3 Monthly aerosol concentration map during (**a**) January, (**b**) February, (**c**) March, (**d**) April, (**e**) May, and (**f**) June in 2020 80
Fig. 4.4 Monthly Sulfur dioxide concentration map during (**a**) January, (**b**) February, (**c**) March, (**d**) April, (**e**) May, and (**f**) June in 2020 81
Fig. 4.5 Monthly Nitrogen dioxide concentration map during (**a**) January; (**b**) February; (**c**) March; (**d**) April; (**e**) May, and (**f**) June in 2020 82
Fig. 4.6 (**a**) India industrial production index growth, 2006–2020; (**b**) India GDP growth rate, 2010–2019; (**c**) Growth in investment in India, 2010–2020; (**d**) India export status, 2010–2020; (**e**) India import status, 2010–2020; (**f**) Growth of govt. revenue in India, 2010–2020 (*Source* CEIC [CEIC Indicator dashboard is available under the URL: https://www.ceicdata.com/en/indicators] for a, b, and c; The Trading Economics Portal [Trading Economics Portal: https://tradingeconomics.com/] and Ministry of Commerce and Industry, Govt. of India for d, e, and f) 84

Fig. 4.7 Gross NPAs Growth in India for the period 2010–2019 (*Source* Prepared by the authors from the Ministry of Commerce and Industry, Govt. of India datasets) 86
Fig. 4.8 (**a**) GDP growth rate from 2000–2020 (**b**) India unemployment rate 2018–2010 (*Source* The Trading Economics Portal & Centre for Monitoring Indian Economy CMIE [CMIE data portal on unemployment, URL: https://unemploymentinindia.cmie.com/]) 88
Fig. 4.9 GDP from (**a**) agriculture; (**b**) manufacturing, and (**c**) construction (*Source* The Trading Economics Portal [Trading Economics Portal: https://tradingecono mics.com/] and Ministry of Statistics and Programme Implementation, Government of India datasets) 89
Fig. 4.10 State-wise number of factories in India (2014–2017) 92
Fig. 4.11 State-wise poverty condition in India (2011–12 MRP Consumption) (*Source* Reserve Bank of India) 93
Fig. 4.12 India state-wise GSDP volume (current price) 2016–17 93
Fig. 4.13 India state-wise GSDP production and state-wise COVID-19 status 94
Fig. 4.14 India state-wise GSDP share from the industry of origin (at 204–05 prices) 2012–13 95
Fig. 5.1 Location of Kolkata Metropolitan City (*Source* Prepared by the authors from KMC, NATMO, Google Earth datasets) ... 104
Fig. 5.2 Monthly average concentration of $PM_{2.5}$ and PM_{10} ($\mu g/m^3$) in Kolkata (*Source* Compiled by author Data: CPCB) 105
Fig. 5.3 Nitrogen dioxide concentration over India during lockdown and pre-lockdown Period (*Source* AURA 2020) 106
Fig. 5.4 Nitrogen dioxide concentration during lockdown and pre-lockdown period in Kolkata (*Source* Computed by author Data: CPCB, W.B.2020) 106
Fig. 5.5 Comparative Map of SO_2 emission in India during pre lockdown (2019) and lockdown period (April 2020) (*Source* ESA, July 2020) 107
Fig. 5.6 Average monthly concentration of SO_2 in $\mu g/m^3$ (*Source* Compiled by author Data: CPCB) 108
Fig. 5.7 Major sources of pollutants (*Source* Compiled by authors based upon data sourced from www.urbanemission.info) 109
Fig. 5.8 Types of indoor patients during 1st Phase of lockdown in Kolkata (as on 30th April) (*Source* Compiled by author Data: Health and Family Welfare Department, West Bengal) ... 111
Fig. 5.9 Disease wise monthly variation of indoor patients in Kolkata (*Source* Compiled by author Data: Health and Family Welfare Department, West Bengal) 112

Fig. 5.10 Trend of indoor patients in Kolkata during pre lockdown and lockdown Phase 1 to 5 (*Source* Compiled by author Data: Health and Family Welfare Department, WB) 112
Fig. 6.1 Structural framework of solid waste management during COVID times (*Source* Compiled by authors) 124
Fig. 6.2 Average daily nitrogen dioxide emission (micrograms/sq metre) in March 2020 in major cities of India (*Data Source* Centre for Research on Clean Air and Energy, Central Pollution Control Board) 127
Fig. 6.3 Clustering of monitoring stations based on the concentration of pollutants (*Source* Central Pollution Control Board 2020) ... 129
Fig. 6.4 Classifying Hand sanitizers (*Source* Compiled by author) 133
Fig. 6.5 Having one's own mix—Pros and Cons of self-manufacturing hand sanitizer (*Source* Compiled by authors) 134
Fig. 7.1 The interrelationship between the three aspects which this chapter deals with 146
Fig. 7.2 COVID-19 and HDI in a single frame (*Data source* World Health Organization) 160
Fig. 7.3 Ratio of doubling time of confirmed COVID cases (3-day period) (*Data source* Observer Research Foundation) 161
Fig. 7.4 Features of human development that are directly enhancing the human capabilities considered as the foundational portion of human development 167
Fig. 9.1 State-wise confirmed COVID-19 cases and death cases as on 20 August 2020 (*Source* Ministry of Health and Family Welfare, Govt. of India) 213
Fig. 9.2 Number of new COVID-19 cases reported on 31 March, 30 April, 31 May, 30 June, 31 July, and 20 August 2020 in different Indian states (Source Ministry of Health and Family Welfare, Govt. of India) 214
Fig. 9.3 Number of new COVID-19 death cases reported on 30 April, 31 May, 30 June, 31 July, and 20 August (*Source* Ministry of Health and Family Welfare, Govt. of India) 215
Fig. 9.4 State-wise unemployment rate for the year 2016, 2017, 2018, and 2019 (*Source* Centre for Monitoring Indian Economy CMIE) 223
Fig. 9.5 State level unemployment rate from January 2020 to July 2020 (*Source* Centre for Monitoring Indian Economy CMIE) 224
Fig. 9.6 Impact of COVID-19 induced lockdowns on livelihood, capital assets, and human vulnerability (*Source* Prepared by the author) 226
Fig. 10.1 Broad channels of pandemic impact on the economy (Drawn by the author based on Evan and Over 2020) 235

Fig. 10.2 Mobility Trends in India during coronavirus 2019 (Drawn by the authors following Radhakrishnan et al. 2020) 236
Fig. 10.3 **a** Daily freight flow; **b** daily passenger non-suburban (intercity) special migrant rail ridership, and **c** daily passenger suburban rail ridership before and during the lockdown in India (Drawn by the authors following Kumar and Sankar 2020) 237
Fig. 10.4 Satellite view of Delhi near India Gate. **a** Before lockdown December 7, 2019 and **b** during lockdown March 28, 2020 (*Source* Google Earth) 238
Fig. 10.5 Trend of passenger volume in Indian railways (as of April 2015–March 2020) (Drawn by the authors with the d*ata from the* Ministry of Railways, Govt. of India) 240
Fig. 10.6 **a** Trend of Foreign Tourist Arrival (FTAs) (*Data source* Ministry of Tourism, Govt. of India 2020) and **b** Foreign Exchange Earnings (FEE) from tourists (*Data source* Bureau of Immigration, Govt. of India 2020) 242
Fig. 10.7 The choice of people regarding the transit for **a** a work trip in post lockdown period and **b** commuting marketplaces in post lockdown period (*Source* Authors) 246
Fig. 10.8 Daily road accidents in India (*Data source* Kumar and Sankar 2020) 247
Fig. 10.9 Commuters are waiting to enter Howrah Station and flouting social distancing norms after lockdown restrictions were eased in Kolkata (*Source* Samir Jana, Hindusthan Times 2020) 251
Fig. 11.1 Spatial pattern of migration in India (*Source* Prepared by authors, based on Census of India 2011 datasets) 260
Fig. 11.2 Dominating reasons for migration in India (*Source* Prepared by the authors based on Census of India 2011 datasets) 261
Fig. 11.3 Two cities witnessing the same scenario when **a** the desperate migrant workers gathered at Anand Vihar bus terminal, Delhi (*Source* ANI [view the Newsroom Post, dated 29 March 2020: https://newsroompost.com/gallery/amid-lockdown-a-sea-of-migrants-in-delhi-try-to-find-their-way-back-homeamid-lockdown-a-sea-of-migrants-in-delhi-try-to-find-their-way-back-home/511214.html. Accessed on 12 August 2020]) and **b** the migrants as thousands gather at Bandra station, Mumbai (*Source* India Today [see the India Today dated 14 April 2020: https://www.indiatoday.in/india/story/lockdown-woes-mumbai-stations-flooded-with-migrant-labourers-hoping-to-get-back-home-1666908-2020-04-14. Accessed on 15 August 2020]) 263

Fig. 11.4 **a** Migrant workers are returning their homeland being piled on a truck (*Source* PTI [see the India Today dated 18 September 2020: https://www.indiatoday.in/india/story/fake-news-triggered-large-scale-migration-during-lockdown-mha-1722102-2020-09-15. Accessed on 20 September 2020]). **b** Migrant workers returning back by walking along the rail track (*Source* ANI [see the Newsroom Post dated 29 March 2020, https://newsroompost.com/gallery/amid-lockdown-a-sea-of-migrants-in-delhi-try-to-find-their-way-back-homeamid-lockdown-a-sea-of-migrants-in-delhi-try-to-find-their-way-back-home/511214.html. Accessed on 20 August 2020]) 266
Fig. 11.5 **a** Villagers barricading the entry to Durdih village in Lakhisarai district of Bihar on March 25, 2020 (*Source* India Spend [see the India Today dated 01 April 2020: https://www.indiaspend.com/bihars-migrants-return-to-face-stigma-under-prepared-medical-facilities/. Accessed on 25 August 2020]). **b** Migrant workers quarantined themselves on the branches of trees (*Source* ANI [see the Times of India dated 28 March 2020, available under the URL: https://timesofindia.indiatimes.com/city/kolkata/west-bengal-labourers-who-recently-returned-to-village-quarantine-themselves-on-tree/articleshow/74866464.cms. Accessed on 21 August 2020]) 268
Fig. 12.1 Share of women health workers in different roles by age group (*Source* Prepared by the author from Boniol et al. 2019 with the LFS Survey datasets [see: https://apps.who.int/iris/bitstream/handle/10665/311314/WHO-HIS-HWF-Gender-WP1-2019.1-%20eng.pdf?ua=1]) 282
Fig. 12.2 Sustainable Development Goals Report 2020 (*Source* UNO 2020 [sustainable development goals report, 2020. UNO. Available under the URL: https://unstats.un.org/sdgs/report/2020/The-Sustainable-Development-Goals-Report-2020.pdf. Accessed on 16 September 2020]) 289
Fig. 12.3 Relation of threat, vulnerability, and risk for any disaster (*Source* Prepared by the author) 290
Fig. 13.1 Disaster Resilience Scorecard for Cities: Public Health System Resilience Addendum: (Reproduced from UNDRR Making Cities Resilient Campaign Scorecard) 298
Fig. 13.2 An example from Essential 6 question (*Source* Reproduced from UNDRR Making Cities Resilient Campaign Scorecard) 303
Fig. 15.1 Dimensions of study of CR (*a.* community differentiation, *b.* spatial differentiation) (*Source* Developed by author) 327

Fig. 15.2 Classified communities useful for studying CR under pandemic condition (*Source* Developed and conceptualized by author) 327
Fig. 15.3 A multilevel framework of research action on community resilience (*Source* Developed by author, based on Naja and Hamadeh 2020) 329
Fig. 15.4 Structure of community resilience and customized index (*Source* Developed by the author *after*, Norris et al. 2008; Qin et al. 2017) 331
Fig. 15.5 Policy suggestion sector-wise prioritization after CSA (*Source* Developed by author) 351

List of Tables

Table 1.1 Health expenditure and mortality rate across WHO regions ... 16
Table 1.2 Quarterly estimates of the Gross Value Added (GVA) at basic prices in the first quarter (April–June) in consecutive three financial years ... 27
Table 2.1 Fifteen measures that must be strengthened under the second level of response ... 38
Table 2.2 Specific prevention and control measures of Beijing to the COVID-19 pandemic ... 39
Table 3.1 Report coronavirus cases in the world in September 18 ... 54
Table 4.1 Economic sectors in India, leading states and their importance ... 90
Table 6.1 Effects of acute and chronic toxicity by active ingredients of hand sanitizers ... 134
Table 6.2 Facing the pandemic head-on: excerpts of ground reality ... 136
Table 7.1 Recommended non-pharmaceutical interventions ... 159
Table 8.1 SDG 4 and human rights violations during the pandemic ... 180
Table 8.2 Rural–urban distribution of educational institutes in India ... 182
Table 8.3 Digital initiatives by the government in India ... 184
Table 8.4 Learners affected in India ... 190
Table 9.1 Confirmed COVID-19 cases and death cases in India ... 212
Table 9.2 Share of formal and informal sectors across broad sectors to GVA (in percentage) ... 220
Table 9.3 Rural–urban variation of unemployment rate (%) in India ... 225
Table 10.1 Gross revenue trends over the years in India ... 237
Table 10.2 Estimated loss in gross value added (sector-wise) due to COVID-19 outbreak ... 239
Table 10.3 Foreign tourist arrival and number of domestic tourist (2018–2019) ... 242
Table 10.4 FTAs on e-Tourist Visa during March 2020 ... 242
Table 10.5 Foreign Exchange Earnings (FEEs) from tourism in rupees terms ... 243
Table 10.6 Characteristics of the respondents ($n = 212$) ... 244

Table 10.7 Mode of transportation used to commute workplaces and market 244
Table 10.8 Online meeting and conferences during and post lockdown period 245
Table 10.9 Use of online education platforms during and post lockdown period 246
Table 11.1 Share of migrants by source states 259
Table 11.2 Relevant legislative measures available that might work 270
Table 12.1 A glimpse on the Sustainable Development Goals 284
Table 12.2 Sustainable Development Goal 3 and associated targets 285
Table 12.3 Functions and Components of the World Health Organization's Health Emergency and Disaster Risk Management Framework 292
Table 14.1 *Atmanirbhar Bharat* Program by the Government of India 314
Table 15.1 Variables composing socioeconomic resilience 332
Table 15.2 Variables composing environmental resilience 335
Table 15.3 Variables, composing the infrastructure and health resilience 337
Table 15.4 Variable composing institutional resilience 341

Part I
Pandemic Trapped World

Chapter 1
COVID-19 Pandemic Trajectory: Challenges and Opportunities for India

R. B. Singh and Mukunda Mishra

Abstract The COVID-19 pandemic exemplified how an urban health risk can rapidly spread to become a global health emergency. Throughout history, cities have been the sites of infectious disease outbreaks that present unique challenges to those leading response, as was evident in Ebola, SARS, MERS, and presently, the COVID-19. As the opening chapter of this present volume this chapter takes the role to present the detailed trajectory of the present pandemic. In the introductory section, we tried to portray the present pandemic as a member in the chain of evidences where the spillover of the virus from the animals to humans invited the health emergency situation. We tried to present a vivid discussion about the propagation of the disease epicenter. And, the last section of the chapter is devoted to explore the opportunities arising out of this pandemic, especially for the developing nations for whom India is a representative.

Keywords Spillover · Urban health emergency · Economic shock · Stigma · WHO regions

1.1 Introduction

The history of the outbreak of virus borne diseases for the last two decades reveals that a series of infectious diseases have been transmitted to the human bodies from the animal kingdoms. These incidences are not unusual from the standpoint of biological sciences. However, the recurring intervals of such incidences are unprecedented in the recorded history of human civilization. The list is more or less, not unknown. Still, we feel that this discussion needs them to mention to understand the sequence

R. B. Singh
Department of Geography, Delhi School of Economics,
University of Delhi, New Delhi, India
e-mail: rbsgeo@hotmail.com

M. Mishra (✉)
Dr. Meghnad Saha College, Itahar, Uttar Dinajpur, West Bengal, India
e-mail: mukundamishra01@gmail.com

M. Mishra and R. B. Singh, *COVID-19 Pandemic Trajectory in the Developing World*,
Advances in Geographical and Environmental Sciences,
https://doi.org/10.1007/978-981-33-6440-0_1

in which the latest one is the *severe acute respiratory syndrome coronavirus 2* (i.e., SARS-CoV-2).

We believe that the readers could memorize the Bird's deadly strain (Avian) Flu H5N1 that continued to spread among poultry in Egypt and certain parts of Asia. Technically, the H5N1 is a *highly pathogenic avian influenza* (HPAI) virus, which is deadly to most birds. Moreover, it is deadly to humans and to other mammals that catch the virus from birds. The first human case was detected in 1997, and, since then, the H5N1 had become fatal to nearly 60% of the people who had been infected (WHO 2006). However, the fact is that unlike the human flu pathogens, H5N1 bird flu does not spread easily from person to person.[1] There are very few human-to-human transmission cases observed, and these rare transmission cases have been reported to have occurred between the peoples with exceptionally close contact, such as a mother who caught the virus while caring for her sick infant.[2] The recurring history is fascinating. The *aviation influenza* was initially called as 'fowl plague'. The beginning of the 'recorded' history of that fowl plague is very often referred to as '1878'. It was the first time when this particular disease was ascribed as 'different' kind from the other diseases that caused high mortality rates in birds (Alexander and Brown 2009). Interestingly, between 1959 and 1995, there were 15 recorded occasions of the emergence of HPAI viruses though they cause minimal losses. However, the same disease outbreaks at least 11 times between 1996 and 2008 in poultry; and, 4 of these outbreaks have infected millions of birds (Alders et al. 2014).

If the damage is concerned, the wake of H5N1 pandemic affected mostly the small-scale commercial farms and backyard poultry producers due to the culling of huge numbers of birds due to HPAI infection and control attempts (Porter 2012). In Vietnam alone, over 50 million domestic birds were killed (Hiromoto 2000). The total economic losses in Southeast Asia were estimated as US$10 billion. It affected the lower incoming strata in Vietnam severely. There was an average loss of 2.3 months of production and US$69–108 for households where most had an income of $2 per day or less (McLeod et al. 2005). The loss of food security for vulnerable households was reflected by the stunting of children under 5 in Egypt and the decreased enrollment of girls in school in Turkey (Alders et al. 2014). Research shows that the women were at risk as in most regions of the world, as the H5N1 pandemic-induced food insecurity stuck the low incoming households (Bagnol 2012).

With a risk of fatality as around 1 in 1,000, the Chikungunya has now been identified in nearly 40 countries in Asia, Africa, Europe, and, most recently, the Americas (Caglioti et al. 2013). It is a viral disease transmitted to humans through the bites

[1] National Institute for Occupational Safety and Health (NIOSH) published this document entitled 'Workplace Safety and Health Topic' which is available in the website of the Centre for Disease Control and Prevention. URL: https://www.cdc.gov/niosh/topics/avianflu/.

[2] The article entitled 'Recommendations for Worker Protection and Use of Personal Protective Equipment (PPE) to Reduce Exposure to Highly Pathogenic Avian Influenza A H5 Viruses: Avian Influenza (Flu)' was published by the Center of Disease Control and Prevention and it is available in their official website under the URL: https://www.cdc.gov/flu/avianflu/h5/worker-protection-ppe.htm.

of *Aedes albopictus* and *Aedes aegypti* mosquitoes infected with the chikungunya virus (CHIKV). The virus may circulate within several animals, including birds and rodents (WHO 2016). The word 'chikungunya' has been derived from the Makonde language (or Kimakonde), which is spoken by the Makonde, an ethnic group in south-east Tanzania and northern Mozambique. It was first described during an outbreak in southern Tanzania in 1952, and Symptoms usually begin 4–8 days after a mosquito bite but can appear anywhere from 2–12 days (CDC 2006). 'Chikungunya' means 'something that compels to bend up', which has been due to the controlled posture of people affected with the severe joint pain and arthritic symptoms associated with this disease (Robinson 1955).

The transmission of the CHIKV is worth mentioning as it brings to fore a typical interaction between the vectors (i.e., mosquitoes), their environments, and human behavior. The adaptation of mosquitoes to the changing climate of North Africa around 5,000 years ago made them seek out environments where humans stored water (Powers et al. 2000). The proximity of the human and mosquito habitation has been making the human bodies susceptible to be used as the reservoirs of CHIKV during periods of epidemics, incubating high amounts of virus in the human blood with acute infection. The virus can be spread from a viremic human to a mosquito, and back to a human (Morrison 2014). During the non-epidemic times, monkeys, birds, and other vertebrates have served as reservoirs (Lee and Hapuarachchi 2010).

The next on the list is the Swine Flu pandemic that lasted from January 2009 to August 2010. Interestingly, it was the second of those two pandemics involving H1N1 influenza virus. The first one being the 1918–1920 Spanish flu pandemic. Trifonov et al. (2009) confirm that the Swine Flu virus appeared to be a new strain of H1N1, which resulted from a previous triple reassortment of bird, swine, and human flu viruses that further combined with a Eurasian pig flu virus. Though, in August 2010, WHO declared the swine flu pandemic officially over,[3] the controversies regarding the casualties still loom large.. The number of lab-confirmed deaths reported to the WHO is 18,449,[4] though the CDC estimates about 284,000 (range from 150,000 to 575,000) deaths.[5] A follow-up study by Roos (2012) under the aegis of the Centre for Infectious Disease Research and Policy (CIDRAP) claims that the risk of serious illness resulting from the 2009 H1N1 flu was no higher than that of the yearly seasonal flu.

The Crimean-Congo hemorrhagic fever (CCHF) is caused by infection with a tick-borne virus (Nairovirus). The disease was first occurred in Crimea in 1944 and later in

[3]Dr. Margaret Chan, Director General, WHO announced on 10 August 2010 that the H1N1 influenza virus has moved into the post-pandemic period. See the official websites under the URL: https://www.who.int/csr/disease/swineflu/en/.

[4]WHO reported in its official website under the headline 'Pandemic (H1N1) 2009—update 112', published on 6 August 2010 which is available under the URL: https://www.who.int/csr/don/2010_08_06/en/.

[5]CDC report entitled 'First Global Estimates of 2009 H1N1 Pandemic Mortality Released by CDC-Led Collaboration' was published on 25 June 2012 in its official website which is available under the URL: https://www.cdc.gov/flu/spotlights/pandemic-global-estimates.htm.

Congo during 1969.[6] The infection causes fever and hemorrhage (i.e., hemorrhagic fevers), which results in the current name of the disease. The CCHF is found in Eastern Europe, particularly in the former Soviet Union, throughout the Mediterranean, in northwestern China, central Asia, southern Europe, Africa, the Middle East, and the Indian subcontinent. Ticks are both 'environmental reservoir' and vector for the virus. It carries the virus from wild animals to domestic animals and humans (Mehravaran et al. 2013). Wild animals and small mammals, mainly European hare, Middle-African hedgehogs, and multimammate rats are the 'amplifying hosts' of the virus. Researchers observe that the birds (except ostriches) are generally resistant to CCHF. Domestic animals and cattle can develop high titers of virus in their blood, but they do not fall ill (Ergönül et al. 2004).

India witnesses sporadic confirmed cases of CCHF, and the first human case was reported in Sanand of Gujarat during January 2011. Unfortunately, that emergence became fatal for four persons, including that index patient, treating physician, and nurse.[7] The re-emergence occurred in Gujarat in July 2013, reaching the fatalities to seven persons in Kariyana village of the Amreli district.[8] Again, in 2014, confirmed cases had been reported from Bhuj, Amreli, Sanand, Idar, and Vadnagar in Gujarat. A doctor and a laborer in north Gujarat were tested positive for the disease in November 2014. In the following weeks, three more people died from CCHF.[9] In March 2015, one more person died of CCHF in the same province.[10] Against this backdrop, the CCHF was recognized as 'widespread' in India, only four years after the first human case had been diagnosed.[11]

The Ebola outbreak during 2014–2016 in West Africa and during 2018 in the Democratic Republic of Congo remains the deadliest attack in terms of mortality. The Ebola virus was first identified in the then Zaire Republic by a team of researchers, including the Belgian microbiologist Peter Piot in 1976. The virus was named after the Ebola River, a tributary of the Congo River in central Africa (Brown 2014). There are six species of Ebola virus, four of which have caused disease in humans—Zaïre ebola virus (EBOV), Sudan ebola virus (SUDV), Tai Forest (TAFV) which was formerly known as Ebola Ivory Coast, and Bundibugyo ebola virus (BDBV). Between 1976

[6]CDC listing in the official website under the URL: https://www.cdc.gov/vhf/crimean-congo/index.html.

[7]Indian daily newspaper, the Indian Express covered the news on 19 January 2011, which is available in their archive under the URL: http://archive.indianexpress.com/news/deadly-virus-makes-first-appearance-in-india-kills-three-in-gujarat/739292.

[8]Indian electronic media India TV telecasted the news on 15 July 2013, which is archived as India TV News under the URL: http://m.indiatvnews.com/news/india/congo-fever-seven-die-in-amreli-in-a-week-25097.html.

[9]Indian daily newspaper, the Times of India reported on 26 January 2015, which is archived under the URL: http://m.timesofindia.com/city/jaipur/Health-officials-confirm-congo-fever-death-of-Jaisalmer-man/articleshow/46017452.cms.

[10]Indian daily newspaper, the Indian Express reported on 29 March 2015, which is achieved under the URL: http://indianexpress.com/article/india/gujarat/kutch-resident-dies-of-confo-fever/.

[11]Reported by the Otbreak New Today on 09 October 2015 which is available under the URL: http://outbreaknewstoday.com/crimean-congo-hemorrhagic-fever-spreads-across-india-82283/.

and 2014, Ebola infections have been primarily reported from the remote villages near tropical rain forests in Central and West Africa. The largest outbreak occurred in West Africa between March 2014 and June 2016, affecting people in Guinea, Liberia, and Sierra Leone, trapping 28,600 people under confirmed infection and causing fatal for 11,325 with an average mortality rate of around 40%.[12]

Experts opine that the 2018's outbreak in Congo was the second-deadliest ever, behind only the 2014–2016 West Africa epidemic. The Democratic Republic of Congo's Ministry of Health reported an Ebola outbreak in North Kivu province on 1 August 2018.[13] This was the 10th Ebola outbreak in the region since the virus was discovered (Feleke and Scutti 2018). As of 17 December 2018, the new cases brought the outbreak total to 539, of which 491 were confirmed, and 48 were probable; and with an utter worry, the official record confirmed 315 deaths with the mortality rate reaching the new record ever (64%) (Soucheray 2018).

2015–2016 Zika virus epidemic has its newer dimension in terms of the rapidly spreading territories of mosquito-borne disease. The Zika virus (ZIKV) belongs to the virus family *Flaviviridae,* which is spread by daytime-active Aedes mosquitoes (Malone et al. 2016). Scientists conducting routine surveillance for yellow fever in the Zika forest of Uganda isolated the Zika virus in samples taken from a captive, sentinel rhesus monkey in 1947, and it got nomenclature accordingly (Dick et al. 1952; Sikka et al. 2016). Where, since the 1950s, Zika infections were reported from the areas surrounding a narrow belt along the equator in Africa and Asia, the virus started spreading eastward across the Pacific Ocean to the Americas between 2007 and 2014 to ultimately burst into the form of 2015's epidemic (Mehrjardi 2017). Zika virus infection appears to have changed in character while expanding its geographical range. The change is witnessed from an endemic, mosquito-borne infection causing mild illness across equatorial Africa and Asia, to an infection causing large outbreaks from 2007 onward. From 2013 onward, the outbreaks are linked with neurological disorders, including Guillain-Barré syndrome[14] and microcephaly[15] across the Pacific region and the Americas (Kindhauser et al. 2016). WHO's factsheet confirms that a total of 86 countries and territories have reported evidence of mosquito-transmitted Zika infection to date.[16] WHO, in its podcast episodes 2017[17] introduced the Zika virus outbreak as:

---

[12]UK Government factsheet available under the URL: https://www.gov.uk/government/publications/ebola-origins-reservoirs-transmission-and-guidelines/ebola-overview-history-origins-and-transmission.

[13]CDC Report entitled '2018 Eastern Democratic Republic of the Congo Outbreak' which is available under the URL: https://www.cdc.gov/vhf/ebola/outbreaks/drc/2018-august.html.

[14]A rapid onset of muscle weakness caused by the immune system damaging the peripheral nervous system.

[15]Microcephaly (my-kroh-SEF-uh-lee) is a rare neurological condition in which an infant's head is significantly smaller than the heads of other children of the same age and sex.

[16]See WHO Factsheet about Zika virus disease which is available under the URL: https://www.who.int/en/news-room/fact-sheets/detail/zika-virus.

[17]The WHO podcast 2017 is available under the URL: https://www.who.int/mediacentre/multimedia/podcasts/2017/en/.

> Zika virus has been reported in dozens of countries around the world from 2015 onwards. WHO's experience over 2016 has shown that the Zika virus and the associated neurological complications represent a long-term public health challenge.

One of the most recent outbreaks involves the Nipah virus (NiV). This virus is a member of the *Paramyxoviridae* family and *Henipavirus* genus. NiV was initially isolated and identified in 1999 during an outbreak of encephalitis and respiratory illness in Malaysia and Singapore. The name 'Nipah' originated from Sungai Nipah, which is a village in the Malaysian Peninsula where the disease was spread among pig farmers and people in close contact with pigs. The relatedness of NiV to Hendra virus directed the virologists to single out *bat* species for investigation. The *flying foxes* of the *Pteropus* genus were subsequently identified as the reservoir for NiV.[18] Nipah virus is a zoonotic virus, which means it is transmitted from animals to humans. It can also be transmitted through contaminated food or directly between people. Nipah virus has been reported to cause only a few known outbreaks in Asia. However, WHO warns that NiV could infect a wide range of animals and cause severe disease and fatalities among people, making it a public health concern.[19]

The Nipah virus outbreak is not a new incidence in India; instead, the 2018's Nipah outbreak in the state of Kerala is the third outbreak, which is technically traced to the fruit bats in the area.[20] The previous outbreaks caused the fatality to 45 (during 2001) and 5 (during 2007), respectively. The 2018's outbreak was localized in Kozhikode and Malappuram districts of Kerala and claimed 17 lives.[21] The outbreak was declared over on 10 June 2018 (Sharma 2018).

The World Health Organisation (WHO) has indicated that 70% of new viruses originate in animals (Khati 2020). The above discussions have probably made it clear that the virus involved in past pandemic outbreaks mostly originated from the animal world. The Novel Coronavirus is the new virus (for which it is named as 'Novel'). It probably complies with the hypothesis of 'animal-to-human' transmission. The scientific communities have tentatively inclined on the fact that the phylogenetic data obtained so far confirms the SARS-CoV-2 is from a species of bat and entered humans from a secondary host animal, most probably pangolins (Cunningham 2020).

Though all things are not clear so far and the search for 'patient zero' (i.e., the 'index case')—the first human COVID-19 infection does matter. It is not because any blame lies with this individual, but because discovering how the pathogen entered the human population (Campbell et al. 2020). The statement of Professor Andrew

---

[18]Centre for Disease Control and Prevention (CDC) factsheet on Nipah virus which is archived under the URL: https://www.cdc.gov/vhf/nipah/index.html.

[19]See WHO factsheet on NiV, available under the URL: https://www.who.int/news-room/fact-sheets/detail/nipah-virus.

[20]Indian English daily the Hindustan Times covered the news under the headline 'Nipah virus outbreak: Death toll rises to 14 in Kerala, two more cases identified' on 27 May 2018 which is available under the URL: https://www.hindustantimes.com/india-news/nipah-virus-outbreak-death-toll-rises-to-14-in-kerala-two-more-cases-identified/story-f4QenFXeOtJnWhw7NrVawJ.html.

[21]Indian print media Frontline covered the outbreak on 20 July 2018 with the headline 'After the outbreak' which is available under the URL: https://frontline.thehindu.com/the-nation/public-health/article24200872.ece.

Cunningham (2020) who is the Deputy Director of Science Deputy Director of Science in the Institute of Zoology of the renowned Zoological Society of London, is pertinent regarding this:

> A paper I co-authored in 2013, for example, identified 137 bat viruses of which 61 were known to be capable of infecting people at that time. The phylogenetic data obtained so far indicate that SARS-CoV-2 (the virus that causes COVID-19) is from a species of bat. The initial progress of the virus is difficult to trace and several other animals – most notably pangolins – have been suggested as intermediary hosts that could have transferred SARS-CoV-2 between bats and humans, although there remains insufficient evidence to support these claims. And while authorities have yet to identify patient zero, the human outbreak most likely began at a wildlife market in the city of Wuhan in China.

It is no coincidence that a city was the origin of its spread. The author of the renowned book *Spillover: Animal Infections and the Next Human Pandemic*, David Quammen (2020) expressed his views as that 'shaking viruses loose from their natural hosts' urging them to find new hosts and transmitting to and between people. The renowned disease ecologist, Peter Daszak (2020) explains how humankind has made the virus by disrupting the natural environment's ecosystems:

> Plagues are not only part of our culture; they are caused by it. The Black Death spread into Europe in the mid-14th century with the growth of trade along the Silk Road. New strains of influenza have emerged from livestock farming. Ebola, SARS, MERS and now Covid-19 have been linked to wildlife. Pandemics usually begin as viruses in animals that jump to people when we make contact with them. These spillovers are increasing exponentially as our ecological footprint brings us closer to wildlife in remote areas and the wildlife trade brings these animals into urban centers. Unprecedented road-building, deforestation, land clearing and agricultural development, as well as globalized travel and trade, make us supremely susceptible to pathogens like coronaviruses.

It is clear from the COVID-19 pandemic that urban health risks can rapidly spread and suddenly become a global health emergency. In 2018, 55% of the world's population lived in urban areas and the 'proportion that is expected to increase to 68% by 2050' (United Nations 2018). Cities are also the places with the highest population densities. Throughout history, cities have been the sites of infectious disease outbreaks (Jones 2020), and outbreaks in cities present unique challenges to those leading response as was evident in Ebola, SARS, MERS, and now Covid-19 (Gatzweiler et al. 2020). However, cities are also the sites of solutions to arrest the spread of infectious diseases and improve population health.

Whatever is the point of origin, the common people from China to Chicago have been getting well acquainted with the terms like 'lockdown', 'social distancing', 'quarantine', 'vaccine', 'immunity', and 'isolation' which are dominating the print and electronic media, scholarly articles, and discussions across the globe. The health crisis is now the 'nightmare' for the developed nations and it is a 'war-like' situation for the developing world.

However, beyond the health, the income, economy, livelihood, education, social structure, power-order—almost all aspects of humankind are on the verge of facing unprecedented challenges. Against this backdrop, the present chapter will make the commentary on the present crisis of the COVID-19 pandemic in the world in general and India in particular. We extend our efforts to exploring the emerging challenges arising out of the present pandemic. However, the prime focus of this chapter is not only to elaborate on the 'dejection'. Instead, the chapter will shed light on the opportunities as well that will pave the way for a sustainable economy and livelihood to fuel the forthcoming development process, emerging as new dawn after the darkness of the pandemic trapped world.

1.2 The Pandemic Trajectory Around the Globe

WHO Member States are grouped into six WHO regions—African Region (AFRO), Region of the Americas (AMRO), South-East Asia Region (SEARO), European Region (EURO), Eastern Mediterranean Region (EMRO), and the Western Pacific Region (WPRO).[22] This grouping is not merely a geographical clustering of countries around the globe (though they are grouped and named on this basis); instead, the regions bring to fore crucial public health differentials as well. The health crisis in the form of COVID-19 exhibits a typical contagious chain of disease transmission across the globe. In this discussion, we will focus on understanding the COVID-19 trajectory, keeping the WHO Regions in the background.

This discussion will visualize the world COVID situation for the span of 24 February to 15 August 2020 (174 days), based on datasets available at WHO's Coronavirus Disease (COVID-19) Dashboard.[23] We have used two basic variables for the visualization of the world scenario—(1) *Numbers of confirmed cases of COVID-19 infection* and (2) *Numbers of death reported due to COVID-19*. The number of confirmed cases of infection in a day within a particular region is one of the key indicators to assess the status of the outbreak in the given region. Involving the WHO Regions in the present discussion brings to fore a fascinating trend.

On 31 December 2019, the cluster of cases of 'pneumonia of unknown cause' was first reported prevailing in Wuhan province in the People's Republic of China. WHO's official website has recorded the information as:

> WHO's Country Office in the People's Republic of China picked up a media statement by the Wuhan Municipal Health Commission from their website on cases of 'viral pneumonia' in Wuhan, People's Republic of China. The Country Office notified the International Health Regulations (IHR) focal point in the WHO Western Pacific Regional Office about the Wuhan Municipal Health Commission media statement of the cases and provided a translation of it.[24]

WHO, on 9 January 2020, reported that the Chinese authorities have determined that the outbreak is caused by a 'novel' coronavirus and the first case of fatality due to

---

[22] WHO website: https://www.who.int/healthinfo/global_burden_disease/definition_regions/en/.

[23] WHO's dedicated COVID-19 dashboard is available under the URL: https://covid19.who.int/.

[24] See WHO website: https://www.who.int/emergencies/diseases/novel-coronavirus-2019.

that 'unknown virus' was reported in the Chinese electronic media on 11th January.[25] It took only two more days for the first case to be reported outside the People's Republic of China. On 13th January, the Ministry of Public Health in Thailand reported 'an imported case of a lab-confirmed novel coronavirus from Wuhan'.[26]

However, the most worrying fact regarding the 'novel coronavirus disease' came under the light of the day when, WHO in a press briefing, on 14th January, stated that, based on experience with respiratory pathogens, the potential for human-to-human transmission existed in the 41 confirmed cases in the People's Republic of China. The statement clarified, 'it is certainly possible that there is limited human-to-human transmission'.[27] Within those acute confusions prevailing due to the mutually contradicting statements about 'to what extent the novel coronavirus causes human-to-human transmission', the first confirmed case in the WHO European Region (EURO) was reported on 24th January. Officials from France informed the WHO of three cases of the novel coronavirus infection, all of whom had traveled from Wuhan.[28] On 30 January 2020, WHO declared the novel coronavirus outbreak a public health emergency of international concern (PHEIC), which is WHO's highest level of alarm for humankind.[29]

The highly contagious nature of the novel coronavirus and particularly its person-to-person spreading efficiency along with a sustained spreading capability has posed havoc for humankind. CDC has recorded their observations[30] about the transmission, and they have pointed out that the virus is thought to spread mainly from person to person along the following channels:

- Between people who are in close contact with one another (within about 6 feet).
- Through respiratory droplets produced when an infected person coughs, sneezes, or talks.
- These droplets can land in the mouths or noses of people who are nearby or possibly be inhaled into the lungs.
- COVID-19 may be spread by people who are not showing symptoms.

---

[25] See WHO statement on 9th January 2020. Available under this URL: https://www.who.int/china/news/detail/09-01-2020-who-statement-regarding-cluster-of-pneumonia-cases-in-wuhan-china.

[26] See WHO statement on 13th January 2020. Available under this URL: https://www.who.int/news-room/detail/13-01-2020-who-statement-on-novel-coronavirus-in-thailand.

[27] See WHO briefing on 14th January 2020 via tweet: https://twitter.com/UNGeneva/status/1217146107957932032.

[28] See WHO's News Released on "2019-nCoV outbreak: first cases confirmed in Europe", available under the URL: https://www.euro.who.int/en/health-topics/health-emergencies/pages/news/news/2020/01/2019-ncov-outbreak-first-cases-confirmed-in-europe.

[29] See the statement released by the Director, WHO on 30th January 2020, entitled "WHO Director-General's statement on IHR Emergency Committee on Novel Coronavirus (2019-nCoV)", available under the URL: https://www.who.int/dg/speeches/detail/who-director-general-s-statement-on-ihr-emergency-committee-on-novel-coronavirus-(2019-ncov.

[30] See CDC newsletter entitled "How COVID-19 Spreads", available under the URL: https://www.cdc.gov/coronavirus/2019-ncov/prevent-getting-sick/how-covid-spreads.html.

However, before the fact about the rapid contagion effect caused by the novel coronavirus confirmed and the international borders could be sealed, and the international flights seized, the migrating humans carried the virus to all the continents through their networks of travels and trades. The world witnessed the contagion effect of the disease when the numbers of confirmed cases of COVID-19 infection took only seven days to reach the figure from 10^3 to 10^4 (i.e., between 25th February and 2nd March, with an average caseload of 1285 per day); 16 days to 10^5 (i.e., between 2nd and 17th March, with an average of 5625 per day); 20 days to 10^6 (i.e., between 17th March and 5th April, with an average of 45,000 per day) and 87 days to 10^7 (i.e., between 5th April to 30th June, with an average caseload of 103,448 per day) (Fig. 1.1a). On 15th August, the total infection caseload in the world was reported by WHO as '297326'.

The percentage of the daily confirmed infection caseload by WHO regions exhibit four distinct phases that speculate the propagation of the disease clusters from the 'index case' or the 'ground zero' (Fig. 1.1b). This observation is valuable from the standpoint of spatial epidemiology. The information on the 'locale' of the 'epicenter' of an outbreak helps in examining a disease and map its geographic variations to analyze its outreach in consideration of the demographic, environmental, behavioral, socioeconomic, genetic, and infection risk factors (Elliott and Wartenberg 2004).

Phase 1 (*The frenzied Wuhan*): The ground zero of the present pandemic is within the WPRO region. Hence the infection caseload in the WPRO region dominates since the beginning of the pandemic due to the outbreak in Wuhan. This phase is considered to have ended when the percentage of the caseload of EMRO region (41.80%) overtakes that of the WPRO region (31.39%) on 2 March 2020. *The daily infection caseload maxima* (1st cycle) for WPRO (WPRO Max_1) occurred within this phase on 29th February (1278 new infections were reported) (Fig. 1.2).

Phase 2 (*The bridging Eastern Mediterranean*): The EMRO region shows the highest share of caseload among the WHO regions for only two days—2 and 3 March 2020. It was overtaken by the EMRO region later to start the epidemic in Europe, which the continent experiences after the Spanish Flu event occurred there from February 1918 to April 1920.

Phase 3 (*The European epidemic*): Worlds most trusted and most sophisticated healthcare system faced the challenge of the devastating manifestation of the COVID-19 as the EURO region started to lead the share of the caseload consistently from 4th March to 5th April. The *daily infection caseload maxima* for EURO (EURO Max) occurred within this phase on 1st April (42,803 new infections were confirmed).

Phase 4 (*The devastation in the New World*): The present pandemic brings the deadliest blows for the Americas, especially the USA, in terms of the infection and the fatality. This pandemic is unprecedented in the 'New World'. The USA has been experiencing a shocking number of caseloads per day and the death toll, which has never been happened in the recorded history of the state. This phase was started with the dominating caseload percentage of the AMRO region (49.10%), which to overtake the EURO region (41.84%) on 6th April. The AMRO region has been experiencing an uncontrolled outbreak of infection in most of its member states, accompanied by a higher mortality rate than any other WHO region (Fig. 1.3).

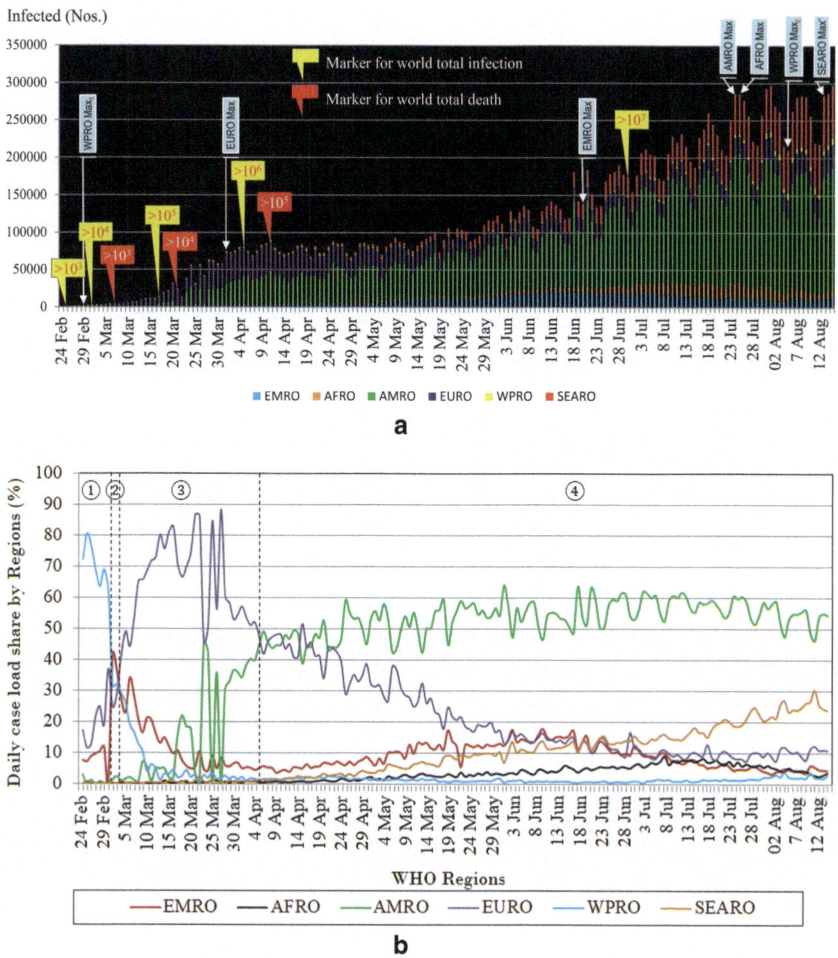

Fig. 1.1 The daily COVID-19 infection caseload sorted by WHO Regions for the period of 24 February to 15 August 2020 as (**a**) absolute number with the markers for total cases of infection and death in the world and (**b**) percentage share by regions

The *daily infection caseload maxima* (2nd cycle) for WPRO (WPRO Max_2) on 5th August (8412 reported new infections) is remarkable. It is an indication of the new cases emerging from the WPRO member states other than the People's Republic of China. Besides, within this present span of observation, the *daily infection caseload maxima* for AMRO (AMRO Max) on 24th July (173,187 new infections), for AFRO (AFRO Max) on the next day, i.e., 25th July (20,614 new infections) and for SEARO (SEARO Max) on 13th August (72,582 new infections) occurred. Since the beginning of August, the caseload share curve shows a slightly recessing trend for AMRO and AFRO (though none of them to be considered as recessing consistently as yet), whereas, the SEARO caseload share is showing a growing trend (Fig. 1.1a). The

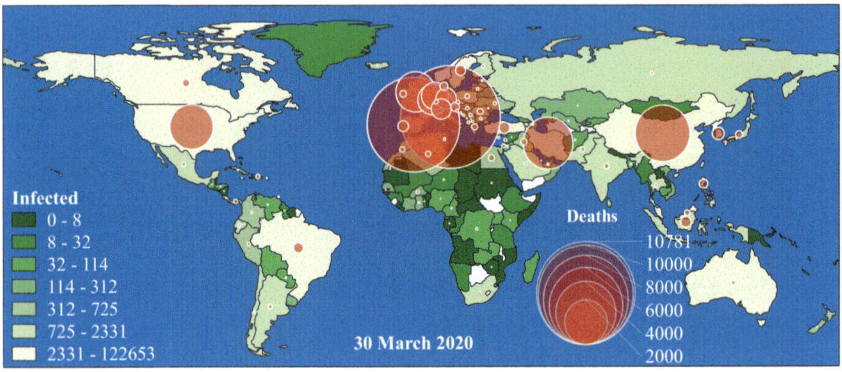

Fig. 1.2 *The European Epidemic*: showing the world distribution of confirmed cases of infection (equal numbers of countries in each class) and death toll due to COVID-19 as on 30 March 2020 (*Data Source* WHO 2020)

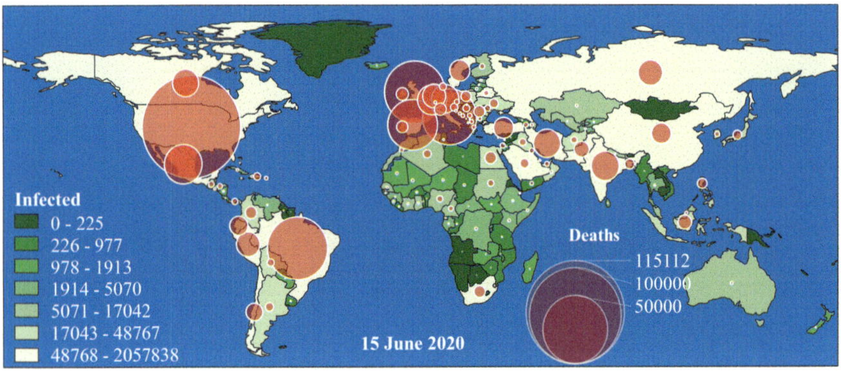

Fig. 1.3 *The devastation in the New World*: World distribution of confirmed cases of infection (equal numbers of countries in each class) and death toll due to COVID-19 as on 15 June 2020 (*Data Source* WHO 2020)

states of Brazil in Latin America and India in Southeast Asia have been experiencing expedited caseloads and growing fatality (Fig. 1.4). The forthcoming surge of health hazard in AFRO and SEARO region has possibly different manifestations. This section will end with that discussion.

As the COVID cases are beginning to spike in the AFRO and SEARO regions, the spatial distribution pattern of population and the share of the urban population are becoming two vital points of concern. Basically, the disease outbreak is in the transition phase, when the positive cases are rising by leaps and bound in Sub-Saharan Africa and the subsistence economic areas of Southeast Asia. Most of the countries in the AFRO and SEARO regions are characterized by more than half of the population living in rural areas, unlike the previous outbreak centers in EMRO, EURO, and AMRO regions (Fig. 1.5). As the COVID-19 infection spreads through person-to-

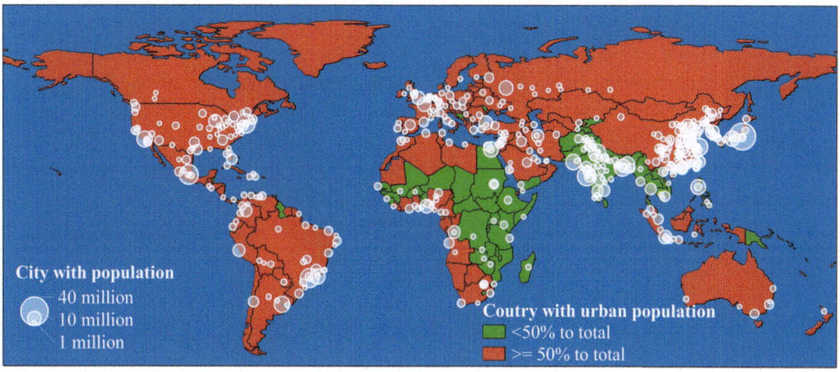

Fig. 1.4 Distribution of the million cities in the world with their population as pie and the level of urbanization as choropleth (*Data Source* United Nations 2018)

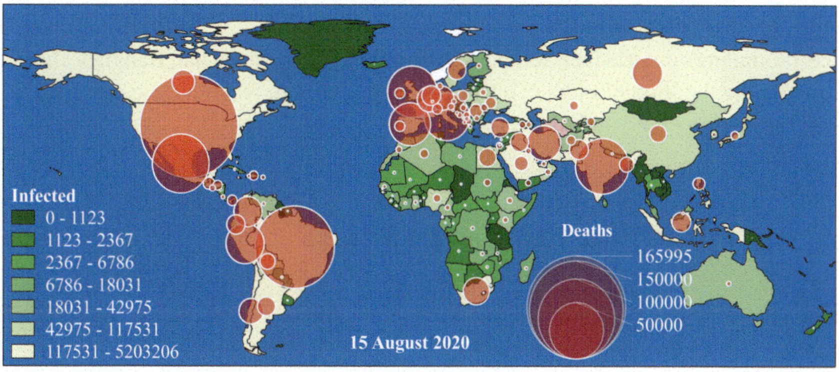

Fig. 1.5 *Shadows of risk over Brazil and India*: confirmed cases of infection and death toll due to COVID-19 as on 15 August 2020 (*Data Source* WHO 2020)

person transmissions, the virus is imported primarily by the airways travelers, flying in a country from the previous outbreak areas. The AFRO region finds ample advantages to control the disease outbreak, despite the poor health infrastructure prevailing in the major part of the continent except for some urban centers. The advantages of the low population density, smaller numbers of million cities, low air-based connectivity with the other continents have overridden the disadvantages of low per capita health financing by the domestic government and paucity of the healthcare facilities in this continent (Table 1.1; Fig. 1.4). In spite of the AFRO region, suffering from a long history of the highest mortality rate among all the regions, the COVID-19 infection and fatality count takes the recessing curve since mid-July. On the other hand, the

Table 1.1 Health expenditure and mortality rate across WHO regions

WHO region	Domestic general government health expenditure (GGHE-D) per capita in PPP int$, in 2017	Domestic general government health expenditure as a percentage of gross domestic product (GDP), in 2017	Probability of dying between 15 and 60 years per 1000 population, in 2016		Mean Population density of the region (Person/Sq. Km)
			Male	Female	
EORO	1986.9	4.9	155	70	110
WPRO	835.3	4.2	104	69	109
AMRO	816.8	4.0	162	89	82
EMRO	956.1	2.6	172	125	112
SEARO	253.3	2.1	206	134	239
AFRO	138.3	1.9	306	248	79

Data Source WHO (2016, 2017), World Bank (2018)

'rice-producing countries' of Southeast Asia, particularly India, and Bangladesh, are experiencing the 'boom' of COVID-19 pandemic, though receiving the surge lately, being at the rear in the outbreak chain (Fig. 1.5).

1.3 COVID-19: The Challenges for India

The COVID-19 has been emerging as a crisis for India. This is obviously a crucial manifestation of the public health care and emergency clinical and medicinal support system under the aegis of the public welfare system of democratic governance. However, this section will not be confined within the discussion of the public health challenges during the pandemic. Instead, it will bring the discussion to explore the multifaceted challenges, incorporating the political, economic, institutional, technological, and sociocultural system to which the surge of the pandemic puts its 'ominous' footprint.

1.3.1 The Acute Health Crisis

India counts the highest numbers of confirmed cases of COVID-19 infection per day and the largest numbers of total cases in Asia (WHO data, as on 15 August 2020). The first COVID-19 case in India was reported on 30th January in Thrissur of Kerala, and two other cases were reported by 3rd February. The initial spreading of the diseases at the multiple clusters at different corners of the country is clearly evidenced by the first 50 COVID-19 cases which were reported in a span of 41 days and were spread in 12 states (Kerala, Tamil Nadu, Telangana, Karnataka, Maharashtra, Uttar

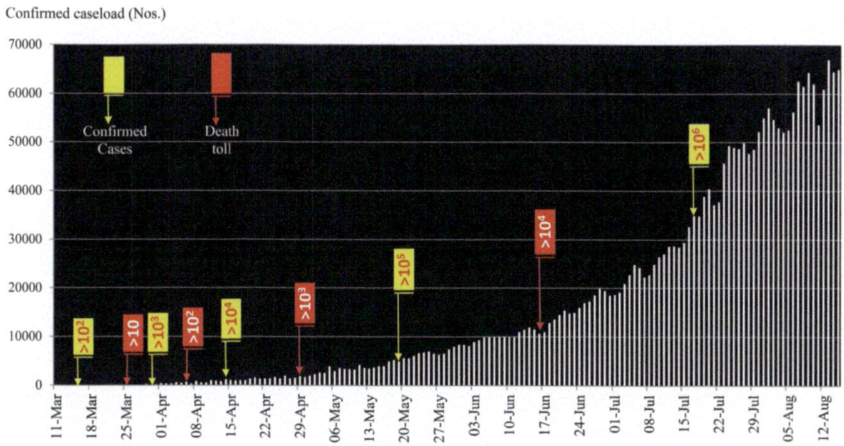

Fig. 1.6 Day-wise COVID-19 caseload in India along with the marker for total confirmed cases and total death toll up to 15 August 2020 (*Data Source* WHO 2020)

Pradesh, Delhi, Haryana, Rajasthan, Punjab, J&K, and Ladakh) and 18 cities or districts (Rawat 2020); and, interestingly, 23 cases out of this first 50 confirmed cases had a travel history to Italy (and, this included the 16 Italian nationals, who tested positive in Jaipur).[31]

India started experiencing a steady growth in the count of COVID-19 infection since mid-April. The number of confirmed cases of COVID-19 infection took only fifteen days to reach the figure from 10^2 to 10^3 (i.e., between 15th and 30th March, with an average caseload of 60 per day); another 16 days to 10^4 (i.e., between 30th March and 14th April, with an average of 580 per day); 35 days to 10^5 (i.e., between 14th April and 19th May, with an average of 2594 per day) and 68 days to 10^6 (i.e., between 19th May and 17th July, with an average caseload of 13,275 per day) (Fig. 1.6). On 15th August, the total infection caseload in the world is reported by WHO as '69239' to make the total confirmed cases '3044940'.

Indian COVID scenario has been exhibiting the general trend that over 60% caseloads are in 5 cities, evidenced by the highest caseloads in the districts of Mumbai, Delhi, Chennai, Ahmadabad, and Thane. Over 75% of caseloads are reported from 10 cities or districts, which includes Pune, Indore, Kolkata, Hyderabad, and Aurangabad in addition to the five above. Furthermore, over 80% of the total infection has occurred in 10 states—Maharashtra, Tamil Nadu, Delhi, Gujarat, Rajasthan, Telangana, Karnataka, West Bengal, Bihar, and Uttar Pradesh (as on 15th August 2020). The National Capital of Delhi, getting threatened by the initial surge, has shown effective responses to mitigate the public health challenges successfully, which is evidenced by the receding trend of the caseload graph of the state (Fig. 1.7). The

[31] See English daily India Today, dated 12th March 2020, available under this URL: https://www.indiatoday.in/india/story/coronavirus-in-india-tracking-country-s-first-50-covid-19-cases-what-numbers-tell-1654468-2020-03-12.

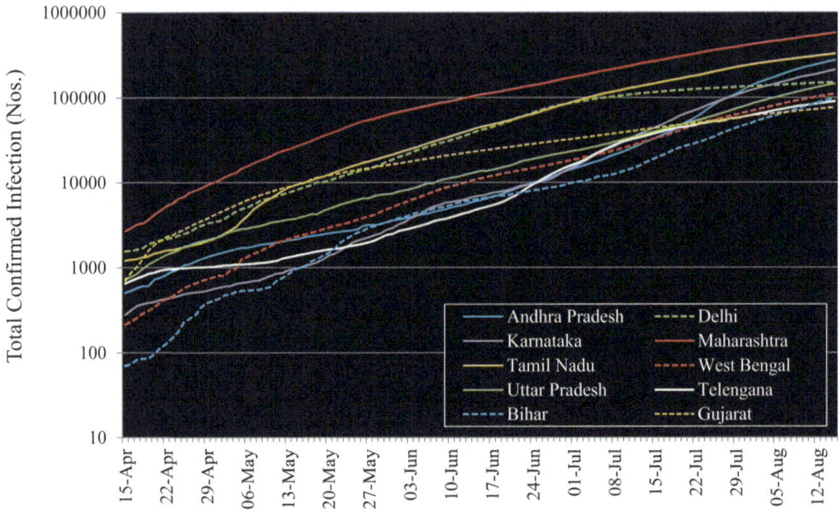

Fig. 1.7 Total confirmed caseloads in ten Indian states with worse condition for the period between 15 April and 15 August 2020 (*Data Source* Prepared by the authors, based on the data from official website of the Ministry of Health & Family Welfare, Govt. of India, i.e., https://www.mohfw.gov.in/)

eastern Indian states of Uttar Pradesh, Bihar, West Bengal, Odisha, and Assam in North East India have been experiencing a higher caseload at a later stage.

1.3.2 The Economic Recession

The sharp nosedive of the global economy is reflected by the declining growth rate and the lowering of GDP in all the countries. 'The year 2020 could see the worst global economic fallout since the Great Depression in the 1930s, involving more than 170 countries likely to experience negative per capita GDP growth due to the COVID-19 pandemic', warned by Kristalina Georgieva (9 April 2020), the Chief of International Monetary Fund (IMF).[32] Parallelly, Oxfam, the UK-based charity organization, issued the alarming statement that the economic fallout could force more than half a billion more people into poverty. According to their assessment, 'by the time the pandemic is over, half of the world's population of 7.8 billion people could be living in poverty'.[33] Simply, India becomes not an exception to this global trend.

[32] See the Time Magazine's coverage on 9th April 2020, available under the URL: https://time.com/5818819/imf-coronavirus-economic-collapse/.

[33] See BBC Report on 9th April 2020, available under the URL: https://www.bbc.com/news/business-52236936.

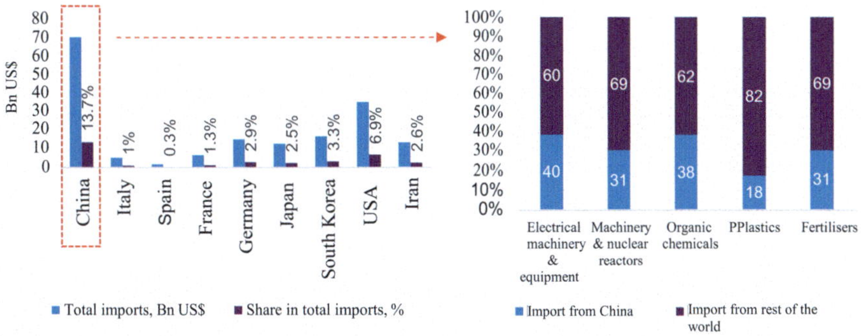

Fig. 1.8 Countries by the import source for India and the percentage of the imported volume of different commodities from the leading import source, i.e., the People's Republic of China (*Data Source* Drawn by the authors, based on official portal of Export Import Data Bank, Department of Commerce, Govt. of India, i.e., https://tradestat.commerce.gov.in/eidb/ergncomq.asp)

With an increasing number of COVID-19 infections, the government locked down transport services, closed all public and private offices, factories, and restricted mobilization. The Government of India announced the nationwide lockdown, starting from 25 March 2020, initially for 21 days, which continued for about two months. In his second address to the nation on the pandemic outbreaking across the world, the Prime Minister of India, while announcing the government's mandate of lockdown, admitted that 'the decision would have an economic cost, but saving people's lives is of the paramount interest to the government'.[34] However, the Government of India stepped forward toward the 'bold decision' of the 'complete lockdown' when this was the most far-reaching measure undertaken by any government in response to the pandemic. It remains the world's biggest lockdown to date, in the context of COVID-19. Travel restrictions and quarantines affecting hundreds of millions of people have left Indian factories short of labor. It poses havoc in the production system.

The disruption in the channels of the import of raw materials remains another issue. Import–export data, given by the Department of Commerce, Government of India, shows that the People's Republic of China remained the largest import source for India for the consecutive years in the last decades. Records show that India imported 13.7% of the total imports in the FY 2018-2019 (Fig. 1.8). China contributes as the leading suppliers of electrical machinery and equipment (40%), machinery and nuclear reactors (31%), organic chemicals (38%), fertilizers (31%), and plastics (18%). The COVID-19-induced shutting down of factories results in the delay in the supply of goods in China. It has caused a shortage of raw materials and intermediate goods for several Indian companies, who depend on China's supply chain.

[34]See the Indian daily, Times of India's news report dated 24th March 2020. URL: https://www.indiatimes.com/news/india/india-put-under-a-complete-lockdown-starting-march-25-to-control-deadly-coronavirus-pandemic-509188.html.

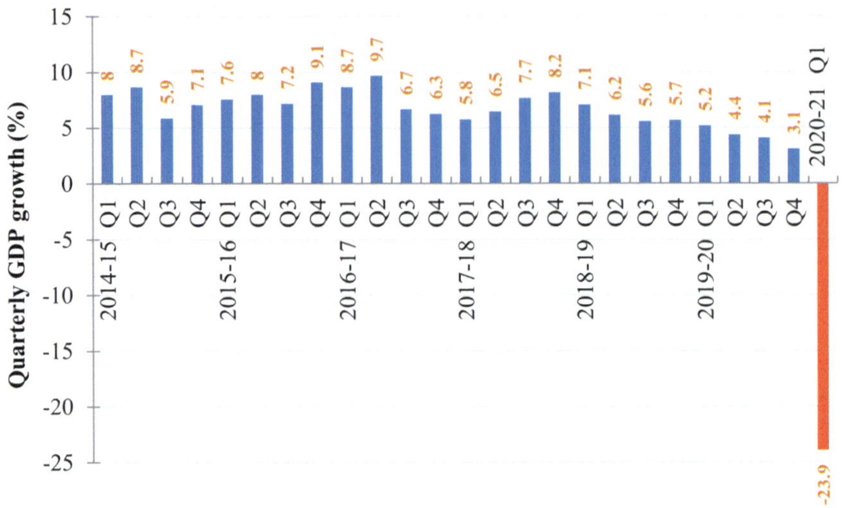

Fig. 1.9 Quarterly GDP growth of India up to the 1st quarter of FY 2020–2021 (*Data Source* Govt. of India)

All these parameters acted cumulatively to cause an unprecedented negative growth of GDP as '−23.9%' during the first quarter of the Financial Year 2020–2021,[35] reflecting the devastation of the COVID-19 pandemic on the country's economy (Fig. 1.9). It highlights an extremely challenging outlook for the Indian economy with only one sector, namely agriculture (+3.4%), showing positive growth on the output side, and only one demand segment, namely, government final consumption expenditure, showing positive growth. The emergence of the agricultural sector as the only sector to witness positive growth can be attributed to an increase in agricultural produce owing to good monsoon rains and targeted government spending.

The manufacturing sector faced a sharp nosedive amidst lockdown, as reflected by the output falling 39.3% in the June quarter after falling 1.4% in the previous consecutive quarter. Moreover, India's fiscal deficit in the four months to the end of July stood at INR 8.21 lakh crore (i.e., $111.7 billion), or 103.1% of the budgeted target for the current fiscal year.[36]

The economic recession directly affected the unorganized sector and semi-skilled jobholders, mostly as they have a greater likelihood of losing their employment. Consequently, there is the possibility for a large volume of the population living just above the poverty line to get drowned below the Poverty Line (BPL). It will

[35]See the Times of India news coverage dated 31 August 2020, available under the URL: https://timesofindia.indiatimes.com/business/india-gdp-data-live-updates-indian-economy-q1-2020/liveblog/77846410.cms.

[36]See the Times of India dated 31 August 2020, available under the URL: https://timesofindia.indiatimes.com/business/india-business/fiscal-deficit-crosses-full-year-budget-target-in-4-months/articleshow/77851470.cms.

cause undue pressure on the national economy due to the public schemes' financial involvement for the newly added population below the poverty line.

The informal workforce significantly contributes to the Indian economy, contributing more than 45% to its overall GDP to place itself as the largest informal economic contribution in the world. Around 37% of regular wage-based employees in urban India are nonagricultural informal workers. Among the Indian States, Rajasthan (54.8%), Punjab (51.8%), Andhra Pradesh (51%), Chattisgarh (49%), and Gujarat (48.4%) involved the higher volume of nonagricultural informal workers in urban areas (Periodic Labour Force Survey 2017–2018).[37] These workers faced uncertain income following the stalling of urban activity due to the lockdown. On the other hand, in rural India, the labor sector under the Mahatma Gandhi Rural Employment Guarantee Act (MGNREGA), 2005,[38] is probably the worst impacted group as they are not provided jobs due to lockdown. As most of the labor sectors are associated with the construction works and daily wage earners, they are to face hardship.

1.3.3 The Issues Concerning the 'Social Stigma'

Besides the economic challenges, the issues associated with provoking stigmatization to COVID-19 infected or patients having similar symptoms (e.g., fever, cough, or breathing trouble which is caused by other diseases) looms large, which unfortunately the Indian citizens are experiencing. Social stigma is 'the negative association between a person or group of people who share certain characteristics and a specific disease. In an epidemic, this may mean people are labeled, stereotyped, discriminated against, treated separately, and experience loss of status because of a perceived link with a disease' (WHO 2020). It is ironical that stigma and fear are commonly reported around communicable diseases and epidemic outbreaks anywhere and anytime globally; however, the magnitude depends on the mass awareness, level of literacy, and the social structure of the area where a disease cluster is located. The consequence is that it hampers the response significantly by driving people to hide the illness to avoid discrimination, preventing them from seeking health care immediately, and discouraging them from adopting healthy behaviors.

The presence of social stigma around the COVID-19 outbreak in India is clear from the statement from the end of the Union Ministry of India.[39]

---

[37]The wage/salaried workers who are not eligible for paid leave and do not have written job contract and enjoy social security benefits are considered as informal workers in this report.

[38]MGNREGA is an Indian labor law and social security measure that aims to enhance livelihood security in rural areas by providing at least 100 days of wage employment in a financial year to every household whose adult members volunteer to do unskilled manual work.

[39]See Zee Media Bureau news coverage on 9 April 2020, available under the URL: https://zeenews.india.com/india/centre-issues-advisory-to-address-social-stigma-associated-with-coronavirus-covid-19-2274974.html.

Cases have been reported of people affected with COVID-19 as well as healthcare workers, sanitary workers, and police, who are in the frontline for management of the outbreak, facing discrimination on account of heightened fear and misinformation about infection. Even those who have recovered from COVID-19 face such discrimination.

The rising incidents of prejudices and social stigma against some communities and areas caused the Union Ministry of Health and Family Welfare (MoHFW), Government of India to seriously intervene, and issue an advisory on 9 April 2020. The advisory says that 'such prejudices and social stigma during the outbreak of communicable diseases may cause fear and anxiety, which could culminate in increased hostility, chaos, and unnecessary social disruptions'.[40]

1.4 The COVID-19 Crisis as the Catalyst for Change

Fighting against a pandemic like the COVID-19 is a war-like situation. It is a symmetric shock with asymmetric impacts that require thoughtful responses and exploring the opportunities arising out of the crisis. The outbreak of war is a cathartic moment that causes suffering, plight, and devastation. However, the historical semantic records confirm that war brings forth societal transformation. Scholarly works confirm World War I as the positive catalyst for progressing women's rights, whereas World War II as the dawn for decolonization. Similarly, the 'fight against COVID-19' has been emerging as the catalyst, particularly for the developing world, toward a positive societal and economic transformation that carries the long-term benefits. In this section, we will discuss all these things concerning India.

1.4.1 An 'Opportunity' to Uplift Health Financing

The public expenditure on health is a vital metric regarding the public health scenario of a nation. WHO's global health dataset on health financing reveals that if the domestic general government health expenditure is concerned, India ranks far lower, even behind some countries, which are classified as the 'poorest' in the world (Fig. 1.10). The Report entitled 'National Health Profile 2019', published by the Ministry of Health and Family Welfare, Government of India,[41] takes a detailed account of socioeconomic status, disease burden, health finance, health infrastructure, and human resources in the sector. It mentions that, in 2016, the public spending on health was just 1.17% of the GDP. In comparison, nations classified as Lower-Income Countries by the World Bank spent 1.57% of their GDP on health that year.

---

[40]See the MoHFW, Govt. of India advisory dated 09 April 2020 under the URL: https://www.mohfw.gov.in/pdf/AddressingSocialStigmaAssociatedwithCOVID19.pdf.

[41]See: http://www.cbhidghs.nic.in/showfile.php?lid=1147.

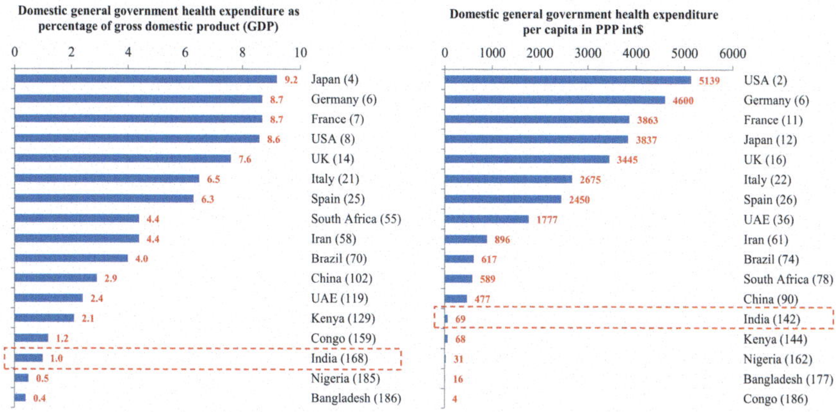

Fig. 1.10 Domestic general government health expenditure for its citizens for selected countries with their world rank within brackets, 2017 (*Data Source* WHO 2018)

The health shock of the COVID-19 in the country has shaken the health system of the country. The 1.28% of GDP on public health in 2019 is found insufficient for the preparedness against a pandemic outbreak. Even the targeted value of '2.5 percent by 2025' may run short for the country with such a large volume of the population depending on the public health and emergency medical system.

The Government of India outlay (2019) for health as INR 623.9 billion (of which INR 317.5 billion for National Health Mission (2019–20) and INR 64 billion for Ayushman Bharat) and the allocation of INR150 billion for healthcare to control COVID-19 outbreak are significant steps toward enhancing the government health financing.[42]

The country boasts of more than two million health workers[43] and a total of 1,159,309 allopathic doctors registered with the state medical councils and the Medical Council of India (MCI) as on 31 March 2019. Assuming an 80% availability, around 9.27 lakh doctors are available for active service. However, it needs serious government initiative regarding the patient to doctor ratio, which is present, 1445:1 (modern medicine) and does not reach the WHO's recommendation (i.e., 1000:1). Parallelly, there are 7.88 lakh ayurveda, unani, and homeopathy (AUH) doctors in the country, practicing traditional systems of medicine. Adding 80% of them (i.e., around 6.30 lakh) as available for the service, together with allopathic doctors, the doctor–patient ratio becomes 1:860.[44]

---

[42] See the press release note at Press Information Bureau dated 9 April 2020, available under the URL: https://pib.gov.in/PressReleaseIframePage.aspx?PRID=1612534.

[43] WHO's report on 'Healthcare Workforce in India', available hunder the URL: https://www.who.int/hrh/resources/16058health_workforce_India.pdf.

[44] See Business Standard news coverage on 29 November 2019. Available under the URL: https://www.business-standard.com/article/pti-stories/doctor-patient-ratio-in-india-less-than-who-prescribed-norm-of-1-1000-govt-119111901421_1.html.

The present public health infrastructure, consisting of 529 Medical colleges (MBBS), 702 Ayush, 313 Dental (BDS), 253 PG Dental, ~158 thousand sub-centers, more than 25,000 Primary Health Centres (PHC), and ~5,624 Community Health Centres (CHC), needs to be equipped with the modern medical facilities and ample human resources to cope up with the national health emergency, not only for the present crisis but for the future as well. COVID-19 outbreak has expedited the process of up-gradation. Presently, the public health facilities dedicated to COVID-19 management are categorized into three categories.

- **Dedicated COVID Hospital (DCH):** Hospitals that offer comprehensive care, primarily for those who have been clinically assigned as severe. These hospitals are accommodated with the fully equipped ICUs, ventilators, and beds with assured oxygen support. All DCHs have separate areas for suspected and confirmed cases. DCHs are designed to serve as referral centers for the Dedicated COVID Health Centres and the COVID Care Centres.
- **Dedicated COVID Health Centre (DCHC):** These are hospitals that serve to offer care for all cases that have been clinically assigned as moderate. The DCHCs have separate areas for the suspected and confirmed cases. There are beds with assured oxygen support, and every DCHC is mapped to one or more Dedicated COVID Hospitals.
- **Dedicated COVID Care Centre (DCCC):** They are designed to offer care only for cases that have been clinically assigned as mild or very mild cases or COVID suspect cases. These are the makeshift facilities that may be set up by the States/UTs in hostels, hotels, schools, stadiums, and lodges, both public and private.

Within the first week of May, 7,740 facilities have been serving in 483 districts in all States/Union Territories that include hospitals and facilities of the State/UT governments as well as the central government. It includes 6,56,769 isolation beds, 3,05,567 beds for confirmed cases, 3,51,204 for suspected cases, 99,492 oxygen-supported beds, 1,696 facilities with oxygen manifold, and 34,076 ICU beds.[45] The country is well focused on the testing of COVID-19 infection and the detection of the cases from the very beginning. Indian Council of Medical Research (ICMR)[46] data shows that a total of 43,324,834 samples have been tested in the laboratories throughout the country up to 31 August 2020. The total number of samples tested on 31 August 2020 only was 1,016,920, which was well above the government's targeted milestone of one million tests daily (attained on 22 August 2020[47]).[48]

---

[45] See One India news coverage, available under the URL: https://www.oneindia.com/india/7-740-dedicated-covid-19-health-facilities-in-483-districts-identified-health-ministry-3086182.html.

[46] See the ICMR website (accessed on 1 September 2020): https://www.icmr.gov.in/.

[47] See The Print dated 22 August, 2020: https://theprint.in/health/india-crosses-10-lakh-daily-covid-tests-milestone-within-month-of-boosting-screening-process/487265/.

[48] See the Ministry statement: https://zeenews.india.com/india/government-is-planning-to-increase-covid-19-testing-capacity-to-one-lakh-per-day-health-minister-harsh-vardhan-2280035.html.

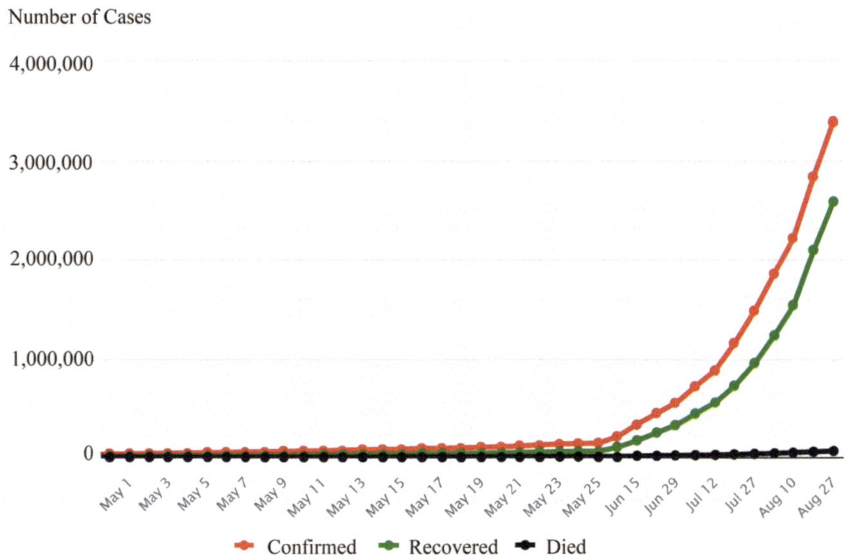

Fig. 1.11 The graph of confirmed, recovered, and fatality relating to COVID-19 in India (*Data Source* Ministry of Health & Family Welfare, Govt of India 2020)

At the end of August, India's COVID-19 recoveries have crossed 26 lakh which has made the recovery rate to reach 76.47% (Fig. 1.11). It has been made possible due to the strategic health policy that includes the 'aggressive' testing, comprehensive tracking, and efficient treating in supervised home-isolation, facility-isolation, and hospitals. The case fatality rate (CFR) is found lower when compared to the global average. The CFR shows a continuous declining trend and reaches 1.81% at the end of August. Despite other parameters, the growing rate of recovered patients is also contributed by the dedicated service of the health workers at the forefront and effective management from the part of the government.

1.4.2 Becoming Self-Reliant in PPE Production

The first item under this heading must be the production of the Personal Protection Equipment (PPEs). PPEs are kinds of protective gear designed to safeguard the health of workers by minimizing the exposure to a biological agent. It includes goggles, face-shield, masks (Surgical/N-95), gloves (surgical/examination), coverall/gowns (with or without aprons), headcover, and shoe cover. The quest for boosting the fight against the novel coronavirus has made the country ramp up the production of COVID-19 protective gear and medical equipment. This effort has made India transform into an 'exporter' of the PPEs where there was not a single domestic manufacturing unit of PPEs before the COVID-19 outbreak in the country.

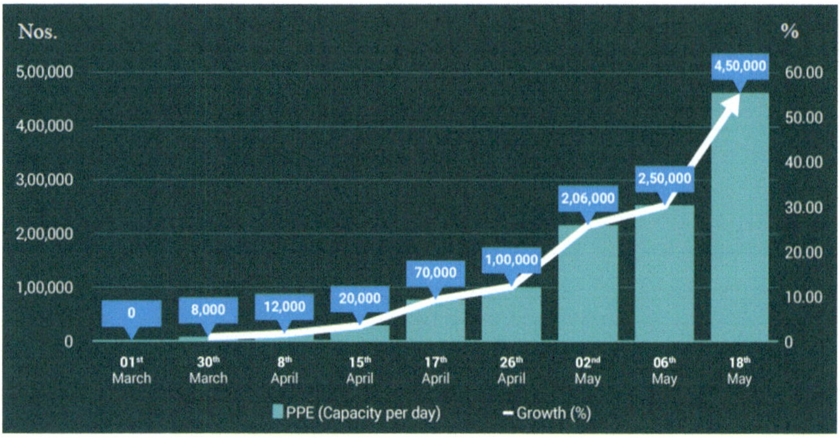

Fig. 1.12 Remarkable growth in PPE production: From 0 on 1st March, India produces nearly 4.5 lakh PPE kits every single day as on 18 May 2020, and in 60 days, the PPE industry in India has witnessed 56 times growth (*Data Source* Ministry of Health & Family Welfare, Govt. of India 2020)

During May, when the infected count has started rising throughout the country by leaps and bound, the Government of India placed a bulk order for PPE. Out of a total of 22 million PPE, 14.2 million were decided to be bought from the domestic manufacturers.[49] Government strategy worked as inertia. A statement from the Union Ministry confirmed that on 18th May 2020, India produced over 4.5 lakh PPE suits by more than 600 companies in the country, certified MSME to manufacture PPEs (Fig 1.12).

The exports of the PPEs started in the month of July. The notification issued on 29 June 2020 by the Director General of Foreign Trade (DGFT) permitted the export of the PPE after satisfying the internal demands. A statement from the government confirmed on 14th August that India had exported 23 lac of PPEs to the USA, UK, UAE, Senegal, and Slovenia to help those countries in their fight against novel coronavirus disease.[50] An economic assessment by the Strategic Investment Research Unit (SIRU) has described the Indian domestic PPE manufacturing as the INR 7000 Crore industry making process, which has more potential ahead than that is exposed so far (Lakshmanan and Nayyar 2020).

[49] See the Times of India news coverage on 2 May 2020: https://timesofindia.indiatimes.com/india/india-ramps-up-production-of-covid-19-protective-gears-medical-equipment/articleshow/75507915.cms.

[50] See Times Now News on 14 August 2020: https://www.timesnownews.com/mirror-now/in-focus/article/self-reliant-for-ppes-india-s-effort-to-fight-covid-19-globally/637055.

1.4.3 Rediscovering the Agricultural Economy

While the Indian economy has taken a massive hit from the coronavirus pandemic and resultant lockdowns, the agricultural sector remains the silver lining for the Financial Year 2020–2021. The estimates of Gross Value Added (GVA) by various sectors in the First Quarter in the FY 2020–2021 show that only Agriculture, Forestry, and Fishing witnessed an increase in income with GVA at 3.4% and rest of the sectors saw a fall in income (Table 1.2). Three factors significantly envisage the positive growth in this primary sector amidst the lockdown—the normal monsoon and the exemption of the agricultural activity from the nationwide lockdown and the added human resources in the rural agricultural fields due to the reverse migration.

The Indian Meteorological Department (IMD) forecasted the monsoon to be 102% of the long-period average. A normal monsoon is always reported well for India, reaching its benefit to a considerable part of the rural workforce who are employed in the agricultural sector. While most economic activities came to a standstill since April due to the lockdown, the farming activities were exempted, which facilitated the uninterrupted harvesting of Rabi crops (i.e., winter crops) and the sowing of Kharif crops (i.e., monsoon crops). The smooth flow of agricultural commodities throughout the lockdown period and across both rural and urban areas ushered the economic benefit for the country.

Table 1.2 Quarterly estimates of the Gross Value Added (GVA) at basic prices in the first quarter (April–June) in consecutive three financial years

Industry	April–June (Q1)				
	2018–2019	2019–2020	2020–2021	Percentage change over the previous year	
				2019–2020	2020–2021
Agriculture, forestry, and fishing	427,177	439,843	454,658	+3.0	+3.4
Mining and quarrying	88,634	92,807	71,209	+4.7	−23.3
Manufacturing	561,875	578,936	351,396	+3.0	-39.3
Electricity, gas, water supply, and other utility services	74,998	81,628	75,877	+8.8	−7.0
Construction	249,913	262,828	130,750	+5.2	−50.3
Trade, hotels, transport, communication, and services related to broadcasting	609,330	630,860	334,284	+3.5	−47.0
Financial, real estates, and professional services	757,850	803,322	760,491	+6.0	−5.3
Public administration, defence, and other services	387,589	417,483	374,659	+7.7	−10.3

Data Source MSPI, Government of India

Hundreds of thousands of migrant workers returned to their native villages across India after the pandemic-induced lockdown left them unemployed in the cities. Reverse migration is seen as a massive threat to the economy, and there are possibilities of raising informality in the economy. However, these human resources, added to the rural agricultural economy, contributed well to the growth of the agricultural economy more than the expectation. Kharif crop, sowing across India in 2020, is about 21% more than last year (2019). The country is witnessing a growth in acreage compared to the last year for some of the crop varieties. Black gram sowing acreage has witnessed a growth of 43.11% this year over last year; pigeon pea (arhar) shot up by 37.74%; finger millet (ragi) by 45%; and groundnut by 56.57%. Official data shows that the acreage of rice cultivation has gone up by almost 19%, oilseed crops by 44.75%, pulses by 32.35%, cotton by 17.28%, and coarse cereals by 12.23%.[51]

The Government of India shows strict supervision on the flow and trade of agricultural commodities and the interests of the farmers that led to the promulgation of three ordinances on 5 June 2020.

- ***The Farmers' Produce Trade and Commerce (Promotion and Facilitation) Ordinance 2020*** seeks to provide for barrier-free trade of farmers' produce outside the markets notified under the various state agricultural produce market laws (state APMC Acts). The Ordinance allows intrastate and interstate trade of farmers' produce outside: (i) the physical premises of market yards run by market committees formed under the state APMC Acts and (ii) other markets notified under the state APMC Acts. Such trade can be conducted in any place of production, collection, and aggregation of farmers' produce, including (i) farm gates, (ii) factory premises, (iii) warehouses, (iv) silos, and (v) cold storages.[52]
- ***The Farmers (Empowerment and Protection) Agreement on Price Assurance and Farm Services Ordinance 2020*** provides a framework for the protection and empowerment of farmers concerning the sale and purchase of farm products. The provisions of the Ordinance will override all state APMC laws. The Ordinance provides for a farming agreement before the production or rearing of any farm produce aimed at facilitating farmers in selling farm products to sponsors.[53]
- ***The Essential Commodities (Amendment) Ordinance 2020*** amends the Essential Commodities Act 1955, allowing the Government of India to delist certain commodities as essential. The Act empowers the government to control the production, supply, distribution, trade, and commerce in certain commodities. The Ordinance seeks to increase competition in the agriculture sector and enhance

[51] See the report posted in the website of the Ministry of Agriculture and Farmers Welfare, GOI: https://pib.gov.in/PressReleasePage.aspx?PRID=1639340.

[52] See the Ordinance in the official website of the Department of Agriculture, Cooperation and Farmers Welfare, GOI: http://agricoop.nic.in/recentinitiatives/farmers-produce-trade-and-commerce-promotion-and-facilitation-ordinance-2020.

[53] See the details in the official website of the Ministry of Agriculture and Farmers Welfare: https://pib.gov.in/PressReleasePage.aspx?PRID=1629750.

farmers' income. It aims to liberalize the regulatory system while protecting the interests of consumers.[54]

Agriculture has been the cornerstone of the Indian economy till date. The growth in the agricultural sector amidst the COVID-19 trapped India provides the 'oxygen' to the 'gasping' national economy. The pandemic brings to fore the practical experiences that give the government and policymakers to nourish the human resources associate with the sector, safeguarding their interests, values, economic, and social security based on which the urban-industrial economy could flourish, carrying with it the long-term benefit for the national economy.

1.4.4 A Shift Toward 'Localization'

The immediate and tangible impact of '*black swan events*' is carried out by the disruption in the supply chain. The geopolitical environment and a globally recessionary climate are likely to lead to greater protectionism and risk aversion. The practical lessons learned from the pandemic outbreak of COVID-19 are expected to lead to direct the policies toward greater localization of supply chains, especially of essential commodities, labor force as well as for sectors that are perceived to be strategically important. The pandemic has pointed out that the transformation in the economy through policymaking is becoming essential. The encouragement in local resource-based enterprises, agro-industrial initiatives, food-processing with local production can be sustainable management for the stream of migrant labor forces. Technology diffusion has a vital role to play for the optimum utilization and processing of agricultural products to a variety of usable goods.

Millet (which is locally named as *Ragi*)-based technology development by the Central Food Technological Research Institute (CFTRI), Mysuru may be taken as an instance. The Millet-based projects undertaken by this institution aims to enhance the visibility and total acceptance of Millet based products in the rural and urban populations. This compilation is brought in order to help in creating a sustainable ecosystem for the growers and processors. The CSIR-CFTRI has developed innovative technologies based on millets, and a large number of these have been successfully transferred to the Small and Medium Enterprises. The utilization of those innovative technologies has been successfully used in producing Ragi Rusk, Ragi Noodles, Ragi Papad, Convenience Flour for Mudde, Instant Beverage From Ragi, Ragi Roti, Ragi Snack, Ragi Flakes, Ragi Murukku Mix, Malted Ragi Flour-Enzyme Rich, Extruded Ragi, Germinated Ragi Drink Mix, Puttu Mix, and many more products.

---

[54]See the Gazztte of India dated 5 June 2020: http://egazette.nic.in/WriteReadData/2020/219748.pdf.

Digital Infrastructure as a Core Utility to Every Citizen	Governance & Services on Demand	Digital Empowerment of Citizens
• Availability of high speed internet as a core utility for delivery of services to citizens • Cradle to grave digital identity that is unique, lifelong, online and authenticable to every citizen • Mobile phone & bank account enabling citizen participation in digital & financial space • Easy access to a Common Service Centre • Shareable private space on a public cloud • Safe and secure cyber-space	• Seamlessly integrated services across departments or jurisdictions • Availability of services in real time from online & mobile platforms • All citizen entitlements to be portable and available on the cloud • Digitally transformed services for improving ease of doing business • Making financial transactions electronic & cashless • Leveraging Geospatial Information Systems (GIS) for decision support systems & development	• Universal digital literacy • Universally accessible digital resources • Availability of digital resources / services in Indian languages • Collaborative digital platforms for participative governance • Citizens not required to physically submit Govt. documents / certificates

Fig. 1.13 The key vision areas, centered by the Digital India Programme (*Source* Ministry of Electronics & Information Technology, Government of India)

1.4.5 A Real Push for Being 'Digital'

Digital India is a flagship program of the Government of India with a vision to transform India into a digitally empowered society and knowledge economy (Fig. 1.13). The core objectives of this project include the broadband internet service for rural areas, integration of all core ICT infrastructure built under the National e-Governance Plan (NeGP), universal access to mobile connectivity, Public Internet Access Programme (PIAP) under the National Rural Internet Mission (NRIM), cradle-to-grave digital identity, digital financing, and many other things to ensure the digital infrastructure as a utility to every citizen.

This pandemic has set the 'being digital' as a compulsion to cope up with the changing sociocultural environment during the post-pandemic world. Institutions, organizations, and public and private offices have been shifted to work remotely, and the employees are now 'online' and working from home. While these trends were already 'in-motion', the entire process will be expedited in the near future. Interestingly, in India, even the most traditional brick and mortar businesses have been forced to experiment with digital channels. This presents a real and immediate opportunity to drive efficiencies through digital media. The crisis has pointed out the importance of government initiative and investment in enabling technologies like cloud, data, and cybersecurity. This will change the way we work with far-reaching implications on online commercial real estate, e-commerce, e-governance, cybersecurity, process automation, data analytics, and self-service capabilities.

1.5 Conclusion

The COVID-19 outbreak manifests its uniqueness in many senses. The global reach of the virus, the ability of human-to-human transmission, the fatality rate in the population in the age group of ≥60 years, the geopolitical reordering, and the economic recession—all these are unprecedented in the recorded history of mankind. More interestingly, the same disease is exposing its malicious effect differently across the individuals, geographical locations, age groups, and working classes. The vaccine is hitherto (i.e., when we are writing) unknown. Humankind is fighting against the pestering virus with the immune system, which exists to protect the host from noxious environmental agents, especially pathogenic organisms. The spillover of the virus, causing the initial consequence as an urban health hazard, has transformed into a global pandemic that is now manifesting the societal and economic transformation throughout the globe. This phase is crucial as the effect is hostile and the impact is diverse. A conventional local grocery shop to a multinational corporate house—all are affected, though the channels of infestation are different.

The practitioners of Medical Science are working at the frontline of the fight against this pandemic. However, beyond this scenario of an acute health hazard, the pandemic has carried with it the other threats for mankind associated with the economy, society, culture, psychology, and politics. The long period of lockdown all over the globe has resulted in environmental reinvigoration and diminishing of the pollution level. On the other hand, the lockdown has affected the world economy to initiate another global economic recession. The fallout of the latter is of grim significance to the entire population of the world. Amidst these multifaceted dimensions of the pandemic, this is high time for global solidarity to save humankind on this only planet in the universe where life exists.

Human society, its ambient environment, the process of socioeconomic development, and politics and power—all drive to set the world order. All these parameters are intimately and integrally related. The interconnections of these three driving forces have a significant bearing on life, space, and time. In parallel, the interrelationship between all these drivers is dynamic, and they are changed drastically with time and space. The social-scientists serve to align the thought, based on which the policymakers need to understand the prevailing equation to project the unforeseen future. The trajectory of the future world helps in planning and policymaking with a scientific direction.

References

Alders R, Awuni JA, Bagnol B, Farrell P, Haan N (2014) Impact of avian influenza on village poultry production globally. EcoHealth 11(1):63–72. https://doi.org/10.1007/s10393-013-0867-x.pmid24136383

Alexander DJ, Brown IH (2009) History of high pathogenic avian influenza. Rev Sci Tech 28(1):19–38. https://doi.org/10.20506/rst.28.1.1856

Bagnol B (2012) Advocate gender issues: a sustainable way to control Newcastle Disease in village chickens. INFPD Good Practices of Family Poultry Production Note No 03. Available via http://www.fao.org/3/aq630e/aq630e.pdf. Accessed 18 July 2020
Brown. R (18 July 2014) The virus detective who discovered Ebola in 1976. BBC Magazine. https://www.bbc.com/news/magazine-28262541. Accessed 22 July 2020
Caglioti C, Lalle E, Castilletti C, Carletti F, Capobianchi MR, Bordi L (July 2013) Chikungunya virus infection: an overview. The New Microbiol 36(3):211–227. PMID 23912863
Campbell C, Yunnan Y, Park A (July 23, 2020) Inside the global quest to trace the origins of COVID-19 and predict where it will go next. Time. https://time.com/5870481/coronavirus-origins/. Accessed 22 July 2020
CDC Centers for Disease Control Prevention (September 2006) Chikungunya fever diagnosed among international travelers–United States, 2005–2006. MMWR. Morb Mort Wkly Rep 55 (38):1040–1042. PMID 17008866
Cunningham, A (9 Apr 2020) As professor of wildlife epidemiology my research explores diseases in wild animal populations. Zoological Survey of London. https://www.zsl.org/blogs/science/covid-19-the-wildlife-facts. Accessed 22 July 2020
Daszak, P (27 February, 2020) We knew disease X was coming—it's here now. New York Times. https://www.nytimes.com/2020/02/27/opinion/coronavirus-pandemics.html. Accessed 16 Aug 2020
Dick GW, Kitchen SF, Haddow AJ (September 1952) Zika virus: isolations and serological specificity. Trans R Soc Trop Med Hyg 46(5):509–520. http://dx.doi.org/10.1016/0035-9203(52)90042-4. PMID:12995440
Elliott P, Wartenberg D (2004) Spatial epidemiology: current approaches and future challenges. Environ Health Perspect 112(9):998–1006. https://doi.org/10.1289/ehp.6735
Ergönül O, Celikbaş A, Dokuzoguz B, Eren S, Baykam N, Esener H (July 2004) Characteristics of patients with Crimean-Congo hemorrhagic fever in a recent outbreak in Turkey and impact of oral ribavirin therapy. Clinical Infectious Diseases. 39 (2): 284–7. https://doi.org/10.1086/422000
Feleke, B and Scutti, S (19 December 2018) Up to 319 people dead as Congo Ebola outbreak worsens. CNN Health. https://edition.cnn.com/2018/12/18/health/ebola-democratic-republic-of-congo-intl/index.html. Accessed 22 July 2020
Gatzweiler Franz, Bojie Fu, Rozenblat Celine et al (2020) COVID-19 reveals the systemic nature of urban health globally. Cities & Health. https://doi.org/10.1080/23748834.2020.1763761
Hiromoto Y (2000) Evolutionary characterization of the six internal genes of H5N1 human influenza A virus. J Gen Virol 81 (Pt 5):1293–1303. https://doi.org/10.1099/0022-1317-81-5-1293
Jones DS (2020) History in a crisis- lessons for Covid-19. N Engl J Med 382(18):1681–1683. https://doi.org/10.1056/nejmp2004361
Khati, D S (Apr 26, 2020) Covid-19 and China's wildlife trade are linked more closely than you think. Hindustam Times. https://www.hindustantimes.com/opinion/covid-19-and-china-s-wildlife-trade-are-linked-more-closely-than-you-think-opinion/story-6sOSmAOctNCnCovZTFPBpJ.html. Accessed 22 July 2020
Kindhauser MK, Allen T, Frank V, Santhana R, Dye C (2016) Zika: the origin and spread of a mosquito-borne virus. Bulletin of the World Health Organization. http://dx.doi.org/10.2471/BLT.16.171082. Accessed 19 July 2020
Lakshmanan R, Nayyar M (25 May 2020) Personal protective equipment in India: An INR 7,000 Cr industry in the making. SIRU. https://www.investindia.gov.in/siru/personal-protective-equipment-india-INR-7000-cr-industry-in-the-making. Accessed 5 Sept 2020
Lee C Ng, Hapuarachchi HC (October 2010) Tracing the path of Chikungunya virus–evolution and adaptation. Infect Genet Evol 10(7):876–885. https://doi.org/10.1016/j.meegid.2010.07.012.pmid20654736
Malone RW, Homan J, Callahan MV, Glasspool-Malone J, Damodaran L, Schneider A, et al. (March 2016) Zika virus: medical countermeasure development challenges. PLOS Neglected Trop Dis 10(3):e0004530. https://doi.org/10.1371/journal.pntd.0004530

McLeod A, Morgan N, Prakash A, Hinrichs J (2005) Economic and social impacts of avian influenza. Food and Agriculture Organisation. http://www.fao.org/avianflu/documents/Economic-and-social-impacts-of-avian-influenza-Geneva.pdf. Accessed 22 July 2020

Mehravaran A, Moradi M, Telmadarraiy Z, Mostafavi E, Moradi AR, Khakifirouz S, et al. (February 2013) Molecular detection of Crimean-Congo haemorrhagic fever (CCHF) virus in ticks from southeastern Iran. Ticks and Tick-Borne Diseases 4(1–2):35–8. https://doi.org/10.1016/j.ttbdis.2012.06.006

Mehrjardi MZ (1 January 2017) Is Zika virus an emerging TORCH agent? An invited commentary. Virology 8:1178122X17708993. https://doi.org/10.1177/1178122x17708993

Morrison TE (October 2014) Reemergence of chikungunya virus. J Virology 88(20):11644–11647. https://doi.org/10.1128/jvi.01432-14.pmc4178719.pmid25078691

Periodic Labour Force Survey (July 2017–June 2018) Ministry of Statistics and Programme Implementation (MoSPI), May 2019. Accessed 24 Mar 2020

Porter N (2012) Risky zoographies: the limits of place in avian flu management. Environ Humanit 1(1):103–121. https://doi.org/10.1215/22011919-3609994

Powers AM, Brault AC, Tesh RB, Weaver SC (February 2000) Re-emergence of Chikungunya and O'nyong-nyong viruses: evidence for distinct geographical lineages and distant evolutionary relationships. J Gen Virol 81(Pt 2):471–479. https://doi.org/10.1099/0022-1317-81-2-471.pmid10644846

Quammen, D (28 January 2020) We made the coronavirus epidemic—it may have started with a bat in a cave, but human activity set it loose. New York Times. https://www.nytimes.com/2020/01/28/opinion/coronavirus-china.html. Accessed 4 Aug 2020

Rawat M (12 March 2020) Coronavirus in India: Tracking country's first 50 COVID-19 cases; what numbers tell. India Today. https://www.indiatoday.in/india/story/coronavirus-in-india-tracking-country-s-first-50-covid-19-cases-what-numbers-tell-1654468-2020-03-12. Accessed 01 Sept 2020

Robinson MC (January 1955) An epidemic of virus disease in Southern Province, Tanganyika Territory, in 1952–53: I. Clinical features. Trans Roy Soc Trop Med Hyg 49(1):28–32. https://doi.org/10.1016/0035-9203(55)90080-8.pmid14373834

Roos R (27 June 2012). CDC estimate of global H1N1 pandemic deaths: 284,000. CIDRAP. https://www.cidrap.umn.edu/news-perspective/2012/06/cdc-estimate-global-h1n1-pandemic-deaths-284000. Accessed 18 July 2020

Sharma NC (22 May 2018) Third outbreak of Nipah Virus in India since 2001. What are the reasons? Mint. https://www.livemint.com/Science/NoQYN3C3YhDUFyjK8nfrjI/Third-outbreak-of-Nipah-Virus-in-India-since-2001-What-are.html. Accessed 22 July 2020

Sikka V, Chattu VK, Popli RK, Galwankar SC, Kelkar D, Sawicki SG, Stawicki SP, Papadimos TJ (11 February 2016) The emergence of Zika virus as a global health security threat: a review and a consensus statement of the INDUSEM Joint working Group (JWG). J Glob Infect Dis 8(1):3–15. https://doi.org/10.4103/0974-777x.176140

Soucheray S (Dec 17, 2018) Ebola hits 539 cases as outreach efforts extend in Beni. CIDARP News. https://www.cidrap.umn.edu/news-perspective/2018/12/ebola-hits-539-cases-outreach-efforts-extend-beni. Accessed 22 July 2020

Trifonov V, Khiabanian H, Rabadan R (July 2009) Geographic dependence, surveillance, and origins of the 2009 influenza A (H1N1) virus. New Engl J Med 361(2):115–119. https://doi.org/10.1056/nejmp0904572

United Nations (2018) World Urbanization Prospects 2018. Department of Economic and Social Affairs. https://population.un.org/wup/. Accessed 26 July 2020

WHO (2006) Avian influenza. https://www.who.int/influenza/resources/avian_influenza/en/. Accessed 16 Apr 2020

WHO (16 April 2016) Chikungunya Fact sheet. WHO. Archived from the original on 27 September 2016. Accessed 18 July 2020

WHO (2017) The global health observatory. World Health Organization. https://www.who.int/data/gho/data/indicators. Accessed 25 Jul 2020

WHO (2018) The global health observatory. World Health Organization. https://www.who.int/data/gho/data/indicators. Accessed 16 Aug 2020
WHO (24 February, 2020) Social stigma associated with COVID-19. https://www.who.int/docs/default-source/coronaviruse/covid19-stigma-guide.pdf. Accessed 2 Sept 2020
World Bank (2018) Population density (people per sq km of land area). https://data.worldbank.org/indicator/EN.POP.DNST. Accessed 18 Aug 2020

R. B. Singh is the Former Professor of Geography at the Delhi School of Economics, University of Delhi; the Secretary General and Treasurer of the International Geographical Union (IGU); the chair of Council for Scientific and Industrial Research (CSIR)–Central Food Technological Research Institute of the Government of India, and a member of the International Council for Science (ICSU) and the scientific committee of Urban Health and Well-Being. He was awarded the prestigious Japan Society for the Promotion of Scientific Research Fellowship and has presented papers and chaired sessions in more than 40 countries. He has published 15 books, 40 edited research volumes, and more than 215 research papers.

Mukunda Mishra is an Assistant Professor and the designated Vice Principal of Dr. Meghnad Saha College in West Bengal, India. Dr. Mishra completed his postgraduate studies in Geography and Environmental Management at Vidyasagar University by securing the top rank in both the B.Sc. and M.Sc. panels of merit. He was awarded a Ph.D. in Geography from the same University. He was selected for the prestigious National Merit Scholarship by the Ministry of Human Resource Development, Government of India. His research chiefly focuses on analyzing unequal human development and on creating multi-criteria predictive models.

Chapter 2
Lessons from the Success Story: The Systematic Approach to Fight Against COVID-19 in Beijing

Yi Han, Jie Li, Kun Jia, Yanxu Liu, and Wenwu Zhao

Abstract Coronavirus disease 2019 (COVID-19) has become a global crisis. Sharing experiences in combating the COVID-19 pandemic and seeking international cooperation is an effective way to facilitate countries' search for appropriate measures to prevent and control the pandemic. The new local confirmed case of COVID-19 suddenly appeared in Beijing on June 11, when there are 56 consecutive days with no new local cases but rapidly controlled in a short period. Its systematic strategy to combat the pandemic is worth sharing and learning from. Beijing's pandemic prevention and control measures can be summarized in the "coordination, classification, and collaboration" approach. Based on this prevention and control strategy, Beijing's anti-pandemic work coordinated the government, society, and individuals' forces. It divided the critical works of pandemic prevention and control, coordinated the conflict between anti-pandemic work and production life, and finally achieved the smooth anti-pandemic work. Beijing's experience in fighting the pandemic can provide one of the solutions to mitigate the global COVID-19 pandemic crisis.

Keywords Pandemic · Containment · Coordination · Classification · Collaboration · Resurgence

Y. Han · Y. Liu · W. Zhao (✉)
State Key Laboratory of Earth Surface Processes and Resource Ecology, Faculty of Geographical Science, Beijing Normal University, Beijing, China
e-mail: zhaoww@bnu.edu.cn

Faculty of Geographical Science, Institute of Land Surface System and Sustainable Development, Beijing Normal University, Beijing, China

J. Li · K. Jia
State Key Laboratory of Remote Sensing Science, Faculty of Geographical Science, Beijing Normal University, Beijing, China

M. Mishra and R. B. Singh, *COVID-19 Pandemic Trajectory in the Developing World*, Advances in Geographical and Environmental Sciences,
https://doi.org/10.1007/978-981-33-6440-0_2

2.1 Introduction

Coronavirus disease 2019 (COVID-19) is a respiratory condition caused by a coronavirus and has been sweeping the world since December 2019 (Zhou et al. 2020). According to the data of Johns Hopkins University in the United States, as of 27 August 2020, the total number of confirmed COVID-19 cases worldwide had reached 24,203,815, with 826,380 deaths (Johns Hopkins University 2020). The pandemic has spread to almost every region of the world, and more than 200 countries, territories, and areas have reported confirmed COVID-19 cases. Despite varying degrees of global response and preparedness, the global pandemic is on the rise, with more than 200,000 new cases diagnosed every day, and the space for transmission is expanding. There is no doubt that the outbreak of the COVID-19 pandemic and its global spread has severely affected the public health and well-being (Yao et al. 2020), which is one of the United Nations Sustainable Development Goals (SDG).

Beijing, one of the world's most populous cities, has experienced a renewed occurrence of COVID-19, even though the COVID-19 pandemic, which broke out in early 2020, was brought under control. However, within only 24 days since the first confirmed case appeared on June 11th, Beijing's COVID-19 pandemic is rapidly brought under control. At present, the prevention and control of the pandemic in Beijing have entered normalcy; students at primary, secondary, and university levels are gradually returning to school, the economy has gradually been recovering, and social operation is on the right track. Behind the rapid control of the COVID-19 Pandemic in Beijing, the methods and experiences in combating the pandemic are worth sharing for global reference (Kokudo and Sugiyama 2020).

The pandemic occurred on 11 June 2020, when the first new confirmed cases being reported in Beijing after 56 consecutive days with no new local confirmed cases.[1] Since the beginning, the cumulative confirmed COVID-19 cases are clustered in the Huaxiang area of Fengtai District, where the Xinfadi market is located, followed by a few cases in Daxing District, Dongcheng District, Xicheng District, and Chaoyang District. Most of those new confirmed cases were appeared between 14 and 21 June, with a steady decline from 28 June to 5 July. The Beijing municipal government, after the pandemic had been reported, immediately took strong preventive and control measures to quickly cut off the chain of transmission of the virus, effectively controlling the further spread of the pandemic in a short period. Although the COVID-19 epidemic from 11 June in Beijing showed a rapid expansion trend from the very beginning, the epidemic did not cause widespread spread thanks to the rapid emergency response and scientific prevention and control measures. After two stages of pandemics, Beijing's experience in dealing with the COVID-19 pandemic has been further validated, based on lessons learned. The experience of COVID-19 prevention and control in Beijing can provide a model for China and the world in the fighting against COVID-19 Pandemic.

---

[1] See: http://wjw.beijing.gov.cn/wjwh/ztzl/xxgzbd/gzbdyqtb/202006/t20200612_1923481.html.

2.2 Response of Beijing to the COVID-19 Pandemic

2.2.1 Specific Prevention and Control Measures

After the reporting of new confirmed local cases in Beijing on 11 June, the government immediately organized professionals to conduct an epidemiological investigation and quickly pinpointed the Xinfadi Agricultural Products wholesale market as the source of the outbreak. The city and county Centers for Disease Control and Prevention organized professional staff to inspect the city's wholesale produce markets and supermarkets on 12 and 13 June, and the Xinfadi market was closed that night to isolate the source of the virus quickly.[2] At the same time, the government quarantined the people associated with the Xinfadi market and closed the community around the Xinfadi market. On 14 June, the government organized medical personnel to conduct nucleic acid tests for people associated with the Xinfadi Market and urged health insurance providers to strictly manage fever outpatient clinics, stressing the need to strengthen infection control in hospitals.[3] In order to fully control the spread of COVID-19 pandemic, on 16 June, Beijing's response level for public health emergencies was adjusted from Level 3 to Level 2,[4] and 15 strengthened prevention and control measures were announced (Table 2.1).

Beijing government suspended university back-to-school and suspended primary and secondary school classes on 17 June to avoid infections in densely populated schools. A certificate of a negative nucleic acid test was notified as a mandatory requirement for personnel leaving Beijing from 18 June to stop the cross-regional transmission of COVID-19 caused by the movement of personnel. Nevertheless, 103,000 couriers had been tested for nucleic acids as of 20 June to control the spread of COVID-19 by service workers in close contact with the public, along with 1.18 million service workers was tested for nucleic acids as of 29 June. In order to ensure the coverage of nucleic acid testing and to screen potential cases where possible, the number of nucleic acid testing facilities was expanded to 124 as of 21 June. The price of a single nucleic acid test at a public nucleic acid testing facility was reduced from ¥180 to ¥120 on 25 June. In light of the pandemic that occurred on 11 June, the government has requested that the city's high-quality medical resources be concentrated on treating patients and further strengthening the prevention and control of hospital-acquired infections. On 29 June, the government stressed that it would strengthen medical services during the normalization of pandemic prevention and control.

Furthermore, the government improved the prevention, control, and supervision of hospital-acquired infections to ensure the quality of medical services and medical safety.[5] Under the unified deployment of the government, Beijing quickly responded

[2]See: http://wjw.beijing.gov.cn/xwzx_20031/wnxw/202006/t20200613_192447-4.html.

[3]See: http://wjw.beijing.gov.cn/xwzx_20031/wnxw/202006/t20200615_1-924869.html.

[4]See: http://wj-w.beijing.gov.cn/xwzx_20031/wnxw/202006/t20200617_1926948.html.

[5]See: http://wjw.beijing.gov.cn/xwzx_20031/wnxw/202006/t20200629_1935075.html.

Table 2.1 Fifteen measures that must be strengthened under the second level of response

1. Strict control measures at farmers' markets, food markets, restaurants, establishments, cafeterias, etc	2. Strict inspection and quarantine of food at ports
3. Strictly control personnel leaving Beijing	4. Restoration of closed community management
5. Adjustment of public transportation ratios to control full load rates	6. Implementation of comprehensive nucleic acid testing of personnel and close contacts in priority areas
7. High school students stop returning to school, primary and secondary schools resume online instruction	8. Body temperature monitoring must be done in public places and construction sites
9. Indoor public places and parks and scenic areas are limited to 30% of the flow, and the flow is limited by appointment for a specific time period	10. Enforcement of strict closed management of custodial, old age, welfare, mental health, and other special institutional premises
11. Strictly control the size of offline meetings and stop mass gathering events	12. Maintaining the outpatient appointment system
13. Cease the opening of the domestic inter-provincial group travel business	14.Wear a mask scientifically and carry a mask at all times
15. Do not congregate if not necessary, and prevent people from bunching up	

Source Xinhuanet News (View details at http://www.bj.xinhuanet.com/2020-06/17/c_1126123440.htm)

with precise prevention and control. It took strong measures at the very initial stage of the outbreak so that the pandemic could be effectively controlled in a short period, and by 5 July, all the new local confirmed cases were cleared. On 20 July, Beijing's emergency response level was resumed from Level 2 to Level 3.

2.2.2 *Effectiveness of Preventive and Control Measures*

Based on summing up the experience and lessons learned in the prevention and control of the pandemic since the second occurrence in Beijing, the relevant departments of the municipal government quickly deployed the correct pandemic prevention and control strategies. It responded to the outbreak in a timely and rapid manner (Table 2.2). The involvement of a series of precise prevention and control measures to efficiently control the further spread of the pandemic achieved remarkable success in pandemic prevention and control. The highest number of new confirmed cases was recorded on 13 and 14 June, after it had become clear on 12 June that the Xinfadi market was the source of infection and the relevant contacts were screened (Fig. 2.1). With the implementation of prevention and control measures, the number of newly diagnosed cases is on the decline. However, the number of new cases fluctuates, and

Table 2.2 Specific prevention and control measures of Beijing to the COVID-19 pandemic

Date	Prevention and control measures	Source
12 June	Identified the source of infection is Xinfadi market	http://wjw.beijing.gov.cn/xwzx_20031/wnxw/202006/t20200613_192447-4.html
13 June	Closed Xinfadi market	http://wjw.beijing.gov.cn/xwzx_20031/wnxw/202006/t20200613_192447-4.html
14 June	Nucleic acid testing for personnel involved in the Xinfadi market	http://wjw.beijing.gov.cn/xwzx_20031/wnxw/202006/t20200615_1-924869.html
16 June	Response level adjusted from level 3 to level 2	http://wjw.beijing.gov.cn/xwzx_20031/wnxw/202006/t20200617_1926948.html
17 June	School closed	http://www.xinhuanet.com/politics/2020-06/16/c_1126122450.htm
19 June	Need a negative nucleic acid test to leave Beijing	http://www.xinhuanet.com/bj/2020-07/20/c_1126260349.htm
20 June	103,000 couriers tested for nucleic acids	http://www.mot.gov.cn/jiaotongyaowen/202006/t20200622_3323204.html
21 June	Nucleic acid testing institutions increased to 124	http://wjw.beijing.gov.cn/xwzx_20031/wnxw/202006/t20200621_1929639.html
25 June	Reduction in nucleic acid testing fees in public health facilities	http://www.xinhuanet.com/local/2020-06/28/c_1126168945.htm
29 June	Nucleic acid testing of 1.18 million service personnel completed	http://www.xinhuanet.com/2020-06/30/c_1126175274.htm
3 July	1,059,000 nucleic acid tests completed in the city	http://www.xinhuanet.com/bj/2020-07/20/c_1126260349.htm

Source Compiled by the authors

on 1 July, for the first time, the number of newly cured cases exceeded the number of newly diagnosed cases in a single day. On 6 July, the COVID-19 in Beijing was cleared of new local confirmed cases. On 8 July, the streets of Nanyuan, Nanyuan Township in Fengtai District, and Guanyinsi in Daxing District were adjusted from medium-risk to low-risk areas. As of 24:00 on 8 July, there was only one high-risk area and 15 intermediate-risk areas in Beijing, virtually eliminating the risk of continued illness in exposed individuals, and high-risk environments and populations have been effectively screened.

As of 20 July, the city had no new confirmed cases for 14 consecutive days, no community cases reported for 16 consecutive days, and all high-risk streets and villages were cleared.[6] As of 6 August, all hospitalized cases of this period had been cleared, and the second occurrence of COVID-19 Pandemic in Beijing was now officially complete.[7] Beijing had learned prevention and control measures of COVID-19 through its experience from the previous round of pandemic, which was effectively applied during the second occurrence. The city authority was able

[6] http://www.gov.cn/xinwen/2020-07/20/content_5528408.htm.

[7] http://wjw.beijing.gov.cn/xwzx_20031/wnxw/202008/t20200807_1975976.html.

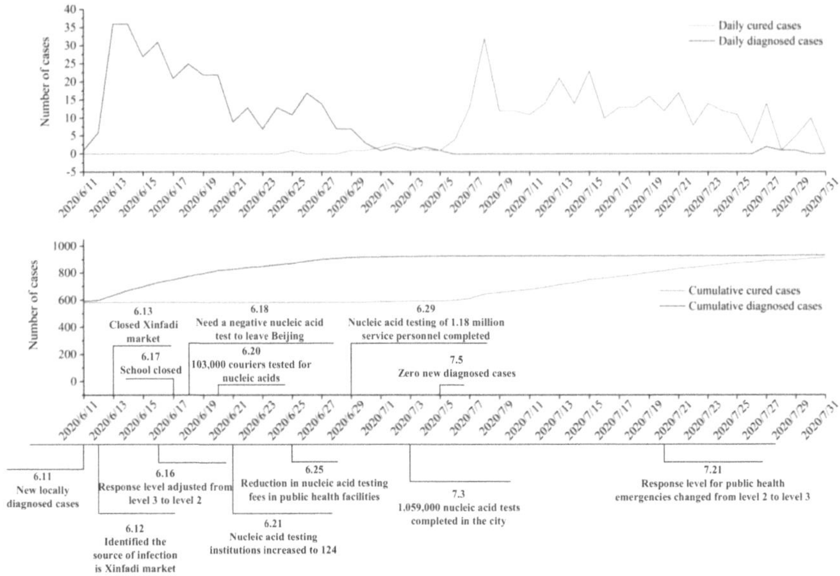

Fig. 2.1 COVID-19 development trend and its countermeasures in Beijing

to control the pandemic within a short period through keeping it contained from spreading widely. Beijing's experience in pandemic prevention is worth considering and learning. It could provide inspiration for countries around the world in dealing with the development of pandemics.

2.3 Lessons of the Prevention and Control of COVID-19 in Beijing

The pandemic prevention and control measures, taken by Beijing, effectively prevented the spread of the COVID-19 pandemic in the city and achieved rapid control of the pandemic. As the global COVID-19 pandemic continues to spread, the pandemic prevention and control measures and ideas were taken by Beijing can provide some reference for other countries to seek their own countermeasure strategies, and thus provide assistance to the global COVID-19 prevention and control.

2.3.1 Cooperation of Government, Society, and Individual

In the face of the resurgence of the COVID-19 pandemic, the concerted efforts of government, society, and individuals are key to bringing the pandemic under control (Johnson et al. 2020). The control of the resurgent of infectious diseases is based on identifying the source of infection. After a new case was diagnosed locally in Beijing on 11 June, the government quickly organized an epidemiological investigation and targeted the source of infection—Xinfadi Market—within 24 h. In the early morning of 13 June, the government issued an announcement that the Xinfadi market was temporarily closed, and 11 communities around the market were closed for management. After that, the government took a series of targeted measures to prevent and control the pandemic. Government departments play a key role in the development of prevention and control policies, the deployment of personnel, and the proper allocation of resources.

The media, businesses and social groups throughout society are responsible for publicizing, mobilizing, and implementing prevention and control measures in the course of pandemic prevention and control (Chan et al. 2020; Van Bavel et al. 2020). Since the occurring of the pandemic, the news media has conducted in-depth and timely interviews and reports on the process of fighting the Pandemic in Beijing. They reported the story of Beijing's fight against the pandemic, disseminated knowledge of pandemic prevention and control, and responded to social and public concerns. Their work has further enhanced the understanding of the whole society about Beijing's fight against the pandemic, as well as the international community's understanding of China's fight against the pandemic. Since the COVID-19 outbreak in the country, public interest from businesses and civil society groups has played a major role. Among them, donations from enterprises, material donations, airlines, and logistics companies to open a green channel for the transport of protective materials for the prevention and control of the pandemic has provided a huge boost. After the resurgence of the Pandemic in Beijing, in order to safeguard the basic needs of residents, wholesale markets, and supermarket chains have continued to supply daily necessities to residents based on strict pandemic prevention, providing a guarantee for social stability.

The active cooperation of the masses is the key to the implementation of pandemic prevention and control. After government departments and the media publicized the scientific prevention and control of COVID-19, the people consciously implemented their own individual prevention and control measures. These measures include home isolation, wearing masks, and attention to personal hygiene. In addition to their own protection, some of the caring people also actively donated money, assisted the hospital in publishing medical masks, protective clothing, and other medical supplies to help information, and provided free meals for medical workers. The active cooperation of patients is also incredibly important in the prevention and control of the pandemic. During the epidemiological investigation, the first confirmed case of the resurgent in Beijing was asked to recall all the places he had visited from 31 May to 10 June. Then, the 38 people he had come into close contact with all recalled, providing essential information for the authorities to target the Xinfadi market and respond quickly.[8]

The combination of government, society, and individuals has formed a working pattern of coordination among all parties and participation by everyone, achieving smooth progress of pandemic prevention and control work from top-level design to policy implementation. The rational utilization of human resources has been maximized, and the rational and orderly work of pandemic prevention and control has been promoted, which has ultimately led to the rapid and effective control of the COVID-19 Pandemic in Beijing.

2.3.2 Coordination of Anti-pandemic Production and Living

Different with the pandemic firstly occurred during the Spring Festival, the resurgent of the pandemic in Beijing taken place after the full resumption of work and production in the city. Consequently, the pandemic prevention and control measures for the new disease clusters of COVID-19 faced a serious problem of coordination of anti-pandemic work, production, and life. In particular, as Beijing is the political and economic center of the country, the negative social impact of pandemic prevention and control was even more serious. So, the fight against the pandemic and the coordination of production and life deserves priority attention.

The pandemic prevention and control in Beijing have adopted hierarchical management of each district according to the three levels of pandemic risk—high, medium, and low—and has adopted a limited closure strategy as far as possible in order to ensure the sustainable operation of the economy. The authorities began with a comprehensive screening, nucleic acid testing, and home observation of all persons in close contact and association with the Xinfadi market to minimize the negative impact of the pandemic. The second is the closure of some of the priority neighborhoods, with nucleic acid testing in priority areas and for priority groups. Unlike the Pandemic in Wuhan, the masses are also relatively free to leave Beijing for production and living needs. As of 18 June, people could leave Beijing with a seven-day "negative" nucleic acid test certificate, a policy that ensures the necessary population movement to the greatest extent possible.

In terms of life, after the closure of Xinfadi Agricultural Products Wholesale Market, a key hub for the supply of agricultural products in Beijing, the National Development and Reform Commission and the Ministry of Commerce urgently transferred vegetables into Beijing in order to stabilize prices in the vegetable market. In order to protect the basic needs of the public and ensure the health of those working in the service industry, on 20 June, Beijing conducted nucleic acid tests on 103,000 courier delivery personnel. By 29 June, Beijing completed 1.18 million nucleic acid tests for delivery and courier service personnel. On 5 July, the nucleic acid testing of more than 70,000 employees in the beauty and hairdressing industry was completed.

[8]See: http://wjw.beijing.gov.cn/wjwh/ztzl/xxgzbd/gzbdyqtb/202006/t20200613_1924443.html.

Prioritizing nucleic acid testing for service industry workers will protect the livelihood needs of the public and prevent service workers from becoming super-spreaders if they contract the virus.

The pandemic prevention and control in Beijing has, on the basis of the experience gained from the previous round, coordinated as far as possible the impact of the anti-pandemic work on production and life. In turn, while maximizing the effectiveness of prevention and control measures, the basic needs of the masses for a living and livelihood have been safeguarded, which is of positive significance for social harmony and stability.

2.3.3 Combination of Coordination, Classification, and Collaboration

The COVID-19 outbreak prevention and control measures in Beijing can be summarized as a systematic approach of "coordination, classification and collaboration" (Fu et al. 2020). Of these, coordination, classification, and collaboration are interrelated (Zhao et al. 2020). The combination of the three promotes mutual cooperation among the government, society, and individuals, which is conducive to the mutual coordination of anti-pandemic work, production, and life.

The coordination is government-led. In the process of pandemic prevention and control, the roles played by the government, social organizations, and the masses are different. Therefore, based on clarifying the functions of district-level governments, social organizations, and the masses, the Beijing municipal government has coordinated the corresponding anti-pandemic work to ensure the reasonable allocation of anti-pandemic resources. The relevant state departments and the Beijing Municipal Government are the decision-makers. As decision-makers, they first quickly set the general level of prevention and control based on the development of the Pandemic in Beijing, in order to allocate human and material resources appropriately. Subsequently, relevant city and district government departments organized relevant professionals to conduct citywide surveys, actively test and track close contacts, and fully control fever clinics and nucleic acid testing sites in hospitals at all levels. By 20 June, the number of nucleic acid testing facilities had expanded to 124, and the maximum daily testing capacity had increased from 100,000 to more than 230,000.[9] At the same time, the process of building the city's nucleic acid testing capacity and the nucleic acid testing capacity of secondary general hospitals continues. In response to the problems in the prevention and control of hospital infections, city leaders first quickly interviewed the person in charge of relevant medical institutions. Strengthening the supervision and inspection of hospital infection prevention and control work, and strict implementation of medical staff, patients, and accompanying staff of the prevention and control requirements were ensured to maintain the capital's pandemic prevention and control situation. In addition, the municipality

[9] Visit: http://wjw.beijing.gov.cn/xwzx_20031/wnxw/202006/t20200621_1929639.html.

strictly manages the number of visitors to Beijing from medium- and high-risk areas to reduce the risk of infection.

The classification work is the basis of the precise prevention and control, limited closure strategy adopted by Beijing. After the resurgent in Beijing, the area of the resurgent was quickly identified and communities and streets were classified according to the severity of the spread of the disease. From 11 June to 1 July, a total of 11 districts and 47 neighborhoods in Beijing found confirmed cases of COVID-19, with the highest peaks in the streets of Huaxiang and Xincun in Fengtai District, Xihongmen Town and Huangcun Town in Daxing District, and Yongding Road in Haidian District. There are five high-risk areas in the street and 39 medium-risk areas in the Xicheng district, including Yuetan Street. Beijing has adopted strict control measures for communities and villages under the jurisdiction of medium- and high-risk cities and towns (Fig. 2.2). Conducting large-scale nucleic acid tests on key populations, key regions, and key areas, such as high-risk individuals in Xinfadi market and surrounding communities, residents of medium- and high-risk streets and villages, and employees in services such as catering, supermarkets, wholesale markets for agricultural products, logistics, and takeaway services.

The orderly collaboration of the government, social organizations, and individuals is a necessary condition for the implementation of Beijing's pandemic prevention

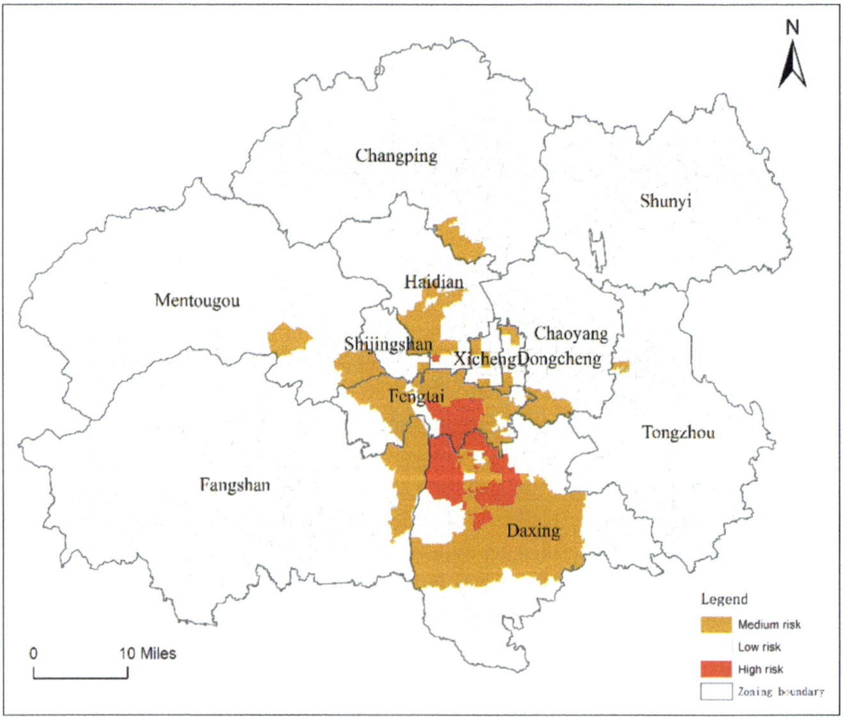

Fig. 2.2 Distribution of risk levels in key areas of Beijing (as of 17:00 on 26 June)

and control policies. In pandemic prevention and control, the government is mainly responsible for overall planning and decision-making; individuals mostly carry out pandemic prevention and control work on their own initiative, and social organizations can participate in pandemic prevention and control in a relatively orderly manner. After the second Pandemic in Beijing, the government quickly mobilized social forces to participate in the prevention and control of the pandemic. Social organizations, on the other hand, are actively involved in the prevention and control of the pandemic on the basis of conducting their own personnel checks. For example, they were assisting community workers with outreach and surveys, delivering household items to people in home isolation, and providing psychosocial support services. The general public, on the other hand, consciously fulfilled their responsibilities, reduced travel, protected themselves, and actively cooperated in the prevention and control of the pandemic. Many people joined the volunteer ranks and did their best to assist in the control of the pandemic. In addition, many cured patients actively donated plasma for clinical treatment. The whole community actively participated and contributed to the victory of this "people's war."

Based on the integration of coordinated, classified, and collaborative responses, Beijing has formed a systemic pandemic prevention and control system. Beijing will focus on solidifying a series of proven institutional measures that have been explored and developed in Beijing's pandemic prevention and control practices, in response to the problems and shortcomings that have been exposed in pandemic prevention and control. In addition, the Beijing Regulations on Emergency Response to Public Health Emergencies is in the process of being formulated, and this upcoming local regulation will provide solid legal safeguards for responding to public health emergencies in the capital.

2.4 Suggestions

Under the systematic anti-pandemic work of Beijing, the COVID-19 Pandemic from 11 June in Beijing was quickly brought under control. In the face of the global COVID-19 crisis, each country needs to explore its own countermeasures on a widely shared basis. The experience gained in Beijing can provide some references for the global fight against the pandemic. Here we highlight five measures that countries can take for reference (Fig. 2.3).

2.4.1 Strengthening the Coordinate Capacity of Government

Government coordination is an effective means of rationally allocating resources for combating epidemics. The basis of government coordination is a sound management system. So it is necessary to focus on strengthening the public health emergency

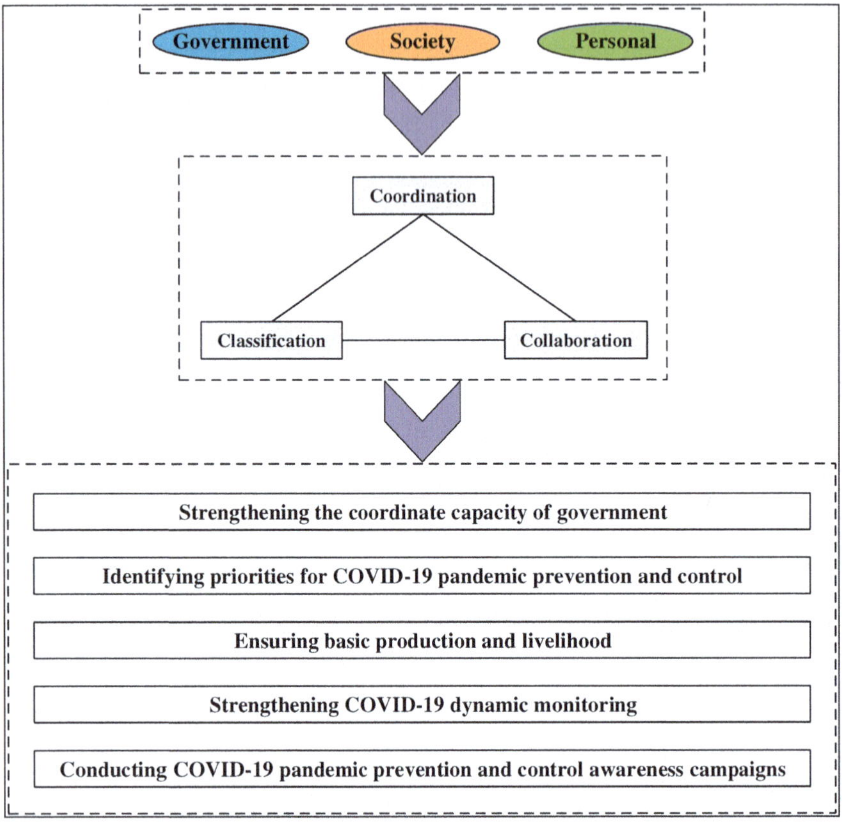

Fig. 2.3 Suggestions based on the systematic approach to combating epidemics of Beijing

management system, coordinating the roles of the government, society, and individuals in the prevention and control of epidemics, and strengthening the ability to deal with public health emergencies, especially epidemics of infectious diseases. Attempts can be made to formulate regulations on emergency management of public health events to clarify the powers and responsibilities of the government in the event of a pandemic of a public health emergency and to clarify the basic procedures for handling public health emergencies.

2.4.2 *Identifying Priorities for COVID-19 Pandemic Prevention and Control*

A clear focus on pandemic prevention and control is the basis for classification and collaboration. The focus of epidemic prevention and control is not limited to identifying key regions and key populations but also includes identifying key environmental factors for epidemic development and key nodes for epidemic prevention and control. In the face of a sudden epidemic outbreak, we can first try to identify the source of the outbreak, after which we need to strengthen the epidemiological investigation, screen, and control the flow of close contacts while identifying the source of infection of the epidemic. It will be followed by the implementation of the responsibilities of the territory, departments, units, and individuals after identifying the key areas and key nodes for epidemic prevention and control according to the environmental conditions and the development of the epidemic.

2.4.3 *Ensuring the Essential Production and Livelihood*

Epidemic control inevitably affects the movement of factors of production, the mobility of people, and the supply of subsistence goods. Prolonged and irrational epidemic prevention and control measures may cause economic stagnation, shortages of subsistence goods, and other problems. Therefore, epidemic prevention and control need to consider the principle of safeguarding production and life. With the trend toward normalization of epidemic prevention and control, prevention and control measures need to take greater account of the impact on production and life in order to ensure normalization of production and life.

2.4.4 *Strengthening COVID-19 Dynamic Monitoring*

COVID-19 cannot be eradicated in the short term, so the world will continue to be threatened by COVID-19 for some time. Although some countries, including China, have achieved better performances in combating COVID-19, the threat of COVID-19 still exists. Therefore, dynamic surveillance of COVID-19 needs to be strengthened regardless of the effectiveness of national outbreak control. This dynamic monitoring is, on the one hand, an overall assessment of localized COVID-19 outbreaks and, on the other hand, a risk assessment of COVID-19 outbreaks based on population movements and environmental factors, with regular sampling of high-risk areas.

2.4.5 Conducting COVID-19 Pandemic Prevention and Control Awareness Campaigns

The masses are the backbone of epidemic prevention and control, and epidemic prevention and control requires everyone in society to protect themselves and hopes that more people will participate in epidemic prevention and control. On the one hand, the dissemination of knowledge on epidemic prevention and control can lead the public to adopt healthy behaviors and actively take personal precautions. On the other hand, it can enhance the acceptance of epidemic prevention and control measures, which is conducive to mobilizing social forces to participate in epidemic prevention and control.

At present, the COVID-19 pandemic has become a global crisis, and the search for reasonable prevention and control measures is a challenge to control the further spread of the COVID-19 pandemic. The success of COVID-19 pandemic prevention and control in Beijing proves that systematic prevention and control measures can be effective in stopping the spread of COVID-19. Therefore, all countries can learn from each other's experience in fighting pandemics and cooperate with each other. This is the way to overcome the pandemic and ensure the sustainable development of human society.

Acknowledgements This work was supported by the National Key R&D Program of China (2020YFC0849100), the Alliance of International Science Organizations (No. ANSO-SBA-2020-01), and the Fundamental Research Funds for the Central Universities of China.

References

Chan A, Nickson CP, Rudolph JW, Lee A, Joynt GM (2020) Social media for rapid knowledge dissemination: early experience from the COVID-19 pandemic. Anaesthesia 75(12):1579–1582

Fu BJ, Zhang JZ, Wang S, Zhao WW (2020) Classification–coordination–collaboration: a systems approach for advancing sustainable development goals. Natl Sci Rev 2020(5):5

Johnson T, Dawes C, Fowler J, Smirnov O (2020) Slowing COVID-19 transmission as a social dilemma: Lessons for government officials from interdisciplinary research on cooperation. J Behav Public Adm 3

Kokudo N, Sugiyama H (2020) Call for international cooperation and collaboration to effectively tackle the COVID-19 pandemic. Glob Health Med 2:60–62

University JH (2020) COVID-19 dashboard by the Center for Systems Science and Engineering (CSSE) at Johns Hopkins University (JHU). https://coronavirus.jhu.edu/map.html. Accessed 27 Aug

Van Bavel JJ, Baicker K, Boggio PS, Capraro V, Cichocka A, Cikara M, et al. 2020. Using social and behavioural science to support COVID-19 pandemic response. Nat Human Behav 4:460–471

Yao H, Zuo X, Zuo D, Lin H, Huang X, Zang C (2020) Study on soybean potential productivity and food security in China under the influence of COVID-19 outbreak. Geogr Sustain 1(2):163–171

Zhao W, Zhang J, Meadows ME, Liu Y, Hua T, Fu B. (2020) A systematic approach is needed to contain COVID-19 globally. Sci Bull 65(11):876–878

Zhou C, Su F, Pei T, Zhang A, Du Y, Luo B, Cao Z, Wang J, Yuan W, Zhu Y (2020) COVID-19: challenges to GIS with big data. Geogr Sustain 1(1):77–87

Yi Han is a Ph.D. student major in the natural resource of the Faculty of Geographical Science, Beijing Normal University. His research focuses on the ecological and environmental effects of land use. He has published 3 theses on *Catena, Science of the Total Environment, Computational Intelligence and Neuroscience,* and 18 theses in Chinese journals such as *Acta Ecologica Sinica, Journal of Plant Nutrition, and Fertilizers, China Land Science*, and *Chinese Journal of Applied Ecology.*

Li Jie is a postgraduate student major in Cartography and Geographic Information System of the Faculty of Geographical Sciences, Beijing Normal University. Research orientation is vegetation remote sensing and ecological remote sensing.

Kun Jia is an associate professor of the State Key Laboratory of Remote Sensing Science, Faculty of Geographical Science, Beijing Normal University. His research focuses on quantitative remote sensing, remote sensing for the ecosystem, and spatial analysis. He is a Tang Scholar at Beijing Normal University. He has published more than 80 theses on international journals such as *Remote Sensing of Environment, IEEE Transactions on Geoscience and Remote Sensing*, and *ISPRS Journal of Photogrammetry and Remote Sensing* and Chinese journals such as *Journal of Remote Sensing, Advances in Earth Science*, and *Resources Science.*

Yanxu Liu is a lecturer of the Faculty of Geographical Science, Beijing Normal University. His research focuses on remote sensing of landscape processes, evaluation of landscape multifunctionality, and Spatial planning for landscape sustainability. He has more than 40 theses on international journals such as *Journal of Cleaner Production, Habitat International, Landscape and Urban Planning, Ecological Indicators, Landscape Ecology*, and Chinese journals such as *Acta Geographica Sinica, Acta Ecologica Sinica, Journal of Natural Resources*, and *Resources Science.*

Wenwu Zhao is a professor of Faculty of Geographical Science, Beijing Normal University. His research focuses on interdisciplinary research, including land use and ecosystem services, human–environmental system, and sustainability. He has published more than 150 papers on international journals such as *Science Bulletin, Journal of Cleaner Production, Catena, Science of the Total Environment, Land Degradation and Development, and Chinese* journals such as *Acta Geographica Sinica, Acta Ecologica Sinica, Progress in Geography, Acta Pedologica Sinica*, and *Journal of Soil and Water Conservation.*

Chapter 3
The Geography of COVID-19 in Rio de Janeiro, Brazil: Conflicts, Tensions, and Challenges

Andrews José de Lucena, Leandro Dias de Oliveira, Pablo Ibanez, Gustavo Mota de Sousa, and André Santos da Rocha

Abstract The frightening capacity for contamination and the impossibility of a medical-scientific response to tackle the coronavirus's spread revealed the fragility of the human race in the face of a powerful invisible agent. To date, there are more than twelve million people infected, and more than half a million deaths worldwide. 2020 has already become a "historic" year when we need to change our relationship with nature and our society's priorities. In addition to the poor historical conditions of basic sanitation and housing and the concentration of income in Brazil, the virus found some critical ally, which the present chapter will focus. Even having a public and universal Unified Health System (UHS), Brazil, unfortunately, became an international highlight in the number of contaminated and killed, becoming a true negative reference in the fight against the pandemic. In Rio de Janeiro, we endeavored to draw a panel of the COVID-19 pandemic impact on its territory. From its entry via global contacts in the more affluent areas, the "virus of globalization" gradually reached the suburban areas and, finally, the metropolitan peripheries which are more impoverished and susceptible to problems. In this sense, the Fluminense Lowland, formed by thirteen cities and where the Federal Rural University of Rio de Janeiro (UFRRJ) is located, has become a favorable space for contamination and dispersion of the pandemic, promoting a dynamic of accelerated expansion and aggressive lethality. Visiting the pandemic's impacts on the metropolitan peripheries is the great research challenge for all researchers committed to the community and social justice.

Keywords COVID-19 · Metropolitan Area of Rio de Janeiro/Brazil · Federal Rural University of Rio de Janeiro · Health policies · Urban periphery

---

A. J. de Lucena (✉) · G. M. de Sousa
Integrated Laboratory of Applied Physical Geography (LIGA), Federal Rural University of Rio de Janeiro (UFRRJ), Seropédica, Brazil
e-mail: lucenageo@gmail.com

L. D. de Oliveira · P. Ibanez · A. S. da Rocha
Laboratory of Economic and Political Geography (LAGEP), Federal Rural University of Rio de Janeiro (UFRRJ), Seropédica, Brazil

M. Mishra and R. B. Singh, *COVID-19 Pandemic Trajectory in the Developing World*, Advances in Geographical and Environmental Sciences,
https://doi.org/10.1007/978-981-33-6440-0_3

3.1 Introduction

The pandemic from the new coronavirus reveals the contradictions in the world-system. This is chiefly because of the precariousness of the accessibility to the health system—both the public and private; and also because of the lousy habitation conditions of the most part the population, especially in Latin America, Asia, and Africa.

Far from what they speak on the coronavirus, the pandemic is not democratic. This had reached the most economic impacts on the poor class in all countries in the world. This asymmetric geography is materialized in the expansion to peripheral spaces. There is real territoriality the conflicts in space when we look to the microscale. In this, we identify the actors, the dynamics, and flows of contaminations in the territory. Analysis of epidemic data is crucial for understanding these aspects of the expansion of disease and our relations with the dynamics of society's realities and the environment.

In Latin American countries, Brazil is an epicenter of contamination of the COVID-19. Studies on the postures the government for combating the pandemic added to Brazilian territory's geographical conditions serve as an example for reflecting the tensions, conflicts, and challenges in the new context. For this, it is essential to understand the process of developing the virus territory to comprehend the dynamic of the noble coronavirus in the country. We thus agree with Santos and Silveira (2001) when they indicate more critical is not the territory but your use. The notion of "the use of territory" means that the process of the formation of the territory is focused. Thus, we cannot ignore the capitalist dynamic of production in the formation of the inequalities, as the pandemic tends to materialize this same unequal geography of use the territory spatially.

In this respect, this paper presents an analysis of the expansion of COVID-19 in Brazil. We have focused on the Rio de Janeiro State to comprehend the conflicts, tensions, and challenges in the new context. For this, the text is divided into three parts. In the first part, entitled "*Pandemic trajectory in Brazil*" we present the information on the COVID-19 on a national territorial scale. We have used official statistics of the governmental and press organizations concerning the expansion of the new coronavirus.

In the second part, "Rio de Janeiro State is a Brazil's representation" we try to show the Rio de Janeiro as the second epicenter of the COVID-19 in Brazil. This section analyzes the historical series using data for by March to September 2020, presenting cartography of the pandemic in the state of Rio de Janeiro.

In the third part, "Metropolitan Area of Rio de Janeiro: a favorable space for contamination and dispersion of the pandemic," we have conducted an impact analysis of the pandemic in the metropolitan area. In this part, we have presented the phenomena of the expansion of COVID-19 in the Baixada Fluminense[1] region, which represents the metropolis peripheral.

Lastly, in conclusion, we point out that the discussions on the process of disease expansion follow the inequality of territorial use and have been growing in direction to the peripheral spaces.

3.2 Pandemic Trajectory in Brazil

On the one hand, the high capacity for contamination and the incapability of the medical-science response to treat the COVID-19 have resulted in the high capacity for contamination. Nowadays, more than thirteen million people are infected, and more than half a million deaths worldwide. 2020 has already become a historic year that we need to change our relationship with nature and our priorities as a societal being.

In Brazil, in addition to the poor historical conditions of basic sanitation and housing and the concentration of income, the virus found an important partner: the Federal Executive, with its inability to promote action together with states and cities in favour of human health. However, it caused misinformation and many conflicts and tensions. Even with a public and universal Unified Health System—UHS, Brazil quickly became an international highlight in the number of contaminations and death toll, to become a valid negative reference in the fight against the pandemic (Table 3.1; Fig. 3.1).

Table 3.1 shows that Brazil is the third country in the world regarding the number of cases, and the second regarding the number of deaths by COVID-19. According to the World Health Organization (WHO), Brazil, India, and the USA are currently the epicenter of the illness in the World.

In Fig. 3.1 the graph shows the Total deaths per day (in yellow bars) and 7-day moving average (in blue line) that best represents the illness's evolution. The disease's highest numbers occur in the months of June, July, and August, with daily death peaks, often above 1400 people. The moving average line defines an accelerated increase in the disease during April and May and stabilizes over the months of June, July, and August. Only at the end of August does the mobile media curve show a slight downward trend, but very far from a "happy ending" and which should not slow down the isolation and care measures, which have been appropriately practiced in most cities in Brazil.

However, Brazil is a continental and complex country, like India, and its particularities must be considered. We can see the complexity and continental peculiarities of Brazil in Figs. 3.2 and 3.3. In Fig. 3.2, Brazilian states are classified according

[1]Fluminense is a gentilic name of the Rio de Janeiro state. All people born in the Rio de Janeiro state call "fluminense". Baixada Fluminense is a political local name formed by cities in Metropolitan Area of Rio de Janeiro (MARJ) or we can translate Baixada Fluminense to "Fluminense Lownland".

Table 3.1 Report coronavirus cases in the world in September 18

Ranking	Country	Total cases	New cases	Total deaths	New deaths	Total recovered	Total tests	Population
	World	30,371,357	+308,206	950,270	+5,568	22,032,021		
1	USA	6,874,596	+46,295	202,213	+879	4,155,039	95,235,022	331,420,450
2	India	5,212,686	+96,793	84,404	+1,174	4,109,828	60,565,728	1,382,900,689
3	**Brazil**	**4,457,443**	**+35,757**	**135,031**	**+857**	**3,753,082**	**14,617,980**	**212,883,816**
4	Russia	1,085,281	+5,762	19,061	+144	895,868	41,748,928	145,948,080
5	Peru	750,098	+5,698	31,146	+95	594,513	3,614,738	33,069,039
6	Colombia	743,945	+7,568	23,665	+187	615,457	3,298,415	50,998,462
7	Mexico	680,931	+4,444	71,978	+300	485,024	1,545,572	129,221,511
8	South Africa	655,572	+2,128	15,772	+67	585,303	3,983,533	59,467,369
9	Spain	654,637	+4,541	30,405	+162	N/A	10,756,835	46,758,719
10	Argentina	601,713	+12,701	12,460	+344	456,347	1,653,616	45,284,429

Data Source https://www.worldometers.info/coronavirus/

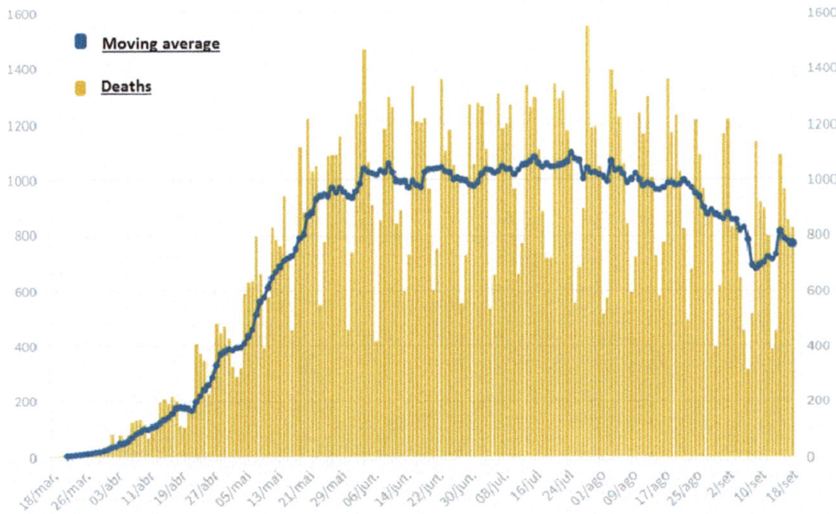

Fig. 3.1 Deaths from COVID-19 per day in Brazil (from 15 March to 18 September 2020) and evolution of the moving average (*Source* Consortium of media outlets in Brazil)

to the number of deaths[2]: falling (green color), stable (yellow color), and high (red color) as on September 18, 2020. Since the beginning of the pandemic, this classification varies daily, showing the spatial differences in Brazil that are due to social, economic standards and public actions in the states in combating COVID-19.

Figure 3.3 shows the number of deaths per city in Brazil. The concentration of deaths is on the coast, where the largest Brazilian cities with a high rate of urbanization developed therein. However, there are frequent cases of the disease in the country's interior, i.e., in the regions of important biomes, such as the Amazon Forest and the Pantanal, which currently, during the months of August and September, suffer from forest fires (BBC 2020; GREENPEACE 2020). Many of these fires are of a criminal nature, and the smoke generated can worsen respiratory diseases.

The first confirmed case of COVID-19 in Brazil was registered on 27 February. Until that moment, little was known about the disease. In March, health authorities and the media began to give greater visibility to the pandemic. The health minister, Henrique Mandetta, with in-depth knowledge in the area, became the main protagonist for the notification and actions to contain the pandemic. Unfortunately, the

[2]The classification of the number of deaths follows the following criteria: researchers indicate the moving average of deaths, which calculates the average of records observed in the last seven days. The technique is best suited to observe the trend of statistics to balance the abrupt changes in numbers over the week. It is possible to talk about a drop in numbers when the decrease is greater than 15% if verified in the last 14 days. If the numbers increase more than 15%, the epidemic accelerates. Intermediate values indicate stability.

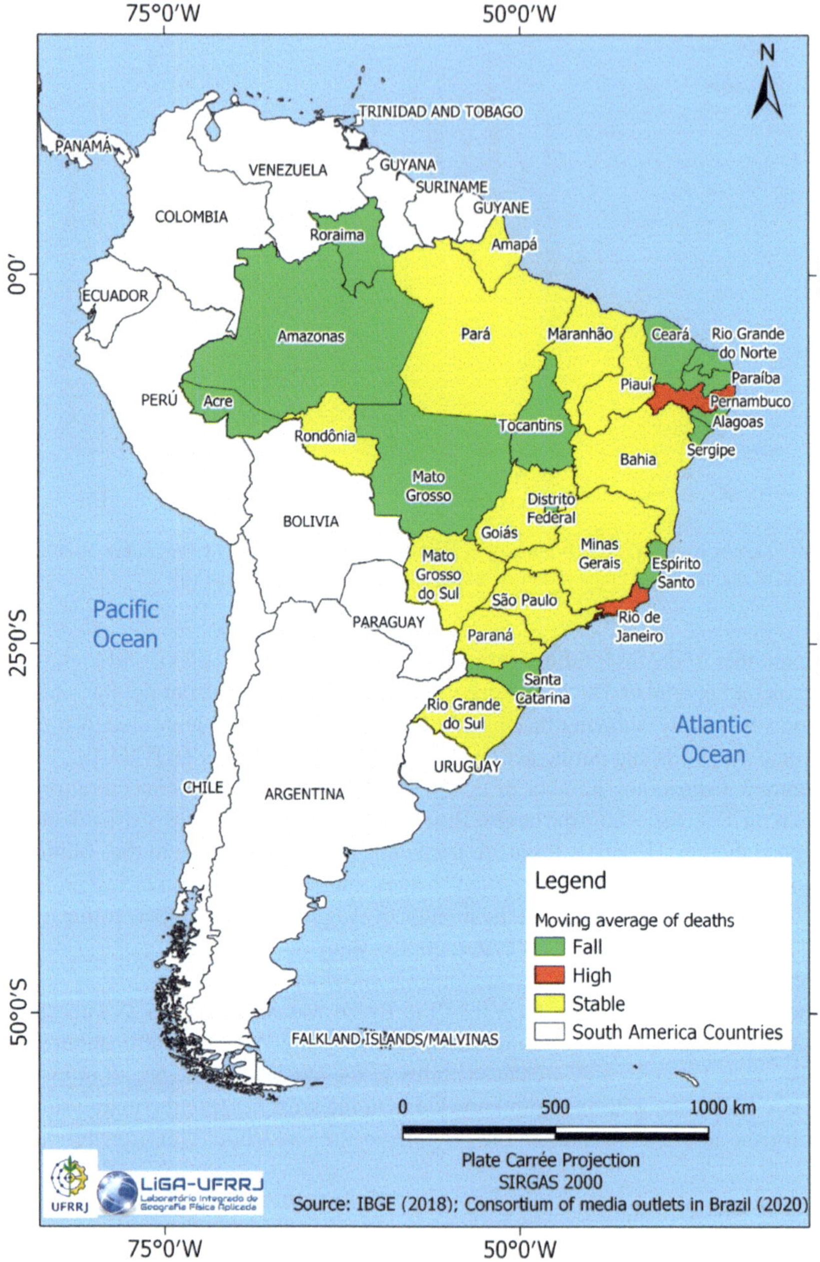

Fig. 3.2 The moving average of deaths in Brazil States on September 18, 2020

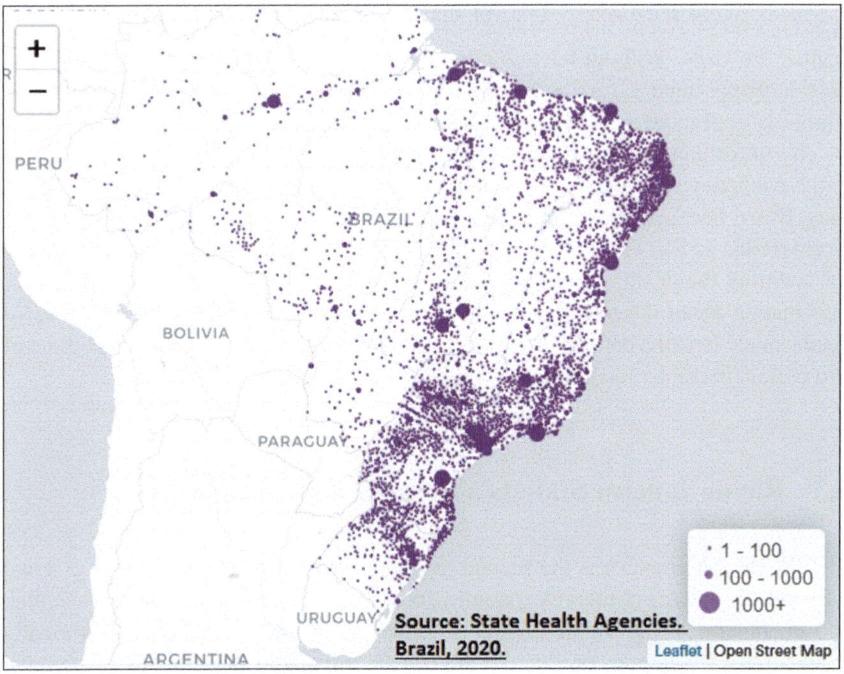

Fig. 3.3 COVID-19 deaths in Brazil by notification city on September 18, 2020

presidency of the republic, President Jair Bolsonaro, began to deny several recommendations from the World Health Organization (WHO) (Estadao 2020) and a political dispute placed Brazil as one of the countries with the expedited spread of the virus, as well as the number of deaths.

Amid political disputes between the health ministry, Mandetta was dismissed from Health Minister on 16 April (BBC 2020). The situation becomes more complicated as the conflict was not only with the guidelines of the Ministry, but also with the Governors and Mayors who started to take more rigid attitudes toward the control of the virus. A debate between containing the virus and preserving the economy started to be the center of attention, while the country witnessed an uncontrolled increase in the number of cases.

Nelson Teich, the newly appointed minister who is also a doctor, was tasked with changing the direction of the Ministry (Folha de São Paulo 2020). Based on the idea that chloroquine would be a medicine capable of curing the disease, in addition to investing in the purchase of the drug from several countries, including India, the federal government started to produce large quantities of the drug. Without proven efficacy, along with enormous pressure from the presidency regarding the use of the drug, Nelson Teich resigned for not agreeing with this protocol (BBC 2020).

Amidst this acute health crisis, health ministry leadership was changed twice within an interval of a month. The number of cases and deaths by COVID-19 was growing every day, and the presidency insisted on chloroquine as an effective remedy against the virus. Without having a doctor to take over the Ministry of Health, the President appointed a Brazilian Army military officer as an interim Health Minister, who lasts until today (CNN Brasil 2020).

This unfortunate political turbulence results in the increase in the number of cases in the pandemic, as well as the denial of a basic protocol: social distancing. On 22 May, Brazil becomes the second place in the number of deaths by COVID-19 in the world (Folha de São Paulo 2020). In June, the Ministry of Health changed the criteria for counting the deaths, and the press consortium became the most used source of information about the pandemic. The spread speed of the virus and the number of deaths made Brazil a country with strict international travel restrictions. The state of Rio de Janeiro ranks second in the number of deaths.

3.3 Rio de Janeiro State Is a Brazil's Representation

The Rio de Janeiro state is the second most important state in Brazil (Figs. 3.4 and 3.5) with significant political, economic, social, and cultural functioning. Its capital, Rio de Janeiro city, was the capital of Brazil for a long time. It has historically concentrated economic, productive, logistical, and social investments of regional

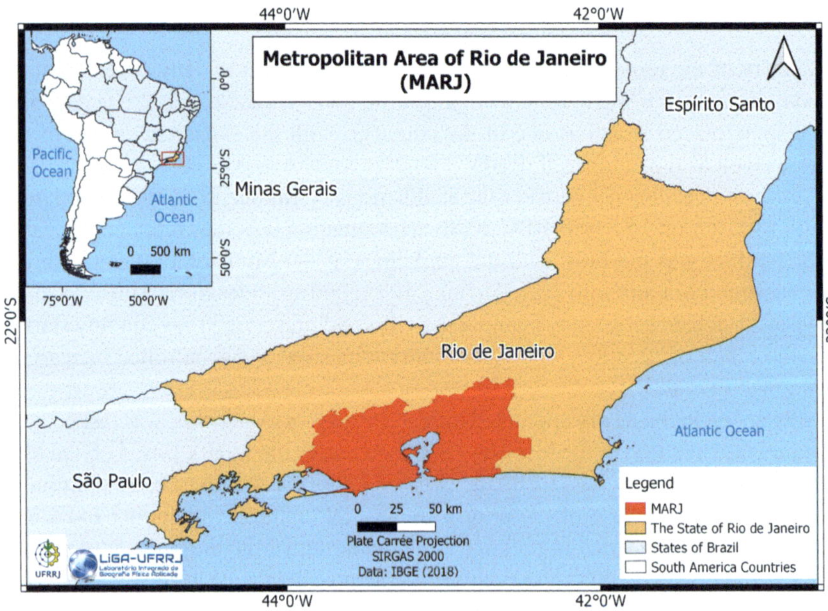

Fig. 3.4 Brazil and Rio de Janeiro State Map: Brazil in grey color; Rio de Janeiro State in orange color; Metropolitan Area in red color

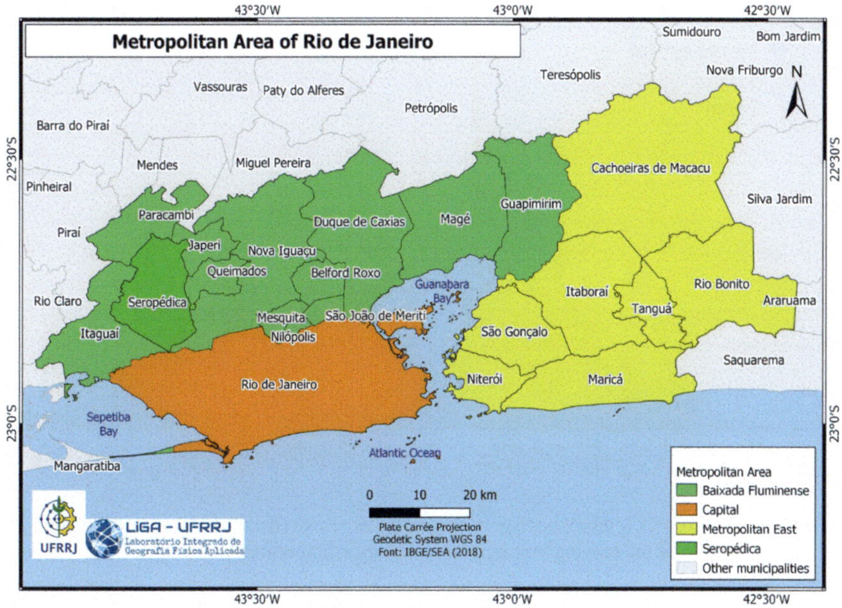

Fig. 3.5 The Metropolitan Area of Rio de Janeiro and its subdivisions: Rio de Janeiro capital (red color); Baixada Fluminense (green color); Seropédica city (dark green color); the cities of the Metropolitan East (yellow color)

integration with the interior. Even with the productive industrial growth of certain interior areas, the city of Rio de Janeiro remains, through its immediate peripheries and with the axes of the metropolitan area, especially with regard to meeting the needs of education and health services.

In the Rio de Janeiro state, the coronavirus arrived through globalized contact networks—international flights. It first contaminated the wealthier citizens who were able to make trips abroad, and their direct employees who, even the poorest, had immediate contact with such travelers. The virus invaded the country through the flow of international flights and reached the city's parts where the middle and upper classes are concentrated.

However, territorial routes of dispersion become robust to fuel community transmission of the disease. Soon, the West Zone districts, especially the overpopulated ones, started to exhibit high numbers, now associated with less favored financial and housing conditions. The myth of the "democratic virus" was lost, as the inhabitants of the most impoverished areas of the city, known as the *favela,*[3] did not achieve minimum housing conditions or access to gel alcohol and other protection items.

[3]It is a space of induced segregation caused by socioeconomic injustice that is very present in the landscape of South American cities. It is marked by many prejudices (for example as a place where all people are criminals). The characteristics are the insufficiency of government investments; the preponderance of poor quality buildings, quite dense and made by the residents themselves; poor education, health, safety, and basic sanitation conditions (Souza e Silva et al. 2009).

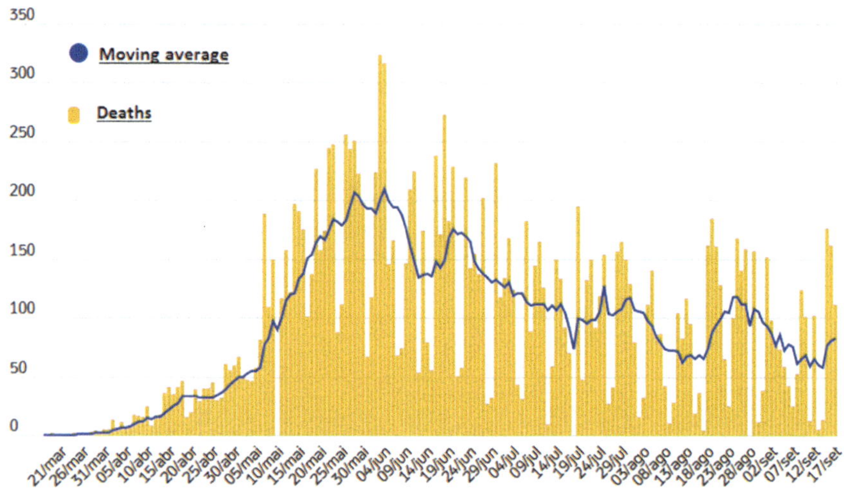

Fig. 3.6 Deaths from COVID-19 per day in Rio de Janeiro State (from 15 March to 18 September 2020) and evolution of the moving average (*Source* Consortium of media outlets in Brazil)

Even though they were aware of the immense underreporting, the numbers started to prove to be more catastrophic and violent, especially with regard to the lethality of the disease.

The main highways of the state also become geographic axes of dispersion of the COVID-19, reaching the Region of the Middle Valley "*Paraíba Fluminense*" (Local Name), the Region of the Lakes (or Coastal Lowlands), and even the North Fluminense. The dispersion of the coronavirus is still a lesson in the economic geography of the metropolis. The same meanings of metropolization in any traditional economic activities also seem to reverberate in the processes of the geographical spread of contamination by COVID-19 in the state (Fig. 3.6).

The graph in Fig. 3.6 shows the total deaths per day (yellow bars) and 7-day moving average (blue line). The disease's peak occurred in the months of May and June, with a daily average of deaths above 200 and peaks over 300. In July, August, and September, the daily deaths did not exceed 200, although there were several days with above 150. The trend line of the moving average shows a high increase during April and May. From June onwards, there is a downward trend until the present moment in September.

However, this fall is not absolute and continuous. It is still necessary to maintain isolation measures with personal health security and social protection to restrain the resumption of the disease's rise. Figure 3.2 illustrates the average daily death by COVID-19 in Brazil on 18 September. The state of Rio de Janeiro maintains a high number of deaths, which are higher than the daily average in Brazil. This fact reveals the need to maintain care and attention with health policies in the fight against COVID-19 and the measures of openness and social interaction that are in progress.

In Rio de Janeiro State, we draw a panel of the COVID-19 pandemic impact on its territory. From its entry via global contacts in the more affluent areas (worthy to say it a virus of globalization), it gradually reached the suburban areas and, finally, the metropolitan peripheries where the population is more impoverished and susceptible to problems. In this sense, the Baixada Fluminense, formed by thirteen cities and where the Federal Rural University is located, has become a favorable space for contamination and dispersion of the pandemic, promoting a dynamic of accelerated expansion and aggressive lethality. Visiting the pandemic's impacts on the metropolitan peripheries is a significant research challenge for all researchers committed to the community and social justice.

Figures 3.7, 3.8, 3.9, 3.10, 3.11, 3.12 and 3.13 will help the readers witnessing the spatial distribution of the evolution of the number of cases, the number of deaths, and the case fatality ratio in the Rio de Janeiro State. In March (Fig. 3.7) just twenty-four cases occurred in the Rio de Janeiro capital and Niterói city, located in the east metropolitan region. In the following months (Figs. 3.8, 3.9, 3.10, 3.11, 3.12 and 3.13) a concentration in the metropolitan area is observed in the number of confirmed cases, deaths, and case fatality ratio. The illness concentration is high in the metropolitan area and spreads, little by little, to other cities in the Rio de Janeiro State.

This situation shows the centralizing role of the metropolitan area of Rio de Janeiro in the state of the same name because it concentrates the density of the

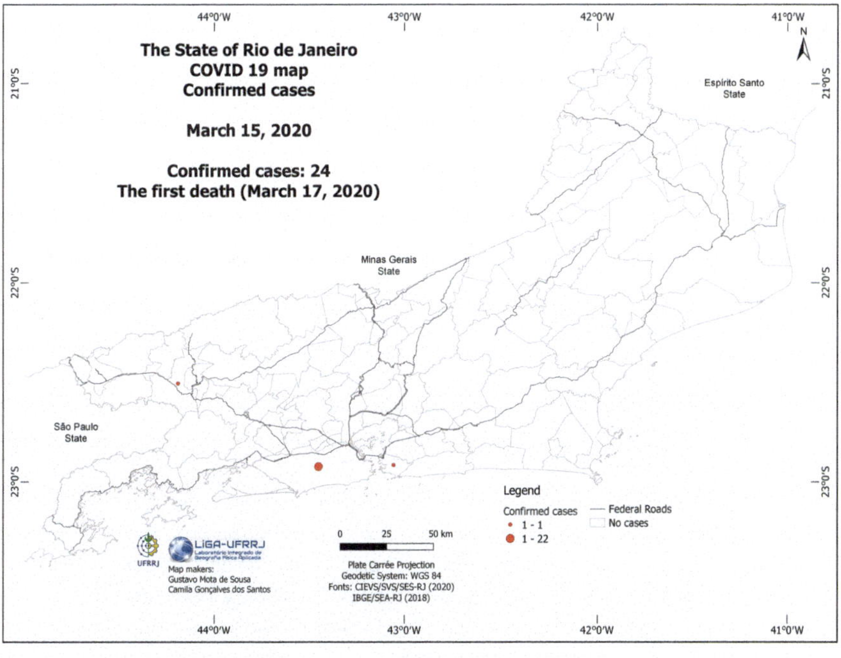

Fig. 3.7 Confirmed cases in Rio de Janeiro State on March 15, 2020

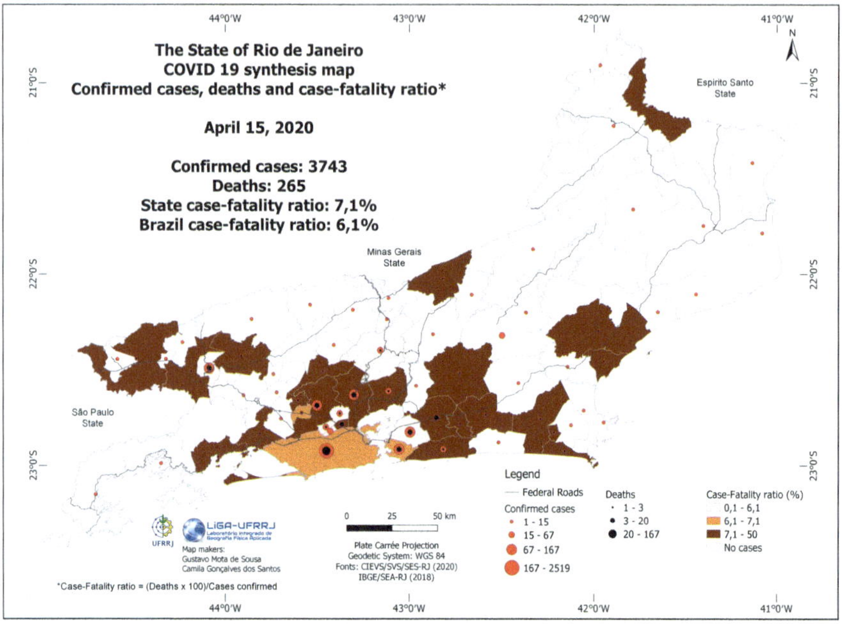

Fig. 3.8 Confirmed cases, deaths numbers, and case fatality (the case fatality ratio is calculated by the number of deaths × 100 over the number of confirmed cases) ratio in Rio de Janeiro State on April 15, 2020

state population, the income, and the main public and private services. Likewise, it concentrates on many problems, conflicts, and challenges that are aggravated in this pandemic moment.

3.4 Metropolitan Area of Rio de Janeiro: A Favorable Space for Contamination and Dispersion of the Pandemic

The Metropolitan Area of Rio de Janeiro has the state capital, the Rio de Janeiro city, Niterói, the second most important city, and the Baixada Fluminense, emphasizing Nova Iguaçu and Duque de Caxias cities. The Baixada Fluminense is an area of great contradictions, where important industrial and logistic companies and other vital investments are located. Besides, this area is also a large contingent of segregated, impoverished spaces, with insufficient basic sanitation and high poverty rates. Baixada Fluminense also has internal differences, especially between the historic Baixada, with denser urbanization, and the portion of the metropolitan west end formed by the Seropédica, Itaguaí, Paracambi, Japeri, and Queimados cities, in addition to Magé and Guapimirim cities, which make up a transition with the Metropolitan

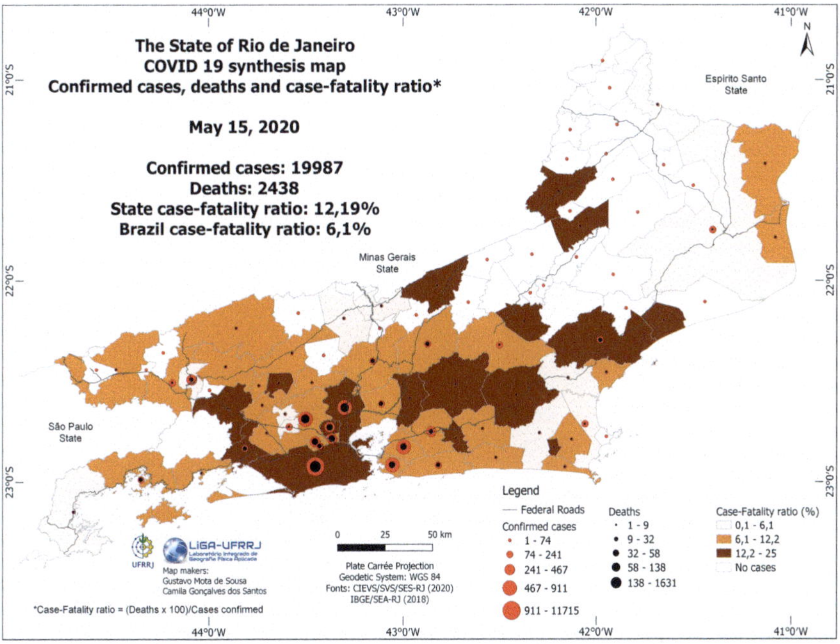

Fig. 3.9 Confirmed cases, deaths numbers, and case fatality ratio in Rio de Janeiro State on May 15, 2020

East. The metropolitan west end has less urban characteristics and a restructuring process supported by recent investments in new logistical and industrial enterprises.

The Metropolitan Area of Rio de Janeiro and Baixada Fluminense present serious structural problems involving lack of basic sanitation (water and sewage), irregular occupations, and deficiency in health equipment, among many others. The Baixada Fluminense is connected to important traffic routes and railway axes destined for the movement of workers and goods, the poor conditions of housing, income, and basic sanitation. Altogether, they placed the Baixada Fluminense as a critical region of COVID-19 expansion in the Rio de Janeiro State.

The host city of the Federal Rural University of Rio de Janeiro (UFRRJ) (Fig. 3.14), Seropédica, is an important site for the analysis. Located at the western end of the Baixada Fluminense, it is an important logistics center. We highlight the Federal Rural University's actions, which isolates its thousands of professors, students, and technical-administrative workers to avoid contamination. The University develops many actions toward its surroundings: the production of masks and alcohol-based sanitizers. It successfully carries out the research, extension, and disseminate information about the pandemic. The UFRRJ plays a very important for the pandemic triggered Baixada Fluminense.

UFRRJ has carried out solidarity actions producing alcohol in gels, masks, and offering organic food to the poorest. The solidarity actions are promoted by teachers,

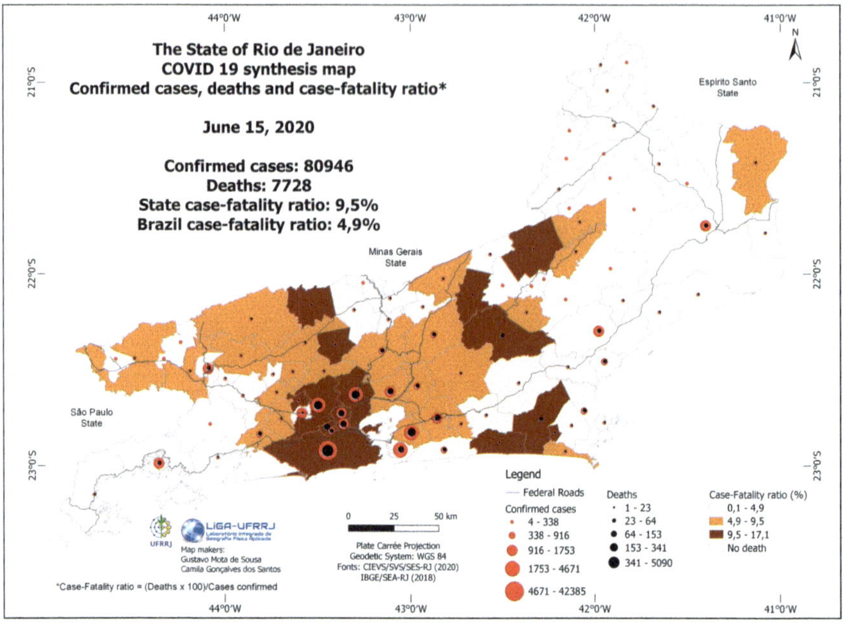

Fig. 3.10 Confirmed cases, deaths numbers, and case fatality ratio in Rio de Janeiro State on June 15, 2020

employees, students, and people from the academic and local communities. Essential hygiene items and food donations bring comfort and hope to those most in need.

3.5 Conclusion

Mike Davis in his work "*The Monster at Our Door: The Global Threat of Avian Flu*" (Davis 2015), makes an important question: how would the almost defenseless cities of the third world react to a pandemic? In the peripheral realities, the concern is revealed: as the confrontation of the pandemic is related to qualified health conditions, availability of hospital spaces and medicines, sanitary conditions for survival, and even the capacity for social isolation, places with greater economic and social poverty become fragile spaces to face the COVID-19. And, Rio de Janeiro is a space of profound contradictions, with slums, a prosperous center, and impoverished periphery, and a series of economic and social problems.

So we share Mike Davis's concern because Baixada Fluminense proves to be a space where the coronavirus pandemic has shown itself to be aggressive. With a dense population, largely with precarious living conditions, with insufficient medical facilities and equipment in its territory, and great difficulty in promoting social distance—either because of the need to build income or because of the very limitations of

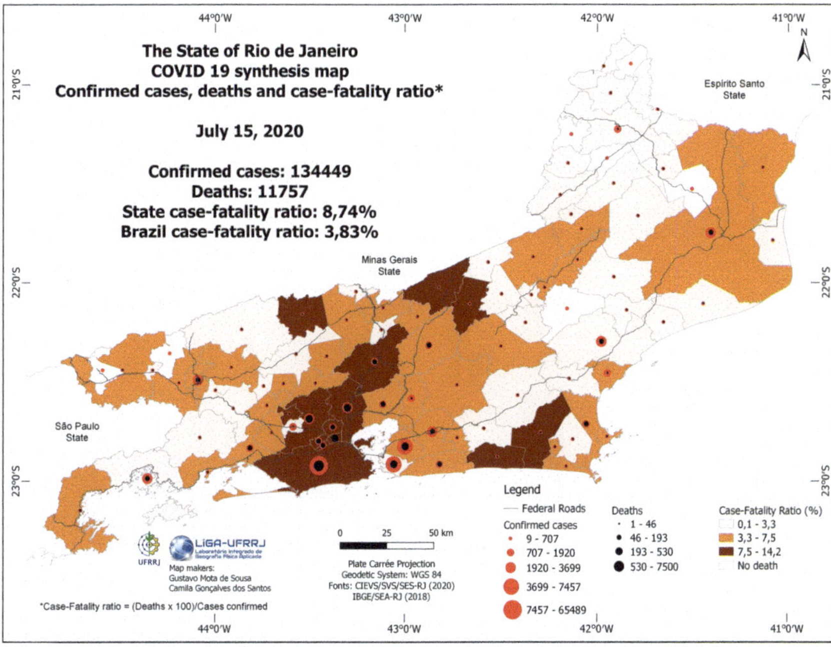

Fig. 3.11 Confirmed cases, deaths numbers, and case fatality ratio in Rio de Janeiro State on July 15, 2020

housing—Baixada Fluminense became a sad laboratory capable of responding to Mike Davis' concerns.

In this scenario, it is essential to highlight the effect of the divestment promoted by an agenda of fiscal adjustments that hit the Brazilian reality more acutely. In fact, we remember the words of David Harvey (2007) on the neoliberalism economy when it indicates economic adjustments in essential areas (such as health) can produce significant impacts in poor's populations. Urban-peripheral areas, such as Baixada Fluminense, are the first to suffer a collapse of their health system, resulting in fewer hospitals, beds, and equipment and reaffirming a framework of territorial injustices. Therefore, the increase in the number of cases of COVID-19 in these cities aggravates a crisis scenario in health. The problematic case of public health is not exclusively for Brazil; As per WHO, all of the countries of the World Periphery exhibits almost the same scenario in responding to the present problems—for a significant portion of the population the access to effective health care during the emergency seems nothing but a reverie (WHO 2018).

Robust public policies are urgently needed for the Baixada Fluminense to meet the living conditions of its populations. Among the actions, we prioritize two:

- A health management model with the expansion of investments in areas with the greatest deficit in the number of beds and improve an integrated policy between cities;

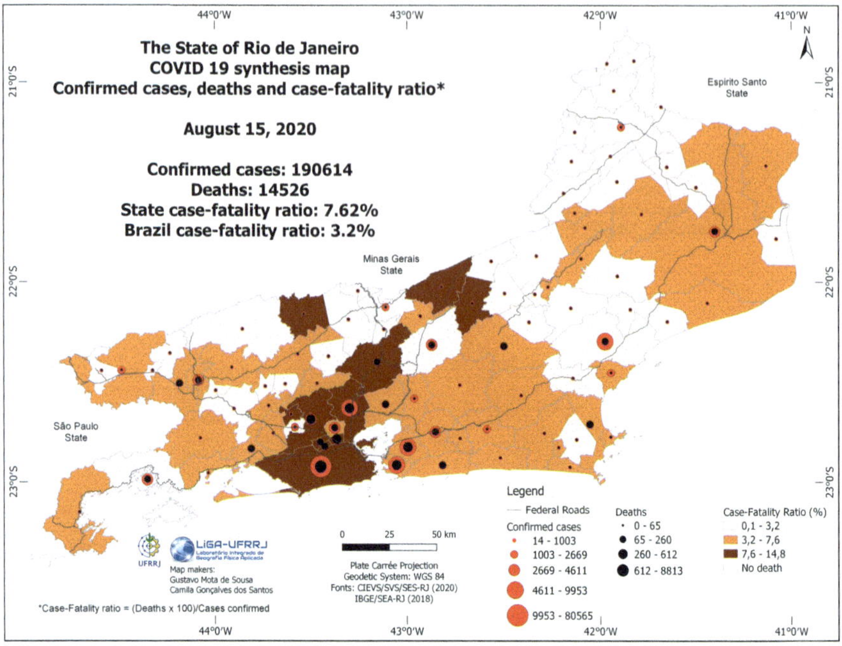

Fig. 3.12 Confirmed cases, deaths numbers and case fatality ratio in Rio de Janeiro State on August 15, 2020

- Investments in health; improvements to urban infrastructure; actions to combat illness, and the implementation of urban infrastructure (sanitation, paving, urban floods mitigation).

This is how we have to overcome the negative representation of the Metropolitan Area of Rio de Janeiro and the Baixada Fluminense, with high levels of inequality and social violence against human life and to face crises like the COVID-19 pandemic.

COVID-19 showed that health is not an issue to be considered in times of pandemic, but a constant effort of the governments to execute social inclusion policies, poverty alleviation, and expansion of cash transfer is the key to combat against a pandemic in the developing world. City management, the right to a more environmentally friendly life, and social justice are urgent demands. The pandemic of COVID-19 is a lesson for social geographers, an alarm for democracy and justice in its true sense.

Baixada Fluminense and the Metropolitan Area of Rio de Janeiro situation may be similar to other regions around the world, such as India, which also needs attention to public policies to face the present crisis and to be prepared for the future.

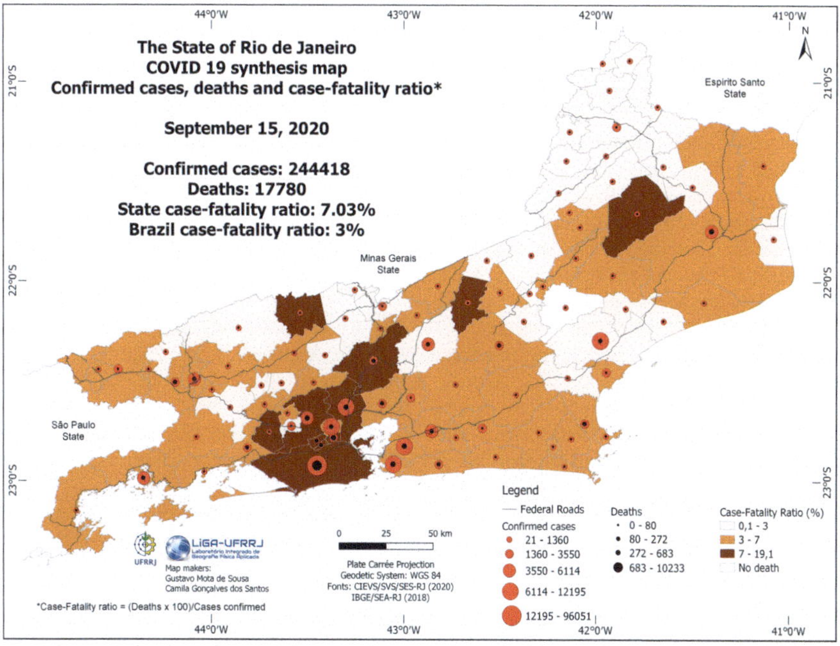

Fig. 3.13 Confirmed cases, deaths numbers, and case fatality ratio in Rio de Janeiro State on September 15, 2020

Fig. 3.14 Area View of Main Building—Federal Rural University of Rio de Janeiro/UFRRJ

References

BBC. Bolsonaro enquadra Teich e diz que Ministério da Saúde mudará protocolo sobre cloroquina (in Portuguese). Available in: https://www1.folha.uol.com.br/equilibrioesaude/2020/05/bolsonaro-enquadra-teich-e-diz-que-ministerio-da-saude-mudara-protocolo-sobre-cloroquina.shtml. Accessed 23 Sept 2020

BBC. Forest fires devastate Brazil's Pantanal tropical wetlands. Available in: https://www.bbc.com/news/in-pictures-54189503. Accessed 19 Sept 2020

BBC. Mandetta é demitido do Ministério da Saúde após um mês de conflito com Bolsonaro: relembre os principais choques (in Portuguese). Available in https://www.bbc.com/portuguese/internacional-52316728. Accessed 23 Sept 2020

BBC. Três temas que expuseram racha entre Nelson Teich e Jair Bolsonaro (in Portuguese). Available in https://www.bbc.com/portuguese/brasil-52684841. Accessed 21 Sept 2020

CNN. Nomeação de Pazuello como interino é publicada 18 dias após saída de Teich (in Portuguese). Available in https://www.cnnbrasil.com.br/politica/2020/06/03/nomeacao-de-pazuello-como-interino-e-publicada-18-dias-apos-saida-de-teich. Accessed 23 Sept 2020

Davis M (2015) The monster at our door: the global threat of avian flu, new edition. London, The New Press

Estadão. Pronunciamento de Bolsonaro contradiz OMS, especialistas e medidas adotadas ao redor do mundo (in Portuguese). Available in https://politica.estadao.com.br/noticias/geral,pronunciamento-de-bolsonaro-contradiz-oms-especialistas-e-medidas-adotadas-ao-redor-do-mundo,70003247141. Accessed 23 Sept 2020

Folha de São Paulo. Brasil passa Reino Unido e se torna o 2° país com mais mortes por Covid-19 (in Portuguese). Available in https://www1.folha.uol.com.br/equilibrioesaude/2020/06/brasil-passa-reino-unido-e-se-torna-o-2o-pais-com-mais-mortes-por-covid-19.shtml. Accessed 23 Sept 2020

Greenpeace. Fires in Brazil's Pantanal wetland and Amazon rainforest worst in a decade. Available in https://unearthed.greenpeace.org/2020/09/04/fires-brazil-pantanal-wetland-amazon-rainforest/. Accessed 19 Sept 2020

Harvey D (2007) Brief history of neoliberalism. Oxford University Press, New York

Santos M, Silveira M L (2001) O Brasil: Território e Sociedade no início do século 21. Record (in Portuguese), Rio de Janeiro

Souza e Silva J; Barbosa J L; Biteti M O; Fernandes F L (eds.) (2009). O que é a favela, afinal? Rio de Janeiro, Observatório de Favelas (in Portuguese). Available in: http://observatoriodefavelas.org.br/wp-content/uploads/2013/09/o-que-%C3%A9-favela-afinal.pdf. Accessed 21 Sept 2020

WHO – Word Health Organization (2018). Public spending on health: a closer look at global trends. WHO/HIS/HGF/HFWorkingPaper/18.3. Available in https://www.who.int/health_financing/documents/health-expenditure-report-2018/en/. Accessed 6 Apr 2020

Andrews José de Lucena was awarded Ph.D. in Atmospheric Sciences from UFRJ in 2012. Since 2004 he has been teaching in higher education and is currently an Associate Professor in Geography at the Federal Rural University of Rio de Janeiro (UFRRJ). He works in Physical Geography (Atmospheric Sciences) with an emphasis on Urban Climatology and Atmosphere Remote Sensing. He is interested in environmental changes in the city(ies) and its various associated phenomena, such as the Urban Heat Island (UHI). He participates and integrates projects and works together with researchers from Geography, Meteorology and Engineering, areas of knowledge where he guides undergraduate and graduate students. He manages the website 'www.climatologia.com.br', a portal with information on the urban climate at RMRJ, especially on the Land Surface Temperature (LST). He coordinates the Integrated Laboratory of Applied Physical Geography (LIGA)/UFRRJ.

Leandro Dias de Oliveira completed the Doctoral in Geography from the State University of Campinas (Unicamp) and Post-Doctoral in Public Policy and Human Training from the University of the State of Rio de Janeiro (UERJ). Presently he is an Associate Professor of the Department of Geography at UFRRJ—Federal Rural University of Rio de Janeiro (DGG-IA), headquarters campus, and teaches in the Graduate Program in Geography (PPGGEO-UFRRJ) and the Graduate Program Interdisciplinary Graduation in Digital Humanities (PPGIHD-UFRRJ). He integrates and coordinates the Laboratory of Economic and Political Geography (LAGEP). He is a researcher in Human Geography, with publications in Economic Geography and Industry, with emphasis on the analysis of Rio de Janeiro's Territorial-Productive Restructuring and on the society-nature relationship, especially in the analysis of the conception and application of sustainable development.

Pablo Ibanez did his Ph.D. in Human Geography from the University of São Paulo. He is currently an Adjunct Professor at the Federal Rural University of Rio de Janeiro and a researcher linked to the research groups: Welfare in Health, Development and Territory, at the Faculty of Medicine of the University of São Paulo; Research Center on Public Policies and Territory, at Fluminense Federal University; and Group of Studies and Research on Politics and Territory (GEOPPOL), from the Federal University of Rio de Janeiro. He was a visiting researcher at Fudan University, Shanghai. His research area is Geopolitics, Political, Economic, and Regional Geography, acting mainly on the themes of science, technology, and innovation; regional development, regionalization and decentralization; analysis of public policies, industrial and innovation policies; and development and territory.

Gustavo Mota de Sousa is a Ph.D. in Geography from the Federal University of Rio de Janeiro—UFRJ. He is currently an Adjunct Professor of the Department of Geography in undergraduate courses in Geography and Geology and of the Postgraduate Course in Geography (PPGGEO) at the Federal Rural University of Rio de Janeiro (UFRRJ). He conducts researches at the Integrated Laboratory of Applied Physical Geography (LiGA/UFRRJ), Laboratory of Cartography (GEOCART/UFRJ), and the Laboratory of Social Dimensions Applied to Physical Activity (LABSAFE/UFRRJ). His teaching, research, and extension projects trace on susceptibility to wildfires, geo-ecological mappings, cartographic representations with studies of tactile cartography, topographic models, Augmented Reality and Digital Fabrication (3D printing and laser router) in addition to geo-technologies applied to participatory mapping.

André Santos da Rocha did his Ph.D. in Geography at the Federal University of Rio de Janeiro – UFRJ. He is a professor in undergraduate and graduate courses (master) at UFRRJ. He works in Human Geography, especially with thematic of political and economic geography. He is also dedicated to studies of International Cooperation with the phase of Brazil-Africa-Latin America relations. He coordinates and participates in research projects with other researchers from UFRRJ and colleagues from Latin American Network Space and Economy (REELE). Dr. Rocha serves as the Editor of the Academic Journal—Continentes (UFRRJ). He is the Coordinator of the Laboratory of Economic and Political Geography (LAGEP).

Part II
Emerging Debates

Chapter 4
Boon for the Environment and Bane for the Economy: Emerging Debate in Pandemic Stuck India

Tanmoy Sarkar and Jayanta Mondal

Abstract The COVID-19 pandemic shock has painted some unprecedented portrayal and presented some exceptional visuals regarding the economy vs. environment debate worldwide. The history of human civilization is impregnated with the exploitation of the environment. The world needs a long and clear vision where economic activities and developmental goals meet environmental sustainability. The fatal coronavirus and Covid-19 pandemic have already affected more than 216 countries worldwide, with more than thirty-six million confirmed cases, including millions of death tolls. Now, India has the second-largest number of Covid-19 affected confirmed cases. The pandemic-driven nationwide lockdown has had a fatal impact on the economy. The supply-side and demand-side for entire economic sectors were disrupted, industrial production has come to a halt, millions of jobs were lost, Gross Domestic Product (GDP) was eroded away at an unprecedented rate, economic anxiety is inevitable in every sphere of life and livelihood. The GDP shows the last 20 years lowest in 2020. The impact of such an economic fallout is supposed to sustain for a long time with its deadly impact directly and indirectly. However, when we come to the environmental aspect, the lockdown exerts a very positive impact. The environmental pollution has been reduced remarkably, and the response in air quality, water quality, wildlife, and vegetation are remarkably flourishing. However, it would not be wise to blame the rampant economic activities and developmental initiatives for the present catastrophic situation. The economy must move on, industrial development would be geared up, and developmental projects must be initialized, but there should be a balance between economy and environment. Human civilization's sustenance needs mutual harmony with the natural environment, and sustainable development is the safest avenue ahead.

T. Sarkar (✉)
Department of Geography, Gazole Mahavidyalaya, Gazole, West Bengal, India
e-mail: srkrtanmoy@gmail.com

J. Mondal
State Aided College Teacher, Department of Geography, Dr. Meghnad Saha College, Itahar, Uttar Dinajpur, Ranipur, West Bengal, India

M. Mishra and R. B. Singh, *COVID-19 Pandemic Trajectory in the Developing World*, Advances in Geographical and Environmental Sciences,
https://doi.org/10.1007/978-981-33-6440-0_4

Keywords COVID-19 · GDP · Sustainable development · Environmental reinvigoration · Pollution · Indian economy

4.1 Introduction

Confusions, contradictions, and controversies have had been accustomed to sensible nexus between environment and economic development. Throughout the history of human civilization, the environment has had an inevitable fate of exploitation. The natural environment has a limited resilience to absorb and sustain against the small-scale exploitations. Every economic principle and mode of production exerts its impact on the environment with varying degrees across time and space. However, the emergence of capitalism and the industrial revolution had revolutionized the mode of environmental exploitation. When every part of nature and natural resources are being evaluated and targeted for capital gains, the exploitation becomes limitless, and the destruction becomes disproportionate. Several International Conventions and strategies are there to protect the environment, e.g., the Stockholm Declaration (1972), the World Conservation Strategy (1980), World Charter for Nature (1983), Brundtland Commission, Earth Summit Resolution, and declaration of Agenda-21 (1992), Malmo Declaration (2000), Rio + 10 and Rio + 20. However, all those global conventions declare the agendas, whereas environmental exploitation is continued in different shapes and various forms. Human civilization seems not in the mood to 'stay.' The 'growth' is felt like an extreme necessity and even at any environmental cost.

However, 2020 is the year when the entire world experiences the most common word of 'lockdown' in every sphere of life due to the fatal novel coronavirus attack. COVID-19 pandemic manifests worldwide. This malicious virus, which causes the World Health Organization (WHO) to declare the global pandemic, was first identified in Central China's Wuhan province (Huang et al. 2020) in December 2019. India has the second-largest number of COVID-19 confirmed cases (6,685,082 confirmed), including 103,569 death as of October 6, 2020, 10:27 am CEST (WHO Coronavirus Disease Dashboard 2020). The first case of COVID-19 in India was reported from Kerala in March, 2020. For combating the rapid transmission of the coronavirus and resist the death toll, the nations worldwide followed a complete shutdown. India was not an exception. Following the voluntary *Janta Curfew* (public curfew) on March 22, 2020, the Indian Government declared a nationwide lockdown from 24 March midnight. The entire nation experienced a complete shutdown for 21 days, and it was extended up to May 3, 2020 in the second phase of the lockdown declared on April 15, 2020; followed by the third phase of lockdown from May 4 to May 17, 2020. With the third phase of the lockdown announcement, India's Government divided all the districts into three zones based on the COVID-19 contamination character—'green' as least vulnerable, 'orange' as moderate, and 'red' as most vulnerable and some relaxations were announced accordingly. The National Disaster Management Authority (NDMA) further extended the lockdown from May 18 to May 31, 2020.

Moreover, the services have been started to resume in a phased manner vide Unlock 1.0 (June 1 to June 30, 2020), Unlock 2.0 (1 to July 31, 2020), Unlock 3.0 (1 to August 31, 2020), and Unlock 4.0 (1 to September 30, 2020). The usual flow of the world's economy, calculation of global politics, sociocultural harmony, academic, environment, people's normal lifestyle, everything has been drastically changed by the grasp of the life-threatening novel coronavirus, i.e., COVID-19 (Bera et al. 2020). Due to the lockdown, a drastic change in the weather pattern and pollutant parameter is witnessed.

4.2 Reinvigorating Environment

The COVID-19 pandemic driven nationwide lockdown heals the earth in a way never seen before. During the lockdown situation, humans are restrained from their routine activities. Nature started to readjust itself during the lockout condition, which had long been hampered due to pollution. India's pollution level exceeds the standard limit, which was determined by national ambient air quality standards. It affects the health condition at a severe rate (Ghose et al. 2004) in the Indian population. A huge number of people have a respiratory problem due to high air pollution (Smith 2000). COVID-19 induced lockdown results a remarkable change in air pollution (Singh et al. 2020) worldwide. If we take the starting date of lockdown as a breakpoint and analyze the environment's health during the lockdown and pre lockdown situation, the trend of pollution will be clear.

The monsoon climate dominates India. India experiences changes in temperature and rainfall trends due to global climate change, though the monsoon predominantly controls it. Different regions of India have faced a negative trend in rainfall during the last few decades. Patra et al. (2012) documented the negative trends in monthly, seasonal, and annual rainfall in the long term over Orissa. Praveen et al. (2020) analyzed the rainfall trend from 1901 to 2015 across India and observed an increasing trend during 1901–1950, while a declining rainfall trend was recorded after 1950 at an appreciable amount.

On the other hand, rising temperature trends have been observed (Fig. 4.1) in India. MERRA v2 data are used for assessing long term temperature trend analysis. The mean annual minimum temperature has been increased at 0.24 °C per 10 years over the Indian subcontinent (Rao et al. 2014). At the same time, the mean annual temperature was increased by about 0.4 degree Celsius per 100 years during 1901–1982 (Hingane et al. 1985).

The rapid urbanization and industrialization enhanced different pollutant levels considerably. The primary emission sources of Nitrogen dioxide (NO_2) are transport, thermal power plants, and industries. Human exposure to NO_2 over a short period causes infuriates asthma. On the other hand, the emission from industries and households contributes to $PM_{2.5}$ and PM_{10} in the lower atmosphere. Such particulate matters do create some serious health hazards, including premature mortality, respiratory symptoms, bronchitis. Sulfur dioxide (SO_2) is emitted mainly from coal-based

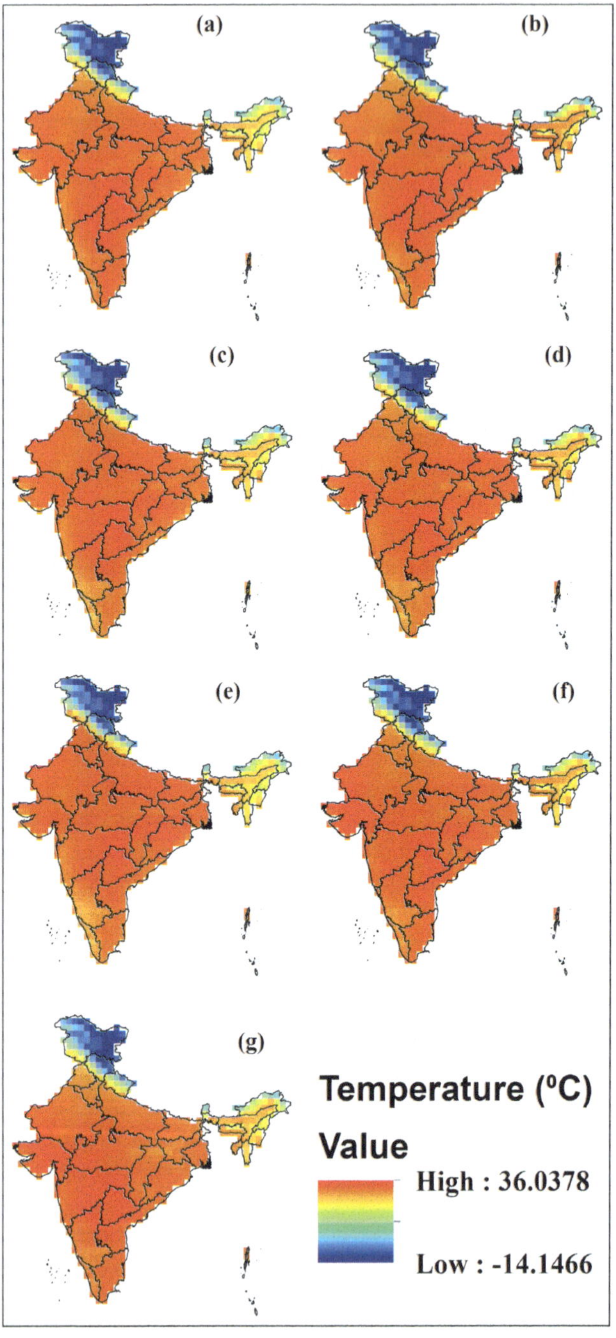

Fig. 4.1 Average temperature in the month of April in **(a)** 1990; **(b)** 1995; **(c)** 2000; **(d)** 2005; **(e)** 2010; **(f)** 2015; and **(g)** 2020

thermal power plants and different industries. SO_2 hampers the lung function and causes irritation to the eyes. Carbon dioxide (CO_2) emission is dominated by the transport sector, biomass burning, and fuel combustion (Sharma and Dikshit 2016).

WHO (2005) reported that 7 million people died throughout the world due to air pollution. Epidemiological research for the last couple of decades pointed out that air pollution plays a crucial role in various respiratory diseases, which is the pivotal cause of premature mortality in developing countries like India. China and India combined contribute to 52% of total $PM_{2.5}$ in the world, causing 1.525 million deaths in 2017 (Health Effects Institute 2019).

In general, the megacities of India used to face severe environmental problems due to heavy air pollution. The pollution levels surpassed the national air quality standards, as mentioned by the Central Pollution Control Board. Aerosols have caused serious human health issues. According to WHO's EPI (Environmental Performance Index), Delhi is regarded as the most polluted megacity in the world. It has been reported that 1 out of every 8 deaths has been caused due to high air pollution in 2017 (ICMR, PHFI, and IHME 2017).

4.2.1 Environmental Resilience Amidst Lockdown

Lockdown as a measure for combating the virus prohibits mass transportation and all non-emergency economic activities. As the commercial, transport, industrial activities have been ceased almost for quite a long time, the environment finds its way to replenish itself and successfully stress out the burden of anthropogenic disturbances (Bera et al. 2020). In this regard, the release of GHG (Greenhouse gases), power generation, vehicular movement, coal combustion, and poisonous particulates emission have been truncated. It results in the quality enrichment of the global climate, pollution-free sky, and a vigorous resurgence of the Ozone (O_3) layer (Chakraborty and Maity 2020). Lockdown becomes an amazing catalyst in reducing the energy consumption level. Fuel combustion drastically dropped down in the first half of 2020 as compared to preceding years. During the period of lockdown, emissions of hazardous gases put a break globally.

In the Yangtze River delta of China, the NO_2, $PM_{2.5}$, and SO_2 have diminished (Li et al. 2020). Megacities of China experience a negligible concentration of pollutant gases amid lockdown (CAMS 2020) due to industrial activities confinement. Tobías et al. (2020) analyzed in detail that in the city of Barcelona (Spain), urban air pollution has diminished drastically just after two weeks of lockdown. PM_{10} concentration has a downfall by 28% for the traffic, and in the urban background stations, it has reduced by 31%. The Italian city of Milan witnesses a cut down of PM_{10} concentration during the partial lockdown phase and experiences a lowering by 13.1% during the total lockdown phase. A huge reduction (−57.6%) has been observed in the case of CO and NO_2 emission in Milan (Collinvignarelli et al. 2020).

The suspension of industrial and commercial activities has lessened the pollution level in 88 cities of India (Sharma et al. 2020). A detailed study by Srivastava et al.

(2020) at 34 monitoring stations in Delhi reveals an appreciable reduction in CO, NO_2, $PM_{2.5}$, and PM_{10} levels. The national air quality index reported an enormous improvement in air quality by 40–50%. Several studies (Maté et al. 2010) reveal that in Delhi, the average concentration of $PM_{2.5}$ has been lowered down to 26 mg/m^3 (on 27 March) from 91 mg/m^3 (on 20 March) within a week of lockdown. An analysis by Mukhopadhyay (2009) spotlighted that in metropolitan Kolkata, 70% of people were victimized by lower atmospheric noxious pollutants. Global Carbon Project 2020 shows that the high emission rate of GHGs has been controlled in this period of lockdown, which is an unexpected phenomenon after World War II.

Worldwide lockdown works as a stimulus for revitalizing world climate. It gives ample opportunity for researchers to work in this unique direction. The present study focuses on the changes of atmospheric components like SO_2, NO_2, CO_2, and aerosols in India during the lockdown period. Therefore, another focus of this study includes the impact of lockdown on the environment. The results of this study will be helpful in determining the fruitfulness of full lockdown as a strategy for diminishing air pollution.

The direct and indirect impacts of lockdown on atmospheric pollutants have been assessed using satellite-based spatio-temporal products related to NO_2, CO_2, SO_2, and Aerosols. Carbon Dioxide, Mole Fraction in Free Troposphere, IR-Only (AIRX3C2Mv005, and AIRS3C2Mv005) products have been used to assess the scenario of CO_2 concentration over India. The concentration of CO_2 varies from 388.102 ppm to 392.737 ppm over India. The concentration of CO_2 hotspots has been increased in the different parts of the Indian subcontinent over the last two decades (Fig. 4.2).

Aerosol optical depth 342.5 nm_0.25 deg product has been used to depict aerosol optical thickness in this study. Long-term trend of aerosol optical depth shows that it has been increasing over the sky of India. 5.2–7.4 unit was the dominant cover of aerosol, observed in India, but interestingly it has been suddenly started to fall in its thickness during the lockdown period (Fig. 4.3).

MERRA v2 product was used to portray the view of SO_2 distribution. 1.80–2.68 kg/m^2 was the long-term range of SO_2 generally found in India. SO_2 concentration in the Indian subcontinent has been remarkably started to decrease during lockdown (Fig. 4.4). OMI product has been used to show the NO_2 distribution. NO_2 is another bold atmospheric pollutant. Normal range of NO_2 was $-2.0921e + 014$ to $1.44568e + 016$ mol/cm^2, generally found in India. However, during the lockdown, it has also been started to confine its arm over India's sky (Fig. 4.5).

4.2.2 Inhale Afresh Amidst a Gasping Economy

Anthropogenic encroachment in the environment teaches us how far we are responsible for our own destruction. The deteriorated air quality hampers the natural rhythm of the environment and obliges to turn down human health. It has been reported that acute air pollution brings death toll across the globe in the form of respiratory diseases

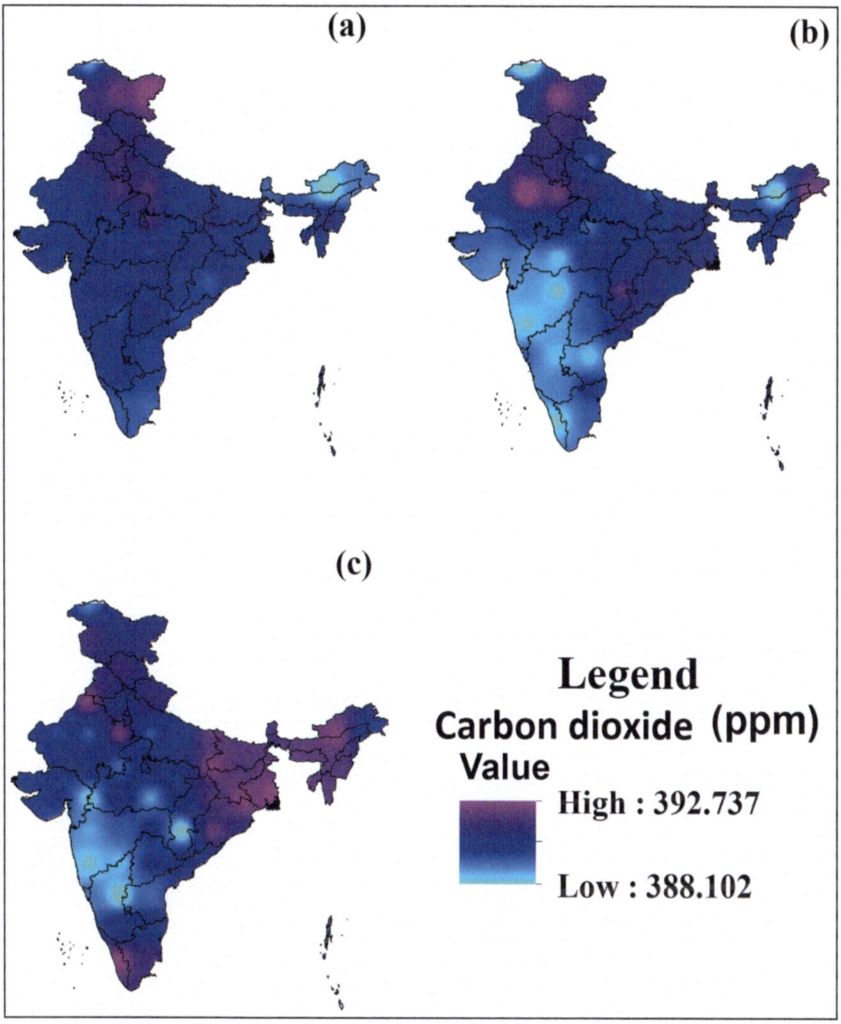

Fig. 4.2 Carbon dioxide concentration in the month of April in (**a**) 2005; (**b**) 2010; and (**c**) 2015

(26%), stroke and ischemic heart (17%), chronic pulmonary disease (WHO 2020). It is documented that 1.1 billion populations worldwide are forced to inhale toxic air (UNEP 2002).

Fortunately (from an environment perspective), global lockdown helps to rejuvenate the air quality. It positively affects the global climate by deducting the concentration of pollutants. The fatal virus threatens our life on one hand and gives the environment a chance to restore itself. So, the global perturbation about air pollution gives ample opportunities to scrutinize over the burning issues amidst the pandemic (Shehzad et al. 2020). Several studies reveal the reduction of pollutants

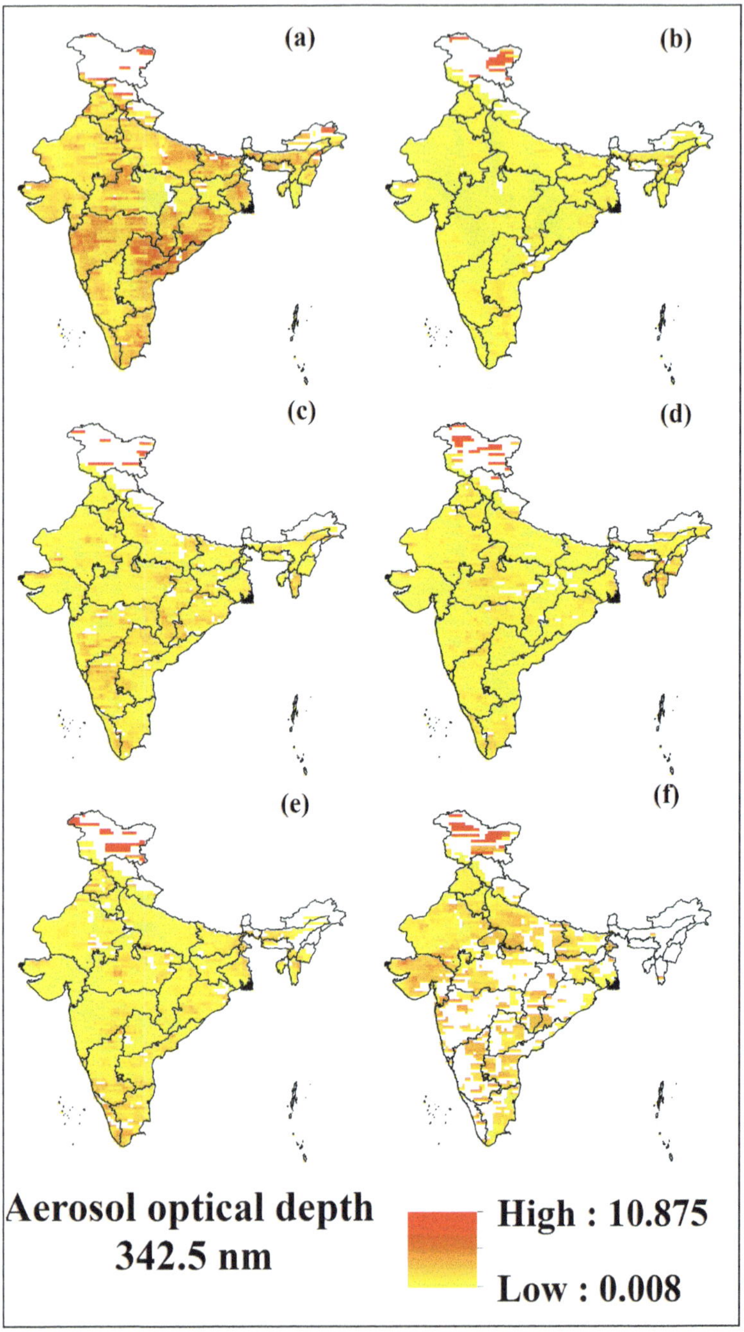

Fig. 4.3 Monthly aerosol concentration map during (**a**) January, (**b**) February, (**c**) March, (**d**) April, (**e**) May, and (**f**) June in 2020

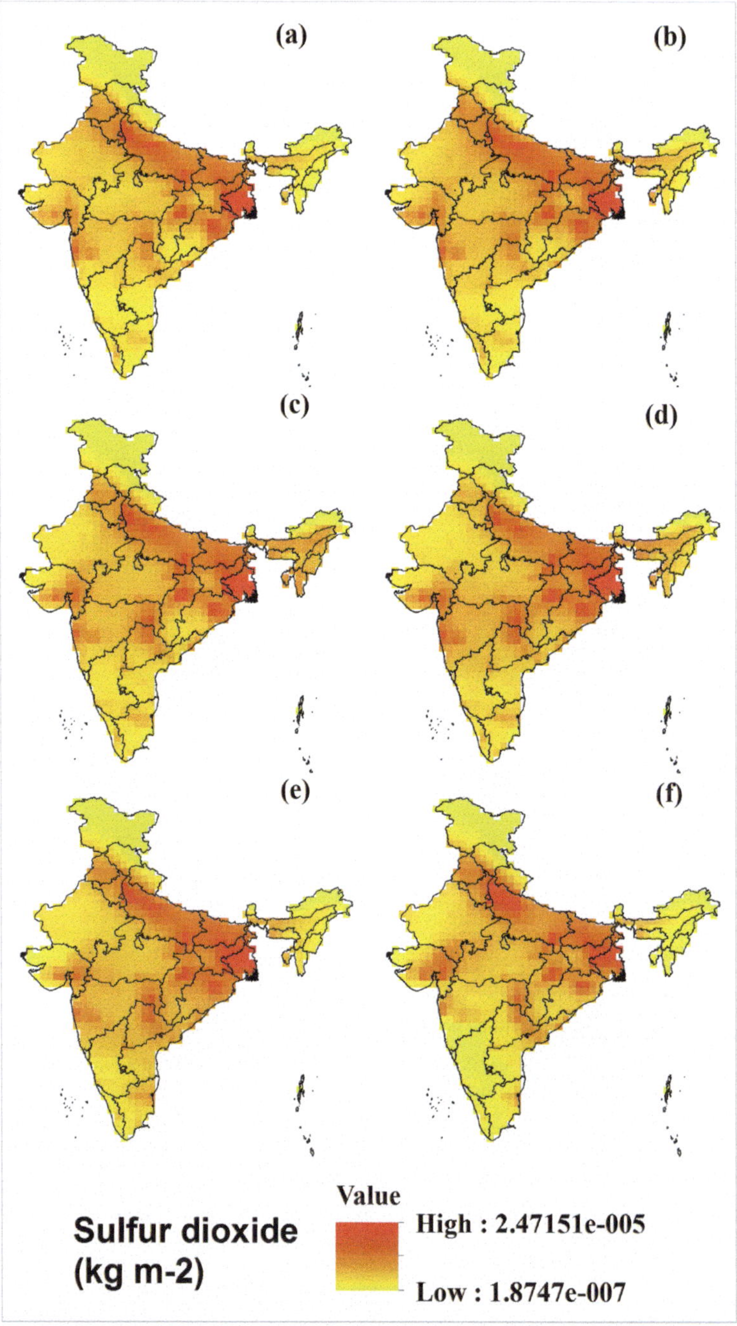

Fig. 4.4 Monthly Sulfur dioxide concentration map during **(a)** January, **(b)** February, **(c)** March, **(d)** April, **(e)** May, and **(f)** June in 2020

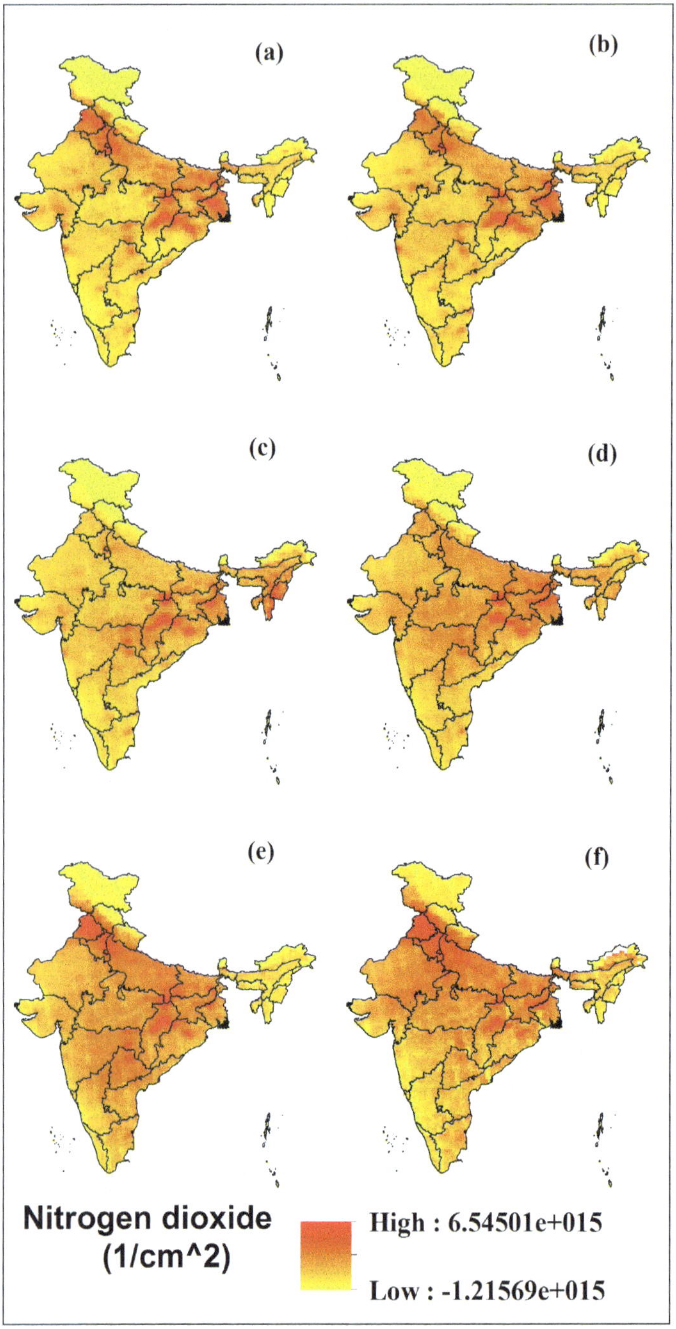

Fig. 4.5 Monthly Nitrogen dioxide concentration map during (**a**) January; (**b**) February; (**c**) March; (**d**) April; (**e**) May, and (**f**) June in 2020

level throughout the globe. A study conducted by Cadotte (2020) reveals the fact that air pollutants level has been diminished in major cities during the lockdown. NO_2 is reduced by 30% and carbon emission by 25% in China (Liu et al. 2015) during this phase. Analysis of Watts and Kommenda (2020), Coccia (2020), and Ogen (2020) also documented the same fact of reduction of air pollutants. The concentration of NO_2 has decreased by 12.1% between 1 and March 21, 2020; while in 2019, an increase was detected of NO_2 concentration by 0.8% during the same period (Biswal et al. 2020). COVID-19 induced lockdown has an awe-inspiring impact on the environment. The restricted human actions have offered nature, wildlife, and the natural ecosystem an excellent opportunity to move, flourish, and replenish. However, for a wider section of the population, the lockdown becomes a nightmare, putting a halt to their jobs. The country is in a perpetual debate, which the next section will focus.

4.3 Turbulence in Economy

COVID-19 pandemic has misstruck the economy with varying degrees of impact across space and sectors. The pandemic's impact on the economy must be addressed considering the pre-COVID pandemic economy, the preparedness and resilience of the economy to fight against the shock of slowing down, and the measures taken during the pandemic situation. The analysis of the Indian economy amidst the pandemic ambiance should be judged with the prelude that the GDP growth has been exhibiting a decelerated trend since 2015–2016 in the country. An inquisitive reader must be knowing that GDP is the final value of the product and services produced within the geographic boundaries of a state for a specified time interval. The nationwide lockdown costs havoc for the Indian economy, including its twenty-nine states and seven union territories. The impact of the COVID-19 pandemic is evident for the entire nation. However, the degree of impact has had varying magnitude across the states, depending on their economic structure, sectoral composition, and workforce.

4.3.1 Pre-lockdown Economy: Was It Tranquil?

Let us concentrate on just a few years back—India's economic scenario clearly tells about the economic slowdown that has been considered a political debate rather than an economic in media houses. India's position as one of the fastest-growing economies from 2014–2017 in the world arena slips in 2018–2019 with reducing GDP growth rate. The indicators of national economic health in the form of GDP, Index for Industrial Production (IIP), investment rates, exports, imports, and government revenues ring the alarm bell (Fig. 4.6).

India's index for industrial production growth (Fig. 4.6a) has been denying a stable growth trend since 2011–2012, and in 2018 it declined sharply. The declining trend in

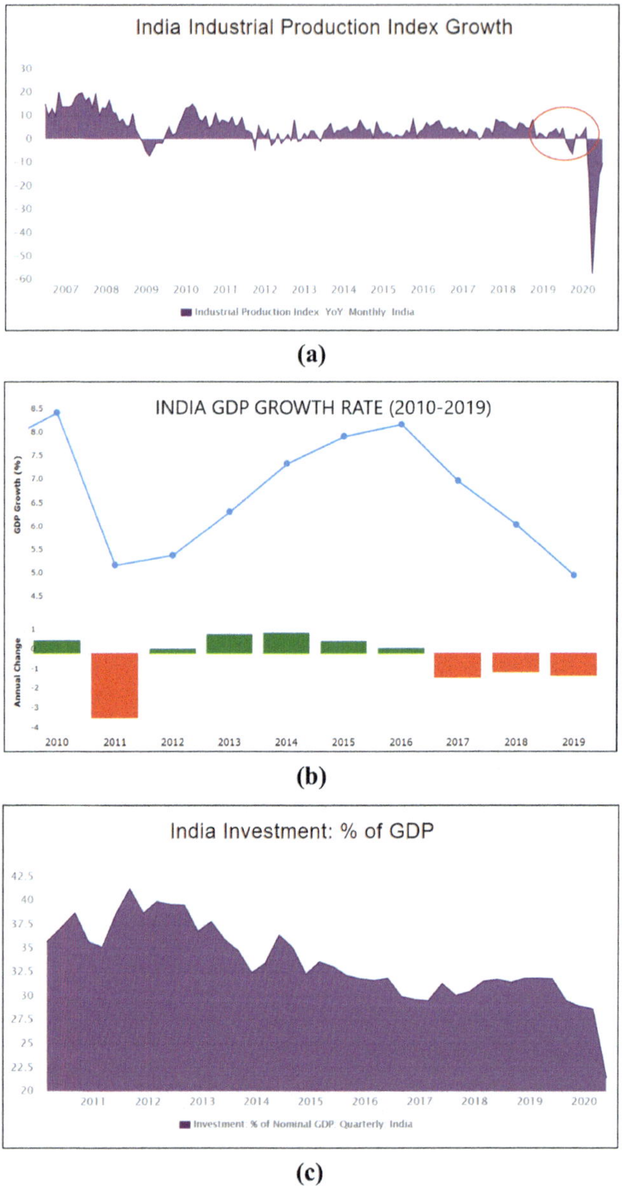

Fig. 4.6 (**a**) India industrial production index growth, 2006–2020; (**b**) India GDP growth rate, 2010–2019; (**c**) Growth in investment in India, 2010–2020; (**d**) India export status, 2010–2020; (**e**) India import status, 2010–2020; (**f**) Growth of govt. revenue in India, 2010–2020 (*Source* CEIC [CEIC Indicator dashboard is available under the URL: https://www.ceicdata.com/en/indicators] for a, b, and c; The Trading Economics Portal [Trading Economics Portal: https://tradingeconomics.com/] and Ministry of Commerce and Industry, Govt. of India for d, e, and f)

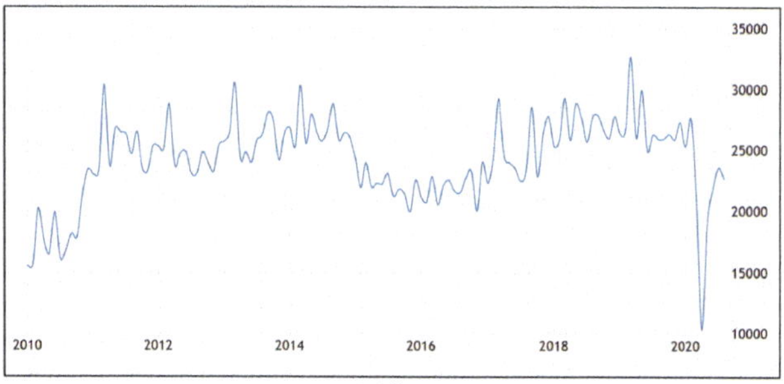

(d)

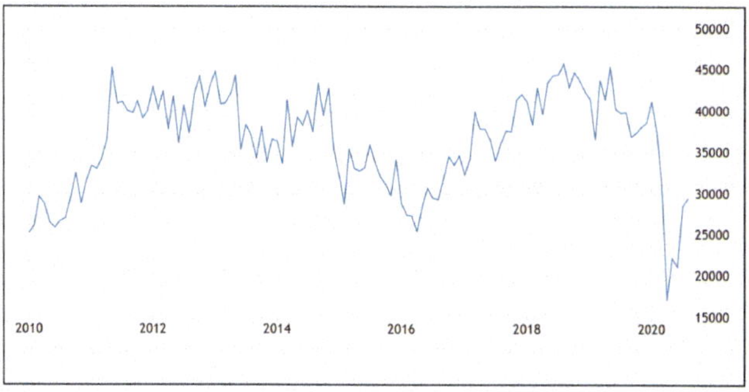

(e)

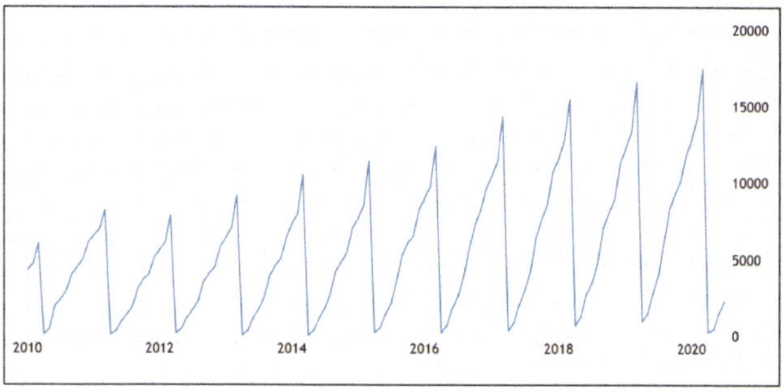

(f)

Fig. 4.6 (continued)

the economy's output and fall in production affected employment and income. India's GDP growth (Fig. 4.6b) has been continuously declining since 2016, whereas the investment rate also shows a reducing trend in investment in terms of the percentage of GDP since 2014. However, the export and import curve (Fig. 4.6d, e) express a healthier scenario comparatively, and the government revenue collection (Fig. 4.6f) gives some hope with its steady growth since 2010. There are several global and structural factors behind this economic turmoil, which include the very debated and controversial 'demonetization,' GST implementation and reforms in the taxation process in India, and so on since 2016. The Global Financial Crisis (GFC) sparked the great recession worldwide and cost the economy worldwide. India was not an exception. The global demand and prices were reduced, which costs the export and investment in India severely. The corporates and small to large-scale companies were experiencing an environment of low demand, high-interest rate, the higher exchange rate of Indian currency, and slow growth that make them very difficult to pay the credit to the bank that was promised during the boom.

Decreasing the repayment capability of corporates leads to India's Twin Balance Sheet problem. Reserve Bank of India (RBI) data shows the steady rise of non-performing assets (NFAs) since 2011 and a steep rise since 2015 with the highest peak in 2018. Due to capital infusion by the Govt. of India the NPAs ratio (Fig. 4.7) has shown a recovery in 2019 but still it stands at around 9% level. High magnitude frauds also contribute to the NFAs that limits the bank's offer to new emerging small enterprises with more ease. The increasing proportion of the non-performing assets (NFAs) severely affects the bank's capacity to grant further credit and lend. In December 2016, RBI introduced the Insolvency and Bankruptcy Code (IBC) as a recovery measure to NPAs, but it raises complexity. Despite this economic volatility and structural issues, the global oil price fall. An increase in the global demand and depreciation of the Indian rupee contributes some ray of hopes to the Indian economy. Diminishing consumer demand and lacking private investments are the key factors that affect economic growth. Moreover, the agricultural sector has been going through a battle against climate change and drought conditions in major parts of the country since 2014. With the demonetization, most of the savings for rural and small-scale enterprises have been taken away. In this overall crisis period with diminishing investment both in formal and informal sectors, the economic slowdown

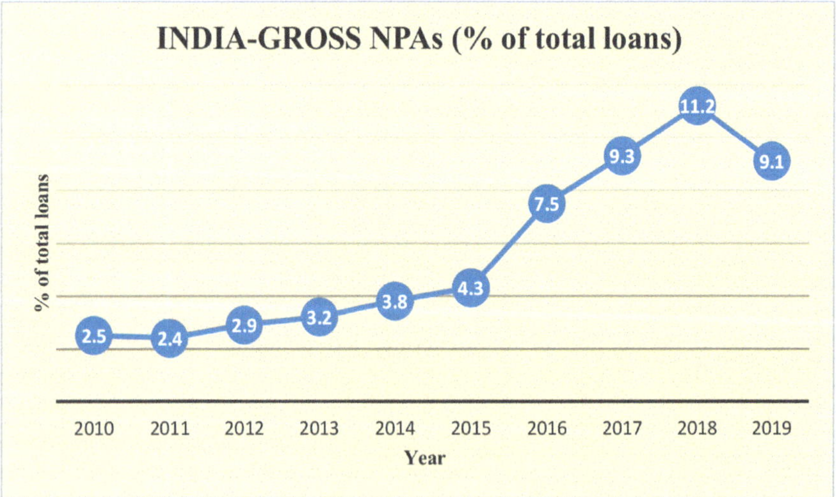

Fig. 4.7 Gross NPAs Growth in India for the period 2010–2019 (*Source* Prepared by the authors from the Ministry of Commerce and Industry, Govt. of India datasets)

was inevitable (Kotwal and Sen 2020). All these things manifest the pre-conditions in the Indian economy on which the surge.

4.3.2 *COVID-19 Pandemic Impact on Economy*

It's lockdown! No movement, no business, no sell! The production exchange, and consumption of all kinds of goods and services, except few extreme essentials, have been restricted due to an unprecedented COVID-19 pandemic situation. The world-wide COVID-19 pandemic affects the world economy badly. India is going through an extremely critical and crisis period in terms of economy and public health. The unemployment rate has sharply escalated during 2020 (Fig. 4.8b). The GDP shows the last 20 years lowest in 2020, slides from 5.02% in 2019 (Fig. 4.8a). The decelerated GDP rate must affect its three contributing components: consumption, investment, and external trade. The GDP share of different economic sectors is unique and unequal (Fig. 4.9) that may lead to variable degrees of pandemic impact on different economic sectors (Table 4.1). The impact of the COVID-19 pandemic is noticeable, but its severity varies across economic sectors and spaces. Indian economy is diverse, and the sectoral structure varies from state to state. The state-wise number of factories (Fig. 4.10) and the poverty condition (Fig. 4.11) has been taken to represent the pre-pandemic economic condition. Apart from this, the Gross State Domestic Production (GSDP) by different Indian states clearly shows the spatial inequality. The Indian states' GSDP share and state-wise COVID-19 status indicate that the 'sound' economy is affected most (Fig. 4.12). The GSDP share from different sectors is hugely different for different states (Fig. 4.13), and the states with greater dependency on industry, manufacture, and services have faced more severe impacts. In contrast, the states with more agriculture dependency are in comparatively stable economic conditions. Various states in India contribute to the economic growth of India in the manufacturing sector. However, five states: Tamil Nadu, Maharashtra, Gujrat, Uttar Pradesh, and Andhra Pradesh collectively hold around 53% of India's total factories (Fig. 4.10).

The state-wise COVID-19 epidemic data reveals that Maharashtra, Tamil Nadu, Delhi, Andhra Pradesh, and Karnataka are mostly affected (Fig. 4.13), followed by Uttar Pradesh, West Bengal, and Kerala. The demand-supply chain was disrupted by prolonged lockdown across the country and restrictions imposed on exchange globally.

4.3.3 *Disruption in the Demand Side*

Prolonged lockdown accelerates a situation of less transaction both in rural and urban areas, and it limits the desire of investment in non-essential goods and services by common people. Due to prolonged lockdown, non-regularity in banking transactions,

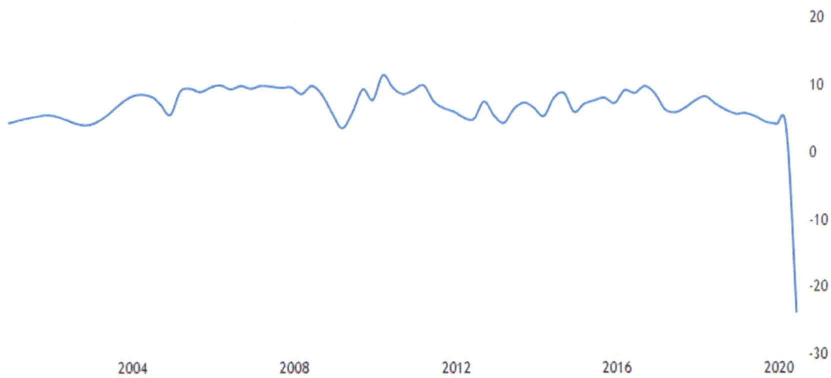

(a)

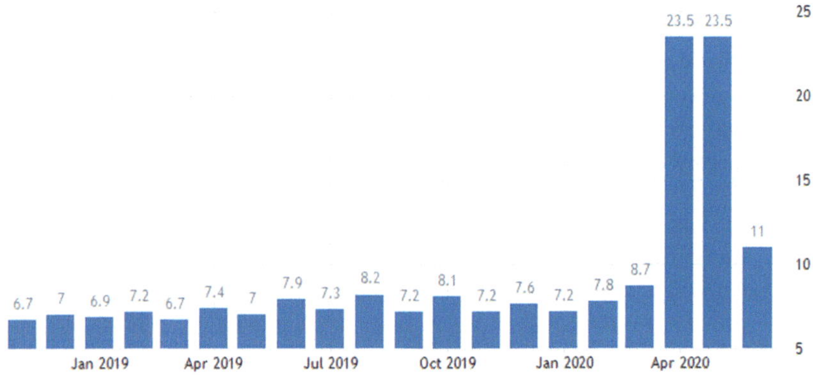

(b)

Fig. 4.8 (**a**) GDP growth rate from 2000–2020 (**b**) India unemployment rate 2018–2010 (*Source* The Trading Economics Portal & Centre for Monitoring Indian Economy CMIE [CMIE data portal on unemployment, URL: https://unemploymentinindia.cmie.com/])

uncertainty in job security across economic sectors limits the available cash in hand for common people. However, there is a decline and limited tendency to move outside in unlocking phases because of restrictions and risk of infection. Despite unlocking phases, there are also the provision of containment zones and a long way to wait for normalcy. This situation could lead to a sharp decline in the demand-side for non-essential investment and everyday life essential.

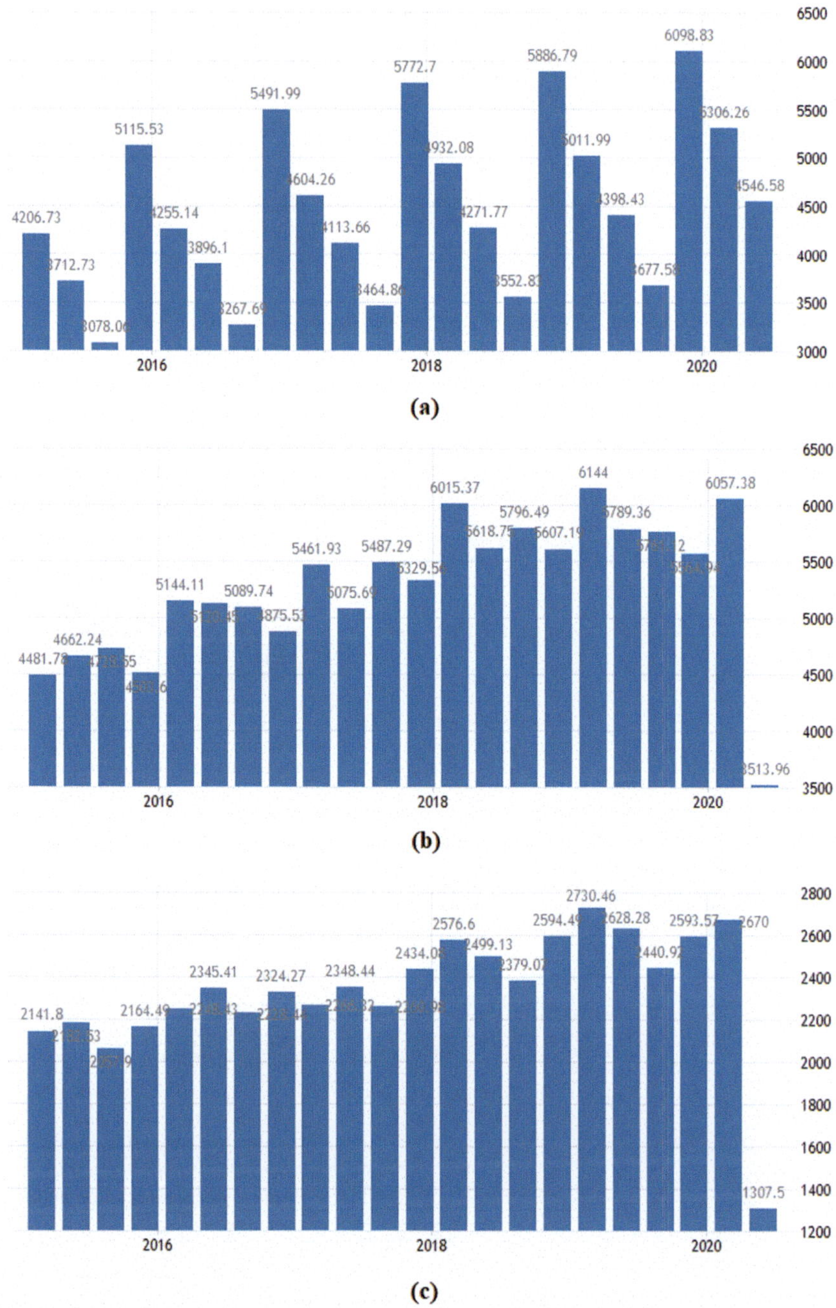

Fig. 4.9 GDP from **(a)** agriculture; **(b)** manufacturing, and **(c)** construction (*Source* The Trading Economics Portal [Trading Economics Portal: https://tradingeconomics.com/] and Ministry of Statistics and Programme Implementation, Government of India datasets)

Table 4.1 Economic sectors in India, leading states and their importance

Sector	Leading States	Importance
Food and agriculture	UP. West Bengal, Gujrat, Madhya Pradesh	The agricultural sector contributes 14.39% of GVA at constant prices and 15.87% at current prices for the service sector in 2018–19[a]; Considered as more than 58% of the Indian population's primary source of livelihood
Micro, Small and Medium Enterprises (MSME)	Maharashtra, UP, Bihar, Tamil Nadu, Madhya Pradesh	Contributes 6.11% of manufacturing GDP and 24.63% of the GDP; the sector is served by more than 114 million people[b]
Apparel and textile	Maharashtra, Gujrat, Rajasthan, Punjab, Karnataka, Tamil Nadu, UP, Delhi, MP, Andhra Pradesh, West Bengal	Shares 2.3% of GDP and 7% of the manufacturing production; Employing more than 45 million direct jobs and a substantial amount of contract labors
Automobile	Maharashtra, Tamil Nadu, Gujrat, UP, Delhi, MP	Contributed 7.1% to the GDP with 49% of the country's manufacturing GDP (FY 2015–2016)
Real estate	Maharashtra, Delhi, Tamil Nadu, UP, Gujrat	Expected to rise in market size of US$1 trillion in 2013 from US$1.72 billion in 2019 with a projected GDP contribution of 13% in 2025[c]
Tourism and Hospitality	Kerala, Rajasthan, Goa, Punjab, Himachal Pradesh, Up, Gujarat, Karnataka, and West Bengal	Contributes 6.8% of the GDP in 2019 and supported 39.80 Mn jobs, which are around 8% of its total employment
Chemicals and Petrochemicals	Gujrat, Andhra Pradesh, Tamil Nadu, Orissa	Ranks 3rd in Asia and 6th in the world by sales value.[d] Huge untapped potential for export and in domestic consumption
Consumer, Retail and Online business	Delhi, Maharashtra, Tamil Nadu, UP, Gujrat	The retail industry shares 10% of the country's GDP and around 8% of employment as per 2018–19 with rank as 5th largest global destination in the retail sector[e]
Education and Allied	All states	India is the 2nd largest E-learning market with an expectation to mark US$1.96 billion by 2021, incorporating around 9.5 million users[f]

(continued)

Table 4.1 (continued)

Sector	Leading States	Importance
Financial service	All states	India has very diversified financial spaces that is undergoing rapid expansion through banking, insurance, co-operatives, stock exchange, and mutual funds, non-banking financial companies
Health service	Kerala, Punjab, Tamil Nadu, and Gujrat	The Healthcare sector in India is expected to achieve a limit of US$372 billion by 2022. GDP share in healthcare service in India is only 1.6% in 2016 and Govt has targeted to reach 3% of GDP share by 2022[g]
Metals and mining	Odisha, Andhra Pradesh, Rajasthan, Chhattisgarh, Jharkhand, Madhya Pradesh and Karnataka	Fast-growing world market, including Asia with GDP share of 2.5% of GVA at 2011–12 prices, for the third quarter in 2019–20.[h] It provides 23 lakh direct employment and 2.3 crores indirect employment[i]
Oil and gas	Assam, Gujarat, Rajasthan, Tamil Nadu, Arunachal Pradesh, Andhra Pradesh	India is the 3rd largest energy consumer after USA and China and among the eight major industries in India[j] with its extremely influential role in other economic sectors
Pharmaceuticals	Andhra Pradesh, Telangana, Maharashtra, Himachal Pradesh, Tamil Nadu, Karnataka, Gujrat, and West Bengal	India is considered the largest producer of generic medicines globally with around 20% share in global supply by volume and shows 10.72% export growth in 2018–19.[k] India supplies 62% of the global demand for vaccines[l]
Power	Dadra and Nagar Haveli, Daman & Diu, Goa, Gujrat, Punjab, Chhattisgarh	India is the 3rd largest producer and 3rd largest consumer of electricity in the world (Tripathi 2018)[m]
Telecommunications	Andhra Pradesh, Maharashtra & Goa, UP, Bihar & Jharkhand, MP, Chhattisgarh, Karnataka, Delhi	India is considered as the 2nd largest telecommunication market in the world with an expectation to reach the limit of 8.2% GDP share in 2020[n]

(continued)

Table 4.1 (continued)

Sector	Leading States	Importance
Transport and logistics	Tamil Nadu, Telangana, Andhra Pradesh, Karnataka, Maharashtra	Rapidly growing sector with > 14% GDP share. And the unorganized sector is mainly accountable for the total sector valued of USD$150 billion with around 99% contribution

[a]http://www.mospi.gov.in/central-statistics-office-cso-1
[b]http://mospi.gov.in/nsso
[c]https://www.ibef.org/
[d]http://vibrantgujarat.com/
[e]http://ficci.in/
[f]www.ibef.org
[g]https://www.ibef.org/
[h]Annual Report, 2019–20, Ministry of Mines
[i]KPMG Report 2020
[j]https://www.ibef.org/
[k]https://www.investindia.gov.in/
[l]https://www.investindia.gov.in/
[m]https://en.wikipedia.org/wiki/Electricity_sector_in_India
[n]https://www.investindia.gov.in/sector/telecom
[o]https://lsc-india.com/content/overview_on_logistics_industry

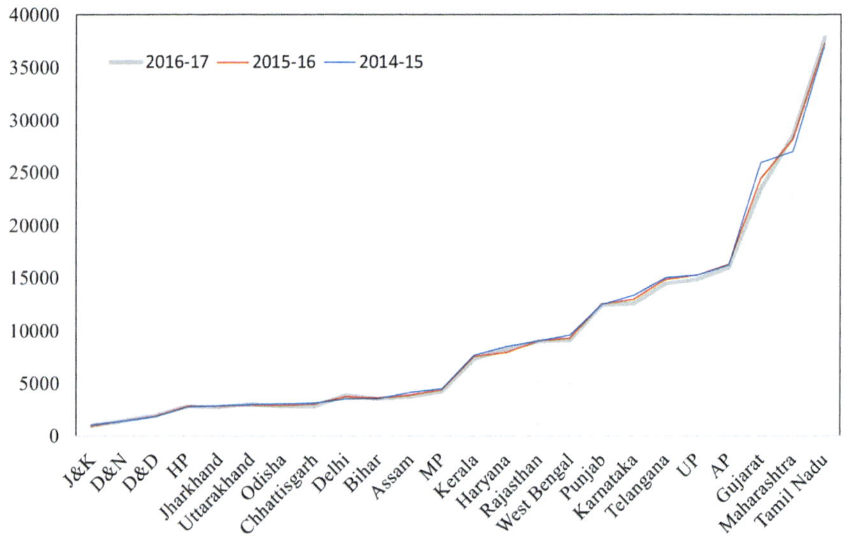

Fig. 4.10 State-wise number of factories in India (2014–2017)

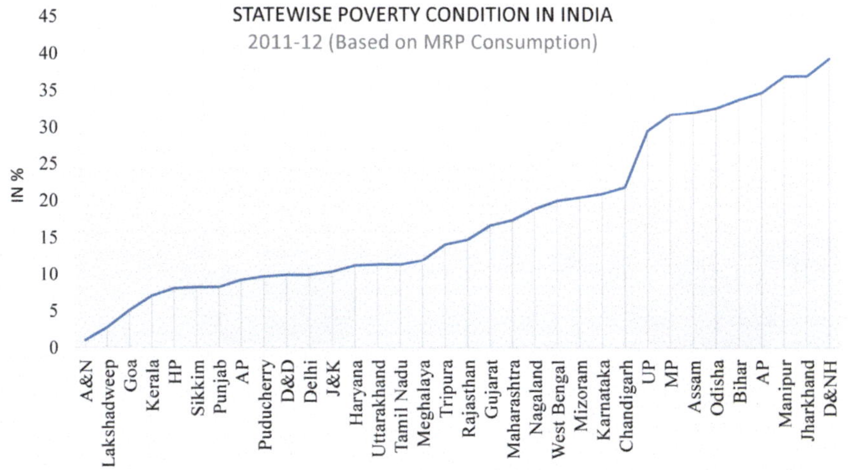

Fig. 4.11 State-wise poverty condition in India (2011–12 MRP Consumption) (*Source* Reserve Bank of India)

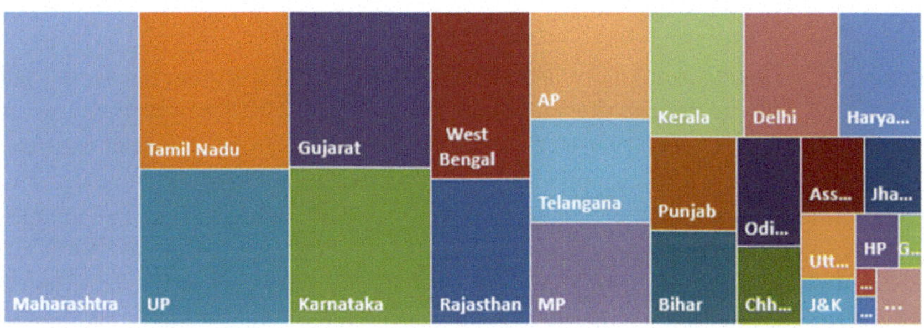

Fig. 4.12 India state-wise GSDP volume (current price) 2016–17

4.3.4 *Stagnation in the Supply-Side*

Major import partners for India are China (16% of total imports), the United States (6% of total imports), United Arab Emirates (6% of total imports), Saudi Arabia (5% of total imports), and Switzerland (5% of total imports).[1] China's Wuhan is the hot spot from where the deadly coronavirus has been outspread to the 188 countries with more than 31.6 million cases as of 23 September 2020. The USA is the worst affected nation with around 2 lakh deaths, and more than 6,800,000 confirmed cases due to COVID-19 pandemic. In this context, entry bans, the imposition of quarantine, restricted international exchange, the supply chain has totally disrupted. Besides, the India–China international border tension contributes more complexities in India–China trade relations. Electrical machinery and equipment, plastic, organic

[1]https://tradingeconomics.com/india/imports.

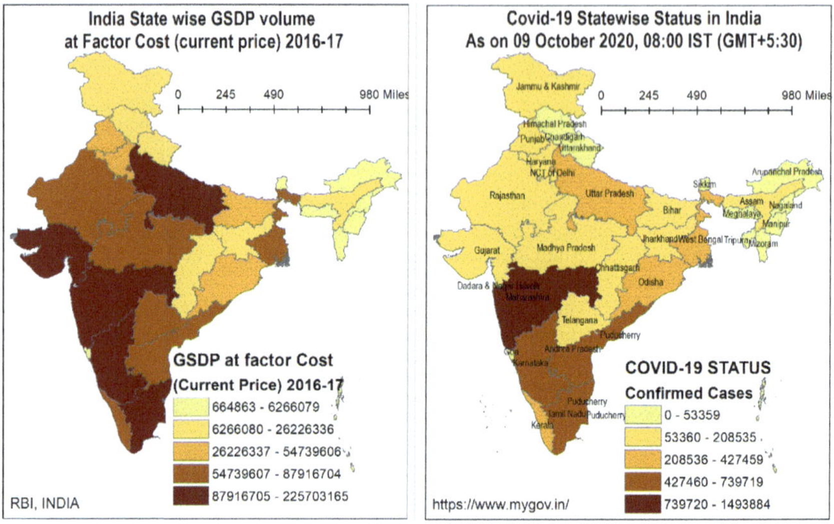

Fig. 4.13 India state-wise GSDP production and state-wise COVID-19 status

chemicals, fertilizers, machinery, and nuclear reactors are the commodities imported from China with the highest share. In 2019 India imported 60.9% of its total import from Asian countries and 15.9% of its purchase by value from European partners, followed by 9.2 and 8.1% supplied by North America and African countries. The sudden collapse of international import would have cost havoc to the Indian manufacturing economy and the supply side. The industries which mostly depend on foreign imported raw material are struggling very hard. The supply chain is also interrupted by huge reverse migration of migrant labourers to their native places, the containment zone restriction within states, irregular transport services, and numbers of masked factors. Labor intensive economic sectors face an unprecedented situation.

4.3.5 Persuading the Policies to Ground Realities

The worldwide economy, including the Indian economy, has been facing an unprecedented situation that has revolutionized the ongoing system in every sphere of life and livelihood. There is an ongoing debate on the advantage and disadvantages between the ethos of 'Capitalism' and 'Socialism'—specifically, their guided mode of production and distribution system to cope with this crisis. However, evidently, both the systems have been affected depending on their effective preparedness, resilience, and exposure.

Analyzing the sector-wise impact of COVID-19 and the pandemic status across the states in India, it might be said that the North-Eastern states in India are in the least degree of risk. This is mostly because of their relative isolation in terms of different

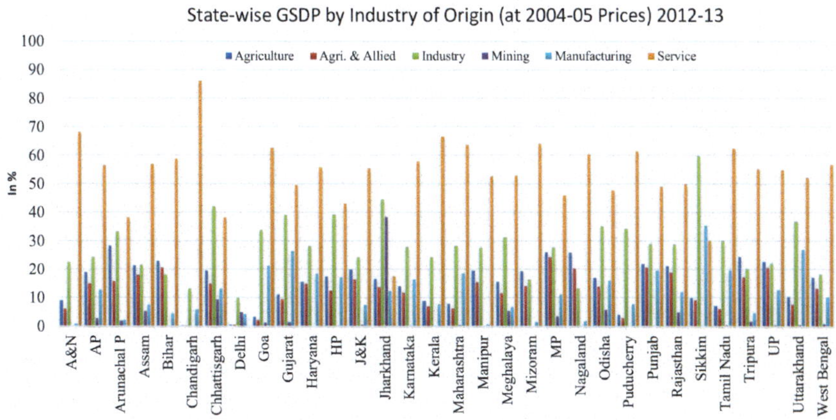

Fig. 4.14 India state-wise GSDP share from the industry of origin (at 204–05 prices) 2012–13

aspects of life and livelihood in comparison to the western and southern states. The lower GSDP share (Fig. 4.13), least market defined economic opportunities, and less exchange and movement have reduced the risk of rapid infection as well as the economy. The people of these states are comparatively more sensitive and conservative towards their society, culture, and traditions and, most importantly, have less exchange with the outside world. These factors contribute to some optimistic scenarios regarding the spread of the coronavirus in these locations. The agricultural sector is more resilient and shows stability with relatively less affected GDP share than industry, manufacture, and service. So, the states with greater dependence on the agricultural sector have a better opportunity to revive.

However, the western and southern part states, such as Maharashtra, Gujrat, Tamil Nadu, Kerala, Karnataka, are more dynamic and produce relatively high GSDP (Fig. 4.14) with more dependence on manufacture, industry, and service sector economy are affected most. Due to the suspension of activities contributing to industry, manufacture, and service, a vast number of migrants who were stranded outside their native places primarily have started to back home in different phases through different modes of transport even the country has witnessed the returning of migrant labor walking thousands of miles. The reverse migration would have a tremendous impact not only on demography but also on the economy. The interstate migration pattern in India varies across states and is mostly guided by their economy. The 2011 census data clearly shows that whereas Maharashtra, Delhi, and Gujrat are the largest receivers of interstate migrants, the state of Uttar Pradesh and Bihar have disproportionately largest number of out-migrants with 37% of total inter-state migrants (Chandramouli et al. 2011). The states of UP, Bihar, Madhya Pradesh, and Rajasthan collectively contribute around 505 of total inter-state migration (Chandramouli et al. 2011). The inter-state migration flow was reported to increase 45% from 2001 to 2011 (Chandramouli et al. 2011). It is obvious that on the verge of 2020, the flow must be increased at a large number. Hence, reverse migration hits havoc the manufacture, industry, and service sector, whereas the in-migration in the recipient states greatly face undue pressure on economic structure.

But the crisis must provoke some opportunities. With the phase-wise unlock periods, the economy has been trying to get back in its path. Indian Govt. has already declared the Self-reliant India Initiative (*Atmanirvar Bharat Abhiyan*) package of twenty thousand crores with some key relief announcements for various economic sectors and practical initiatives to boost up the global supply chain in various sectors, reforms in several sectors, and relaxation in foreign investment rules. Most importantly, to make the country self-reliant, the Centre–state relation has to be redefined to make the state as self-sustaining economic territories in terms of energy, economy, food production, water, and environment. All the policy decisions must focus on more distribution principles rather than a concentrated focus toward the Centre. The way ahead of the pandemic era demands the federal economy in real sense along with good governance and smart fiscal discipline.

4.4 Conclusion

The COVID-19 pandemic has actually painted and presented some impossible scenarios in various aspects of life and livelihood. Whereas the natural environment has got a very positive impetus, but the economy and livelihood have been affected very badly. Millions of jobs have been lost; employment generation has been slowed down, shortage of available cash in hand has made serious constraints on the demand side, and the total production and distribution system has been paralyzed, leading to an unprecedented recession since the 1930s' great depression in the world economy. However, long-time mass confinement due to the nationwide lockdown has also reduced the atmosphere's pollutant concentration. A restricted mass movement, closure of industries, and drastic reduction of fossil fuel consumption make the natural environment purer to live, breathe, and sustain not only for humans but also for every species and wildlife on the planetary earth. The economy must move on; development must be targeted, but human civilization has to make some balance between environment and development, and the only way is 'sustainability'. Unfortunately, there is an immense probability of environmental exploitation in the post-COVID-19 pandemic situation. It shall not be wise to adulterate environmental safeguards and protection regulations to revive the economy and gear up the developmental works. The PARIVESH[2] portal data clearly shows that the Ministry of Environment, Forest, and Climate Change (MoEFCC) of India has approved 2604 proposals (85.6%) out of 3042 proposals submitted between July and October (October 7, 2020) for environment clearance, and it is highly alarming. Interference in bio-diversity hotspot areas around protected areas and wildlife sanctuaries should not be tolerated at any cost or for the sake of the developmental project. Being a developing nation, India has a huge burden and responsibility to flourish immense economic and developmental

[2] See: https://parivesh.nic.in/.

opportunities through significant investment in industry, manufacture, and service. However, at the same time, India also has the responsibility to balance competing needs. The very principles of sustainable development are the only way to follow and reestablish the man–environment nexus to sustain human civilization.

References

Annual Report, 2019–20, Ministry of Mines, Govt. of India. https://mines.gov.in/ViewData/index?mid=1385. Accessed 26 Sept 2020

Bera B, Bhattacharjee S, Shit PK, Sengupta N, Saha S (2020) Significant impacts of COVID-19 lockdown on urban air pollution in Kolkata (India) and amelioration of environmental health. Environ Develop Sustain:1–28

Biswal A, Singh T, Singh V, Ravindra K, Mor S (2020) COVID-19 lockdown and its impact on tropospheric NO_2 concentrations over India using satellite-based data. Heliyon 6(9):e04764

Cadotte M (2020) Early evidence that COVID-19 government policies reduce urban air pollution

CAMS (2020) https://atmosphere.copernicus.eu/amid-coronavirus-outbreak-copernicus-monitors-reduction-particulate-matter-pm25-over-china

Chakraborty I, Maity P (2020) COVID-19 outbreak: migration, effects on society, global environment and prevention. Sci Total Environ:138882

Chandramouli C, General R (2011) Census of india 2011. Provisional Population Totals. Government of India, New Delhi, pp 409–413

Coccia M (2020) The effects of atmospheric stability with low wind speed and of air pollution on the accelerated transmission dynamics of COVID-19. Int J Environ Stud:1–27

Collivignarelli MC, Abbà A, Bertanza G, Pedrazzani R, Ricciardi P, Miino MC (2020) Lockdown for CoViD-2019 in Milan: what are the effects on air quality? Sci Total Environ 732:139280

Ghose MK, Paul R, Banerjee SK (2004) Assessment of the impacts of vehicular emissions on urban air quality and its management in Indian context: the case of Kolkata (Calcutta). Environ Sci Policy 7(4):345–351

Health Effects Institute (2019) State of Global Air 2019. Special report. Health Effects Institute, Boston

Hingane LS, Rupa Kumar K, Ramana Murty BV (1985) Long-term trends of surface air temperature in India. J Climatol 5(5):521–528

Huang C, Wang Y, Li X, Ren L, Zhao J, Hu Y et al (2020) Clinical features of patients infected with 2019 novel coronavirus in Wuhan, China. Lancet 395:497–506. https://doi.org/10.1016/s0140-6736(20)30183-5

ICMR, PHFI, IHME (2017) India: health of the nation's states-the India state-level disease burden initiative. Indian Council of Medical Research, Public Health Foundation of India, Institute for Health Metrics and Evaluation, New Delhi

Kotwal A, Sen P (2020). What should we do about the indian economy? Accessed from https://www.theindiaforum.in/article/what-should-we-do-about-indian-economy

KPMG Report on Impact of COVID-19 on the mining sector in India, May (2020) https://home.kpmg/

Li L, Li Q, Huang L, Wang Q, Zhu A, Xu J, ... Chan A (2020) Air quality changes during the COVID-19 lockdown over the Yangtze River Delta Region: an insight into the impact of human activity pattern changes on air pollution variation. Sci Total Environ 732:139282

Liu C, Hsu PC, Lee HW, Ye M, Zheng G, Liu N, … Cui Y (2015). Transparent air filter for high-efficiency PM 2.5 capture. Nat Commun 6(1):1–9

Maté T, Guaita R, Pichiule M, Linares C, Díaz J (2010) Short-term effect of fine particulate matter ($PM_{2.5}$) on daily mortality due to diseases of the circulatory system in Madrid (Spain). Sci Total Environ 408(23):5750–5757

Mukhopadhyay K (2009) Air pollution in India and its impact on the health of different income groups. Nova Science Publishers, New York

NSSO Report (2016) 73rd round of national sample survey. http://www.mospi.nic.in/https://www.ibef.org/industry/oil-gas-india.aspx#:~:text=The%20oil%20and%20gas%20sector%20is%20among%20the,adopted%20several%20policies%20to%20fulfil%20the%20increasing%20demand. Accessed 27 Sept 2020

Ogen Y (2020) Assessing nitrogen dioxide (NO_2) levels as a contributing factor to the coronavirus (COVID-19) fatality rate. Sci Total Environ:138605

Patra JP, Mishra A, Singh R, Raghuwanshi NS (2012) Detecting rainfall trends in twentieth century (1871–2006) over Orissa State. India Clim Change 111:801–817

Praveen B, Talukdar S, Mahato S, Mondal J, Sharma P, Islam ARMT, Rahman A (2020) Analyzing trend and forecasting of rainfall changes in india using non-parametrical and machine learning approaches. Sci Rep 10(1):1–21

Rao BB, Chowdary PS, Sandeep VM, Rao VUM, Venkateswarlu B (2014) Rising minimum temperature trends over India in recent decades: implications for agricultural production. Global Planet Change 117:1–8

Sharma M, Dikshit O (2016) Comprehensive study on air pollution and green house gases (GHGs) in Delhi. A report submitted to Government of NCT Delhi and DPCC Delhi, pp 1–334

Sharma M, Jain S, Lamba BY (2020) Epigrammatic study on the effect of lockdown amid Covid-19 pandemic on air quality of most polluted cities of Rajasthan (India). Air Qual Atmos Health:1–9

Shehzad K, Sarfraz M, Shah SGM (2020) The impact of COVID-19 as a necessary evil on air pollution in India during the lockdown. Environ Pollut 266:115080

Singh V, Singh S, Biswal A, Kesarkar AP, Mor S, Ravindra K (2020) Diurnal and temporal changes in air pollution during COVID-19 strict lockdown over different regions of India. Environ Pollut. https://doi.org/10.1016/j.envpol.2020.115368

Smith KR (2000) National burden of disease in India from indoor air pollution. Proc Natl Acad Sci 97(24):13286–13293

Srivastava S, Kumar A, Bauddh K, Gautam AS, Kumar S (2020) 21-Day lockdown in India dramatically reduced air pollution indices in Lucknow and New Delhi, India. Bull Environ Contam Toxicol 105:9–17

Tobías A, Carnerero C, Reche C, Massagué J, Via M, Minguillón MC, … Querol X (2020) Changes in air quality during the lockdown in Barcelona (Spain) one month into the SARS-CoV-2 epidemic. Sci Total Environ:138540

Tourism and Hospitality sector overview, Invest India website. Accessed 24 Sept 2020. https://www.investindia.gov.in/

Tripathi B (2018) Now, India is the third largest electricity producer ahead of Russia, Japan. Business Standard India, 26 March. Accessed 27 Sept 2019

UNEP (2002) Environmental threats to children: children in the New Millennium. United Nations Environmental Programme, UNICEF, WHO. https://www.who.int/watersanitationhealth/hygiene/settings/ChildrenNM7.pdf?ua=1

Watts J, Kommenda N (2020) Coronavirus pandemic leading to huge drop in air pollution

WHO (2005) WHO Air quality guidelines for particulate matter, ozone, nitrogen dioxide and sulphur dioxide: global update 2005: summary of risk assessment (No. WHO/SDE/PHE/OEH/06.02)

WHO (2020) Coronavirus disease (COVID-19) situation reports. https://www.who.int/emergencies/diseases/novel-coronavirus-2019/situation-reports. Accessed 30 Apr 2020

Web Sources

http://vibrantgujarat.com/writereaddata/images/pdf/chemical-and-petrochemical-sector.pdf. Accessed 25 Sept 2020

http://ficci.in/sector.asp?sectorid=33. Retail Sector overview. Accessed 25 Sept 2020
https://www.ibef.org/industry/education-sector-india.aspx. Accessed 26 Sept 2020
https://www.ibef.org/industry/agriculture-india.aspx Accessed 26 Sept 2020
https://www.ibef.org/industry/healthcare-india.aspx. Accessed 26 Sept 2020
http://www.social.niti.gov.in/health-index. Accessed 26 Sept 2020
https://www.investindia.gov.in/sector/pharmaceuticals. Accessed 27 Sept 2020
https://en.wikipedia.org/wiki/Electricity_sector_in_India. Accessed 27 Sept 2020
https://www.investindia.gov.in/sector/telecom. Accessed 28 Sept 2020
https://parivesh.nic.in/. Accessed 7 Oct 2020

Tanmoy Sarkar is an Assistant Professor in the Department of Geography of Gazole Mahavidyalaya (College) in Malda, West Bengal, India. The college is affiliated to the University of Gour Banga. Mr. Sarkar completed his postgraduate studies in Geography and Environment Management with a specialization in Remote Sensing & GIS at Vidyasagar University. His research chiefly focuses on soil erosion modeling, geospatial tools, and techniques in physical and human geography, and multi-criteria predictive models. He has been teaching Geographical Science at the undergraduate level for more than ten years. He has the working experience as investigators in different research projects, sponsored by DST and ICSSR. Mr. Sarkar has published three research papers in reputed journals, published by Springer Nature.

Jayanta Mondal is a State Aided College Teacher in the Department of Geography of Dr. Meghnad Saha College. The college is affiliated to the University of Gour Banga. Mr. Mondal completed his postgraduate studies in Geography at the University of Gour Banga. Further, he was assigned as a Research Assistant in ICSSR major project under the same department. His research interest is on environmental health monitoring and remedial simulations. He has hands-on experiences for four years in dealing with remote sensing on the geostatistical modeling.

Chapter 5
Blessing in Disguise in the Megacities: Environmental Co-benefits in Air Quality Amid Covid-19 Lockdown in Kolkata

Kushagra Rajendra, Namita Yadav, and Dipanjana Chakraborty

Abstract Kolkata is the third-most populous metropolitan area in India after Delhi and Mumbai, known as eastern India's cultural, educational, and commercial centre. It is the 2nd furthermost polluted city in India with a yearly average of 59.8 $\mu g/m^3$ $PM_{2.5}$ concentrations (IQAir 2019). Kolkata recorded the highest AQI (191) from 12–19 March 2020 & the lowermost (33) on 23–24 March 2020. This is happened due to the government imposed a nationwide lockdown. As a result of the lockdowns, air and noise pollution decreased for a short period, and air and water quality, biodiversity improved due to reduced industrial and human activities. On this backdrop, the present study endeavours to explore the reason and facts of sudden decrement in the concentration of air pollutants during lockdown due to the COVID-19 pandemic. Due to fewer vehicles and the absence of industrial activities and subsequent heavy rainfall made by cyclone Amphan landfall around Kolkata and adjacent areas, it improved air quality by its cleansing effect. It also highlights the interconnections between the air quality and socio-economics fabrics of the city and the collapse of economic activities. The impact of improved air quality on people's social and health status is positive. The paper also prescribes how these positive impacts can be carried out for the wellbeing of society.

Keywords AQI · Air pollution · Pandemic · Metropolitan · Economic activities

5.1 Introduction

Globally, air pollution is responsible for the death of millions of peoples. According to the report from Global Alliance on Health and Pollution (GAHP Report 2019), air

---

K. Rajendra · N. Yadav
School of Earth and Environmental Sciences, Amity University Haryana, Gurgaon, Haryana, India

D. Chakraborty (✉)
Department of Geography, Sitalkuchi College, Sitalkuchi, West Bengal, India
e-mail: cdipanjana@yahoo.com

M. Mishra and R. B. Singh, *COVID-19 Pandemic Trajectory in the Developing World*, Advances in Geographical and Environmental Sciences,
https://doi.org/10.1007/978-981-33-6440-0_5

pollution is the main reason for the premature deaths, causing 15% of all deaths—some 8.3 million people. India registered around 2.3 million deaths in 2017 because of pollution (India Suffers 2020). Another study on the risk assessment, Global Burden of Disease comparative for 2015, reveals that exposure to air pollution adds around 1.8 million premature deaths and 49 million disability-adjusted life-years (DALYs) lost. Worryingly, India ranked among the top risk factors for India's health problems (PHFI and CEH 2017). The air quality status has significantly degraded in the last few years, particularly in urban centres, including major metro cities in India, compared to rural set-up. The Air quality is complex as far as its control is concerned, having several linear and non-linear relationships with many parameters including land use, meteorology, sources, location, and many more (Zhang et al. 2014; Wang et al. 2019). Despite several measures taken by agencies including the National Clean Air Programme (NCAP) and Graded Response Action Plan (GRAP), the urban centres maintained their flip side of Air Pollution throughout the year (IQAir 2019; MoEFC 2019). Indian cities have been among the top-most polluted cities globally (Kota et al. 2018).

Different measures have been taken globally to cope with the COVID-19 Pandemic spread: the different lockdown regime is the most common. The government has taken measures to control the spread of this virus. The implementation of complete lockdown had restrained people from moving out of their homes. World Air Quality Index exhibits that countries like the United States, Spain, Italy, China, France, Mexico, and the UK witnessed a fall in the concentration of air pollutants during lockdown due to the restriction of vehicular movement and industries' closure due to deficient workforce. In India, cities like Delhi, Mumbai, Kolkata, and Hyderabad, which comes under pollution city ranking, have shown improved AQI (IQAir 2019). According to the latest Report on Air Quality of 10 major cities in the world—Delhi, London, Los Angeles, Milan, Mumbai, New York City, Rome, São Paulo, Seoul, and Wuhan have shown a substantial drop in the concentration of fine particulate matter ($PM_{2.5}$). Delhi has witnessed a 60% fall in the concentration of $PM_{2.5}$ as compared to 2019, Seoul by 54%, Wuhan by 44%, and rest other cities by 9–35%. This significant decline in PM 2.5 was observed because there was a decline in the burning of fossil fuels and reduced road dust due to restricted vehicular movements and construction works during lockdown (IQAir 2020).

There is a noticeable result in the concentration of Nitrogen Dioxide in China, Italy, and England envisaged by the data from European Space Agency (ESA) comparing 2019 and 2020 (Chen et al. 2020). During the lockdown, NO_2 concentration was dropped by 22.8 $\mu g/m^3$ and 12.9 $\mu g/m^3$ in Wuhan and China, respectively. PM2.5 was dropped by 1.4 $\mu g/m^3$ in Wuhan but decreased by 18.9 $\mu g/m^3$ across 367 cities (Chen et al. 2020). The concentration of Carbon Dioxide was reduced in New York by 5–10% (Watts et al. 2020) and in China by 25% (Carbon Brief 2020) compared to the 2019 level. This substantial decline was possible because there was a reduction in vehicular and industrial emission during the lockdown.

In India, the nationwide lockdown was imposed starting from 24 March midnight to reduce Corona infections. It caused to suddenly cease down the transport, industries, infrastructure development, which further resulted in emission reduction significantly (Sharma et al. 2020). Significant air quality improvement has been reported across the country, including Kolkata. The Central Pollution Control Board in India informed that out of the 103 cities, air quality for 23 cities was enlisted as 'acceptable' air quality, while 65 cities were registered as 'satisfactory' air quality during the initial weeks of lockdown (Air Quality 2020).

The decline in air pollution during the complete lockdown, when there was a substantial decline in air pollutant emission from across sources, can give hope to the achievability's extent to control air pollution targets in city clusters. City wise analysis will further help regulators to come up with better plans to curb air pollution as every city has distinct air pollution characteristics in terms of source, meteorology, location, economic activities the city of Kolkata has.

5.2 On Exploring the Air Quality of the 'City of Joy'

Kolkata: the city served as the British capital, with the name 'Calcutt,', till 1912, before Delhi took the crown. It is also the second-largest city in India by area after the present national capital city of Delhi. Kolkata had significant importance to the British Empire. The city has earned the nickname 'city of joy' for its soulful embodiment of culture, enthusiasm, and festivity and the traditions of amazing sweet delicacies. It is a city that upholds a perfect juxtaposition between the old world and the modern one. The Metropolitan Area of Kolkata[1] is located in Hooghly River's left bank, which provides a very gentle slope towards the east and southeast to the city. The average elevation of Kolkata is 7 feet above sea level. Marshy wetlands characterize the Eastern part of Kolkata with swamps spreading over an area of about 12,500 hectares, creating a unique urban eco-system.

As per the Census of 2011, Kolkata city has a population of around 4.5 million and covers an area of 200.71 square km, divided into 16 Boroughs comprising 144 Wards and governed by the Kolkata Municipal Corporation (KMC). The city is densely populated (24306 persons/km^2). The Census 2001 enumerated Kolkata's population as 4,572,876 in 2001 (Census of India 2001, 2011). Among all the West Bengal districts, Kolkata recorded a negative Decadal growth rate (-1.7%) between 2001 and 2011 (Fig. 5.1).

The city's air was degraded due to rapid population growth, increasing vehicles, commercial, industrial activities, which reveal adverse anthropogenic effects on the

[1]Kolkata is the Metropolitan City of West Bengal located between 22°37′ N to 22°30′ N and 88°23′ E to 88°18′ E, and it is bounded by the North 24-Parganas district on the North and East, by South 24-Pargana district on the south, and by river Hooghly on the West (District Census handbook, West Bengal, 2001 and 2011, Series 20, Part XII-A & B, https://censusindia.gov.in/2011census/dchb/1916_PART_B_DCHB_KOLKATA.pdf) and, also view: https://www.censusindia.gov.in/Metadata/Metada.htm.

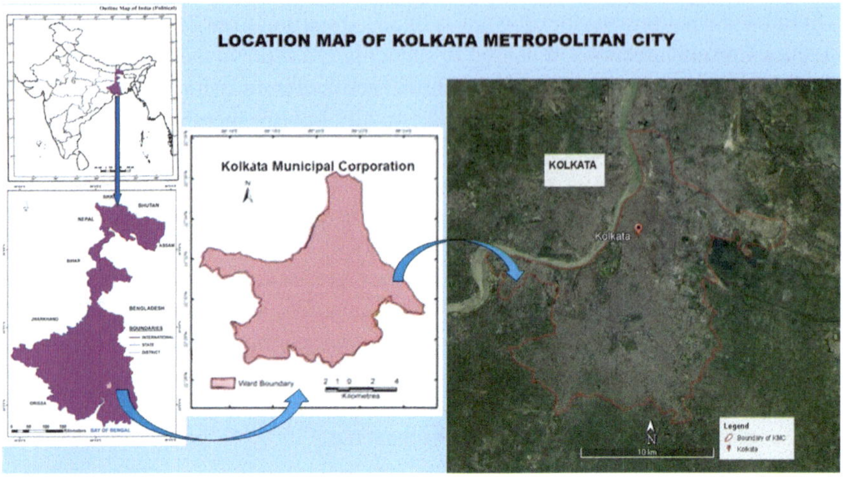

Fig. 5.1 Location of Kolkata Metropolitan City (*Source* Prepared by the authors from KMC, NATMO, Google Earth datasets)

city's environment. Kolkata is the second most polluted city in India, with an annual average of 59.8 $\mu g/m^3$ $PM_{2.5}$ concentration (IQAir 2019). The pollution level of Kolkata city was much higher than the WHO's guideline for a clean air level, i.e., 10 $\mu g/m^3$ annual mean of $PM_{2.5}$ concentration (WHO Air Quality Guidelines 2005).

The present study analyzes Kolkata's city's air pollution status during different stages of lockdown (from mid-March to the end of June 2020) based on air pollutants data collected across the city from a network of automated monitoring stations of CPCB during the study period. Analysis and understanding of data supported with few space-based observations help to pinpoint the impact of stopping major potential sources of air pollutants during the lockdown. This study also gives a glimpse of Kolkata's socio-economic landscape to understand its linkages of air pollution scenarios. Several mass-media information (Newspaper reports) are also considered for the present investigation.

To study on the Air pollution/Air Quality across the city of Kolkata during the lockdown, eight major air pollutants, namely Particulate Matter (PM_{10}, $PM_{2.5}$), NO_2, NO_x, NH_3, Ozone, CO, and SO_2, are collected from the CPCB network of air pollution monitoring stations.[2] Data gathered from seven monitoring stations, namely Jadavpur, Rabindra Bharati University, Rabindra Sarobar, Victoria, Ballygunge, Bidhannagar, and Fort William spread across the city. Their respective air pollution data give the representative status of the city as a whole. Concentrations of above mentioned eight pollutants for the time period mid-March to June end, 2020 is analyzed.

[2]See for details: https://app.cpcbccr.com/ccr/#/caaqm-dashboard-all/caaqm-landing/caaqm-data-availability.

The daily average concentration of all eight pollutants has been computed station-wise to understand the overall decline in Kolkata's air quality status. Station wise average gives an average concentration of pollutants as well as the Air Quality Index (AQI) on a daily basis. With the help of computed representative daily concentration of all selected pollutants, several comprehensive graphs have been prepared, which reflect the gradual decline of air pollution in Kolkata during the lockdown. The pollution data-based finding is further supplemented with space-based observations of NASA and some socio-economic findings of few general studies.

5.3 Impact on City's Air Quality

During the COVID19 lockdown environment, Kolkata's air became healthier, which was reflected through a clear blue sky, clean air, and noise pollution-free environment. $PM_{2.5}$, which had been 60.5 μg/m^3 in pre lockdown period, decreased to 34.3 μg/m^3, 17.2 μg/m^3, and 18.6 μg/m^3 during the 1st, 2nd, and 3rd Phases of lockdown accordingly (Fig. 5.2). It may be linked to the closing of industrial areas of Kolkata like Khidirpure and Garden Reach areas, as unlike other metro cities, Kolkata and Chennai air pollution loads are mainly contributed by industrial sectors (Multi-pollutant 2020). Similarly, PM_{10} also follows a similar trend like $PM_{2.5}$ from March to June end, which indicates less production of transport, domestic and industrial dust particles to a larger extent.

The highest NO_2 levels are in eastern India and are primarily associated with fossil fuel and coal-based electricity generation (Aura 2020). NO_2 emission is mainly

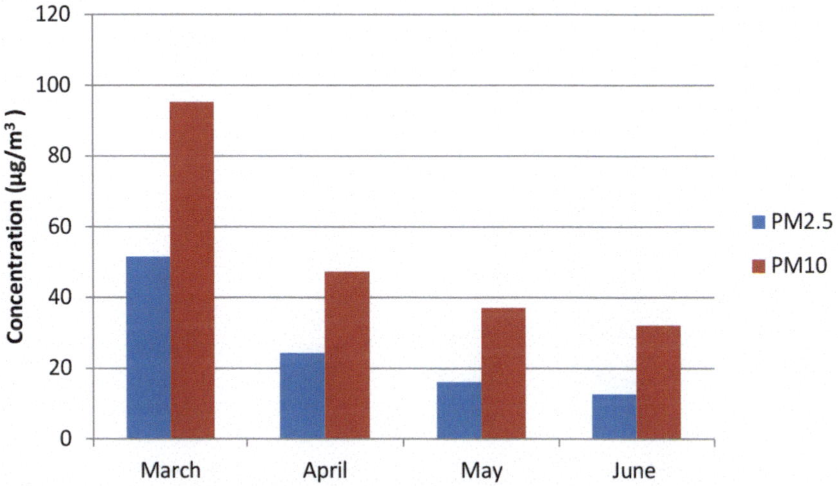

Fig. 5.2 Monthly average concentration of $PM_{2.5}$ and PM_{10} (μg/m^3) in Kolkata (*Source* Compiled by author Data: CPCB)

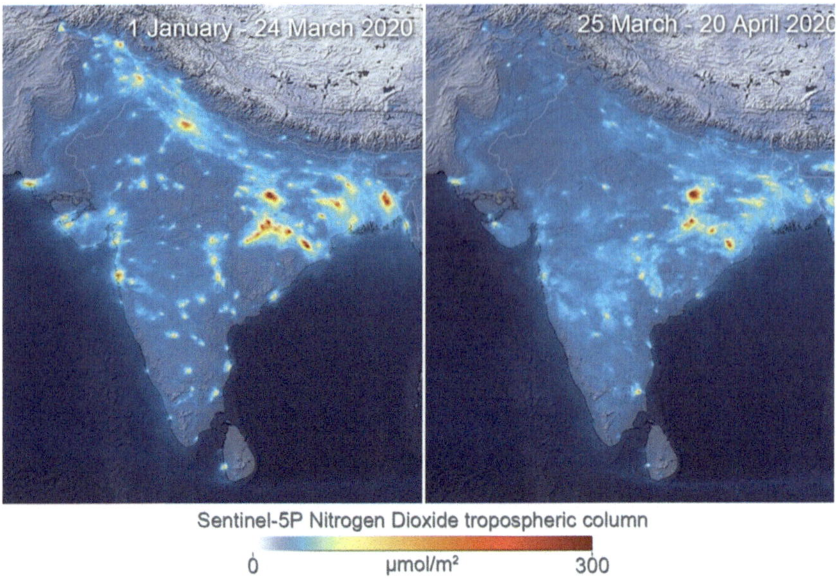

Fig. 5.3 Nitrogen dioxide concentration over India during lockdown and pre-lockdown Period (*Source* AURA 2020)

associated with coal-powered thermal plants situated near Kolkata, such as Bandel Thermal Power Station, Hugli, New Cossipore Generating Station, Kolkata, and CESC South, Kolkata. Some other plants are situated near the metropolitan area includes Budge Budge, Kolaghat, and Durgapur. The NO_2 level in Kolkata declined gradually from more than 30 $\mu g/m^3$ to its one-third value by June (Fig. 5.4). The significant decline of NO_2 levels initially can be further substantiated by finding

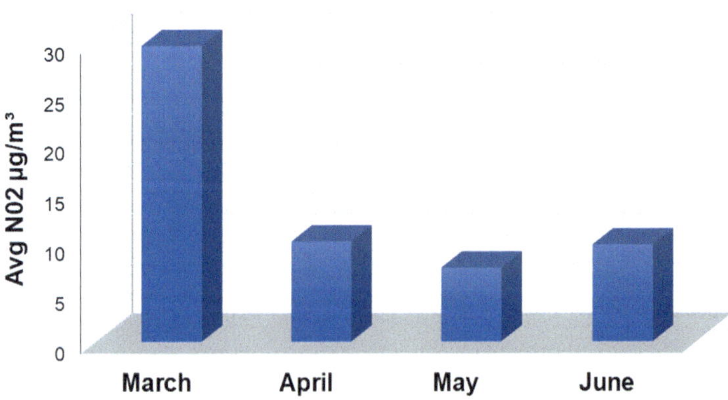

Fig. 5.4 Nitrogen dioxide concentration during lockdown and pre-lockdown period in Kolkata (*Source* Computed by author Data: CPCB, W.B.2020)

NASA's space-based comparative assessment image of NO_2 levels of pre-lockdown to April end (Fig. 5.3, source: AURA 2020).

Figure 5.3 represents the month-wise variation of NO_2 in Kolkata from March to 5 June, which shows a sudden decrease in NO_2 concentration during the 1st to 4th Phase of lockdown with a slight variation at the beginning of June. This is due to the reduction of power generation in thermal power plants located in and around Kolkata.

The emission of SO_2 is a significant contributor to urban air pollution, an indication of biomass burning. The most considerable amount of pollutants (SO_2) comes from fossil fuel burning in power plants, industries, and most significantly from urban transport. ESA's finding on comparative SO_2 distribution during lockdown and 2019 also indicates a substantial decrease in spread and concentration (Fig. 5.5). This comparative image of the average concentration of SO_2 emission over India during the Pre-lockdown period (2019) and lockdown in April 2020 is based on measurements gathered by the Copernicus Sentinel-5P satellite (ESA, July 2020).

The finding of the present study on average SO_2 concentration shows a declining trend by the 4th Phase of lockdown with explained reasons for the decline in transport and industry-related emission in line of other pollutants like particulate matter, NOx, and others (Fig. 5.6). During June, increasing SO_2 concentration is associ-

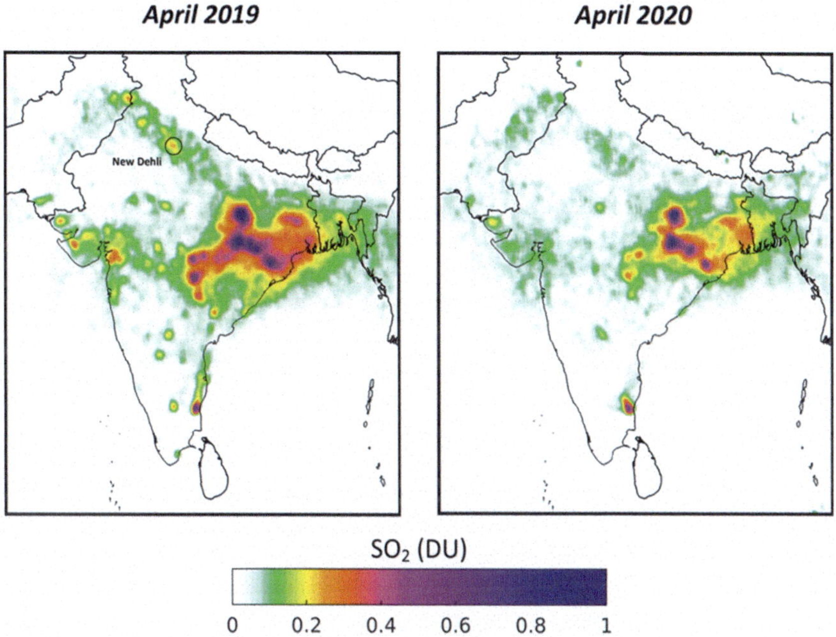

Fig. 5.5 Comparative Map of SO_2 emission in India during pre lockdown (2019) and lockdown period (April 2020) (*Source* ESA, July 2020)

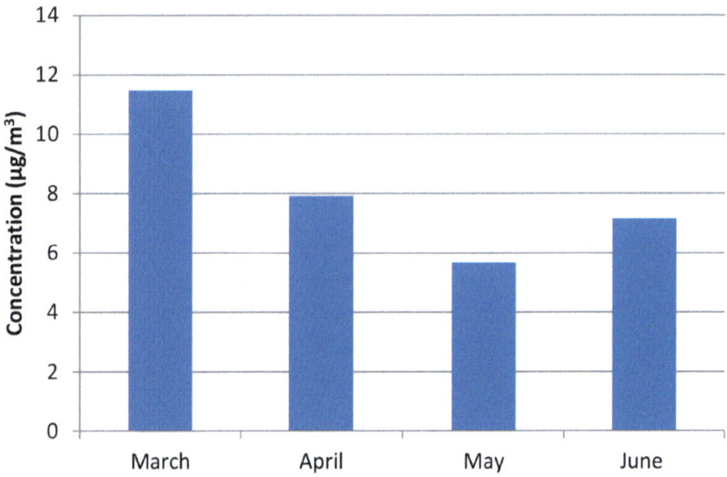

Fig. 5.6 Average monthly concentration of SO_2 in µg/m^3 (*Source* Compiled by author Data: CPCB)

ated with the resumption of transport and other activities gradually. The percentage of 24 hours average of SO_2 is under the acceptable limit of 20 µg/m^3 (WHO Air Quality Guidelines 2005) in Kolkata. However, as a reactive pollutant, it converts easily into sulfates and generates $PM_{2.5}$ as a secondary pollutant.

Like other urban cities reported to have significant improvements in their Air Quality across the country during the lockdown, Kolkata observed cleaning of their air space. Air quality analysis of Kolkata is based on data collected from eight automated stations of the CPCB network during several phases of lockdown, which shows spatial variation over the urban fabric. Among all stations, only two (Victoria, Bullygung WBPCB Stations) have shown moderate (AQI value 101–200) air quality, whereas the rest of four stations indicated satisfying (51–100) air quality. The city areas with a substantial green cover, comprising housing complexes (i.e., Jadavpur, Golf Green), showed the lowest AQI as the Jadavpur station's average concentration. In contrast, areas with mixed land use like that of Rabindra Bharati University and Victoria stations indicated a slight increase in pollution lead from both residential and commercial areas with restricted activities.[3]

Data based analysis of pollutants shows average decline throughout the lockdown period having synchronized minimum concentration (for all studies pollutants) during Phase second onward (3rd week of April onward). This may be associated with pollutants' residence time in the air and meteorological events in the city (good rainfall) during the 3rd Phase of lockdown. The ferocious wind and rains of cyclone Amphan further flushed the air pollutants from the city sky, which is reflected in low pollution concentration across the average daily concentrations across the city (BBC 2020).

[3]For details, view: https://app.cpcbccr.com/AQI_India/.

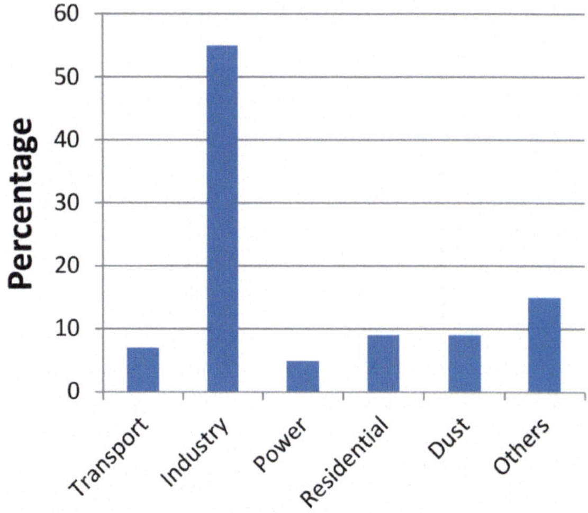

Fig. 5.7 Major sources of pollutants (*Source* Compiled by authors based upon data sourced from www.urbanemission.info)

5.4 Impact on Socio-Economic Environment

5.4.1 Economic Sector

Kolkata is the 3rd productive metropolitan of India, which depends on medium-sized industries, heavy industries, and services, and these medium-sized and heavy industries are the major source of pollutants in Kolkata (Fig. 5.7). GDP of Kolkata is US$170 billion during 2018–2019, with a growth rate of 14.6%, which indicates a steady economic growth of Kolkata City.[4] During COVID-19 pandemic, travel restriction affected Kolkata's tourism industry, which had a direct impact on the GDP of Kolkata. Labour statistics released by CMIE indicate loss of jobs of some portion of salaried employees and labourers during the 1st Phase of lockdown, which negatively impacted the economy of the state (*Economic Outlook* 2020). The jobless migrated labourers, who returned back to West Bengal, made the situation much more complicated. According to news sources, about 75% of workers returned to UP, Bihar, Orissa, Jharkhand at the beginning of the lockdown in March that created a labour shortage in industries, wholesale trade, manufacturing, construction, and informal sectors (Javed 2020). However, as soon as they returned to Kolkata's commercial hub, economic activities geared up from the 1st week of June (*The Times of India*, 13 July 2020).

[4]For details, view: https://www.india-briefing.com/regional-intelligence/kolkata.html.

5.4.2 Industrial and Energy Sectors

Global emission of CO_2 has increased substantially during the last few decades. Surprisingly lockdown brings down the pace of daily emission of CO_2 and other GHGs. The Daily global fossil CO_2 emission is decreased 17% during early April 2020 compared to the mean daily estimates in 2019 level (United in Science 2020; Le Quere et al. 2020). According to the analysis made by the Centre for Research on Energy and Clean Air, CO_2 emission of India fell by 15% during March and 10% in April during nationwide lockdown (Carbon Brief 2020). A similar decreasing trend has been observed in Coal powered thermal power electricity generation for India by 10 and 25% in March and April 2020, respectively.

Drastic fall in pollution levels is observed during lockdown due to the absence of emissions from vehicular sources, industries, constructions, and brick kilns. $PM_{2.5}$, and Nitrogen dioxide level decreased in Kolkata may be linked to suspension of work in factories, thermal power plants, and suspension of public, private transports. At the end of April, IT Sectors and Jute industries provided selective relaxation to reopen with 25 and 15% workforce, respectively. The Government of West Bengal permitted to deploy a cent per cent workforce in Jute industries from 1st June to overcome the problem of job loss and prevent the migration of labourers (*The New Indian Express* 2020). Pollution level increased slightly at the late stage of the 4th and 5th Phase of lockdown when industrial activities started partially. The sudden suspension of industrial activities had a negative impact on the finance and investment sectors also.

5.4.3 Transport, Construction, and Household Sectors

Congested and polluted roads of Kolkata have changed entirely due to the absence of petrol-fueled cars, buses, and other public and private vehicles on the roads. As a result, a sudden drop in pollutants like NOx, SOx, and fine particulate matter changed air quality. During the 1st Phase of lockdown not only the roadway, but each and every mode of transport such as railways, waterways, and airways also stopped functioning, which reduces the emission of CO, CO_2, and ultimately enhances the quality of air of Kolkata. Public transport was restarted from 18th of May 2020 (*Times of India* 23 March; 18 May, 2020). The emission of dust particles was also reduced due to the temporary halting of construction work during the lockdown. The emission of pollutants from household activities increased (Fig. 5.7).

5.4.4 Impact on Health

Ambient air pollution is a significant cause of death and disease globally. Global data shows that respiratory diseases cause 26%, chronic pulmonary diseases cause 25%, and ischaemic heart diseases cause 17% death globally (GHO, WHO 2016). Pollutants can make COVID-19 Pandemic more potent because air pollution aggravates diseases like Diabetes, Hypertension, Cardiovascular diseases, which makes people vulnerable to COVID-19 fatalities (Bandyopadhyay K 2020). On the other hand, pollution-free clean air enables one to fight with diseases, especially lung diseases, which will help to flatten the curve of COVID-19.

The patient-statistics of three major hospitals in Kolkata, namely, Seth Sukhlal Karnani Memorial Hospital (SSKM), R.G. Kar Medical College & Hospital, and Calcutta Medical College (CMC), are taken into consideration to analyze the impact of air pollution on disease patterns during several phases of major lockdown. During the 1st Phase of lockdown, patients admitted in those hospitals in Kolkata with respiratory diseases (13%), Acute Myocardial (10%), Invasive diseases (20%), and Non-Invasive (14%) diseases, and other types of diseases, including COVID-19 (40%) (Fig. 5.8). It reveals that diseases caused due to air pollution shared fewer patients (CCMIS, Health and Family Welfare Department, Govt. of W. B. 2020).

Figures 5.9 and 5.10 represent the trend of patients admitted to public hospitals in Kolkata, which reveals a sudden decreasing trend in respiratory and invasive diseases from April to July (i.e., from the 1st Phase to 5th Phase of Lockdown). On the other hand, it was high in the Pre Lockdown stage (i.e., from January to March) (Health and Family Welfare Department, Govt. of W.B. 2020). The sudden decrease in air pollutants' emission helped reduce invasive diseases like Pneumonia, Epiglottitis, infectious Arthritis, and specifically the respiratory diseases in Kolkata.

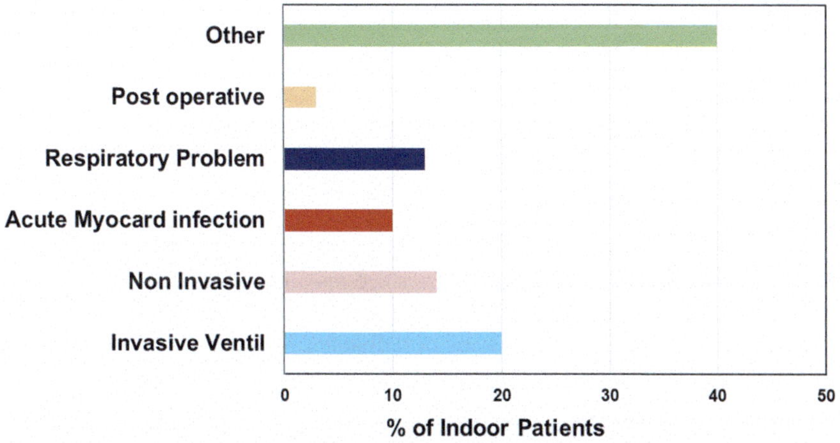

Fig. 5.8 Types of indoor patients during 1st Phase of lockdown in Kolkata (as on 30th April) (*Source* Compiled by author Data: Health and Family Welfare Department, West Bengal)

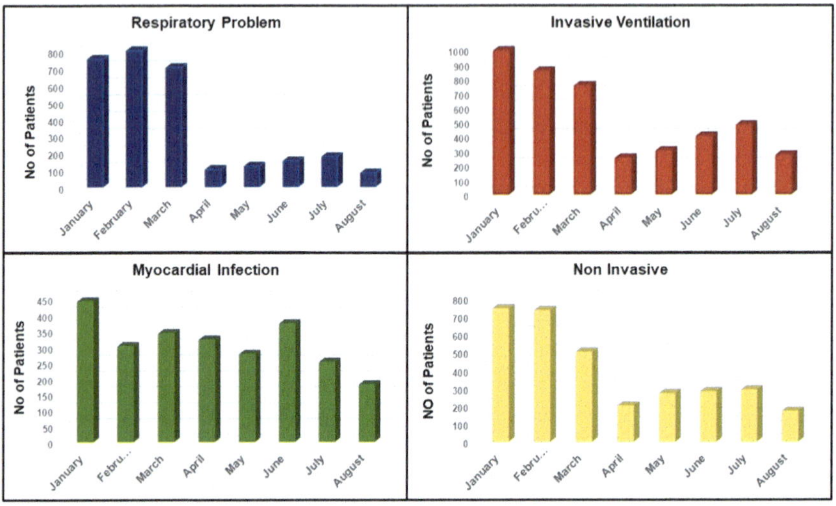

Fig. 5.9 Disease wise monthly variation of indoor patients in Kolkata (*Source* Compiled by author Data: Health and Family Welfare Department, West Bengal)

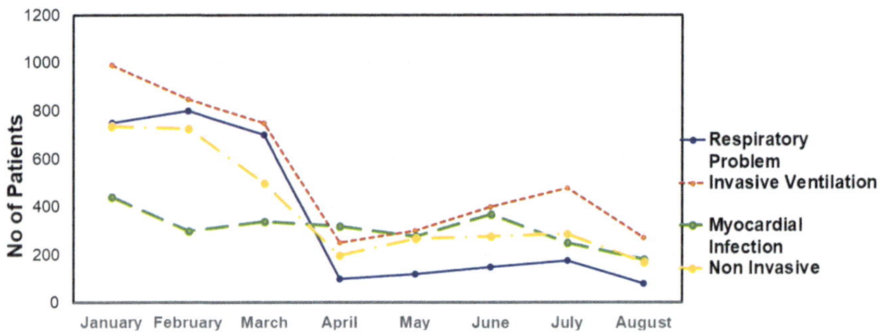

Fig. 5.10 Trend of indoor patients in Kolkata during pre lockdown and lockdown Phase 1 to 5 (*Source* Compiled by author Data: Health and Family Welfare Department, WB)

5.5 Conclusion

Change in air quality and its consequences is eye-opening to all. Nature is changing unexpectedly and reclaiming itself. But it is a short-term improvement of the air quality due to the reduction of emission of pollutants. So, the main concern is how to maintain long-lasting positive environmental change after the lockdown.

- It is possible to reduce air pollutant particles' emission from industries by introducing improved technology such as Spark Ignition engine equipment to maintain

Bharat stage emission standards or BS-VI norm.[5] Along with these old pollutants, vehicles should be phased out.

- To improve the city air quality Calcutta State Transport Corporation will deploy 5000 electric buses and fully electric ferries on the Ganges river by 2030. As of 2019, 80 electric buses have been introduced to the city, with another 100 planned for 2020. These 180 electric buses will lead to an annual reduction of 14086 tonnes of CO_2 emission (Calcutta State Transport Corporation, WBTC 2019).
- Some species of plants are good absorbers of suspended particulate matters. The city environment can be changed by plantation of pollution resistant species with the additional benefit of maintaining city ecology and aesthetic values. The plantation can be done in Bidhannagar, Salt Lake area, Maidan, and densely populated residential areas and in and around the industrial belt along river Hooghly.
- Green infrastructure will be initiated massively mostly in highly congested residential areas where the new plantation is not possible. The green roof, green wall, the green building should be initiated on a priority basis.
- Urban waste management should be run properly.
- The country should move towards non-conventional energy instead of using coal or non-renewable energy. In recent years, India has made progress in renewable electricity deployment and successfully coped up increasing electricity demand and clean cooking. Investing more in renewable energy under the National Clean Air Programme (2019) will help in achieving clean air.
- Emergency Relief Fund should be developed to provide liquid funds to small scale entrepreneurs to manage financial uncertainties.
- Mass awareness through social media can be a key element of environmental movements across India's megacities.
- A sustainable environmental management plan is required to improve the environment of Kolkata city.

Indian megacities experienced the blissful blue sky and cleaner air during the nationwide lockdown. It gave opportunities to rethink and properly implement eco-friendly technologies in India's energy and industrial, commercial, and transport sectors. A large portion of pollution is locally generated and associated with emissions from the power plants, industries, vehicles, road dust, burning of waste materials, cooking fuel, etc. It needs area-specific effective planning and mass awareness on a priority basis. Due to restricted activities during confinement, the present environmental co-benefits scenario strengthened the positive relationship between man and its natural environment and raised hope for a pollution-free city.

[5]The Bharat Stage (BS) are emission standards instituted by the Government of India to regulate the output of air pollutants from motor vehicles. The central government has mandated that vehicle makers must manufacture, sell, and register only BS-VI (BS6) vehicles from April 1, 2020.

References

Air quality improving in India as 130 bn people stay home amid lockdown (2020) Business Standard, 22 April. Accessed from https://www.business-standard.com/article/current-affairs/air-quality-improves-significantly-across-india-amid-lockdown-cpcb-120040200831_1.html

AURA (2020) Reductions in sulfur dioxide & nitrogen dioxide air pollution over South Asia associated with efforts to control the spread of COVID-19. Accessed from https://aura.gsfc.nasa.gov/science/feature-20200501.html

Bandyopadhyay K (2020) Kolkata: Transport system gears up for Monday test as offices reopen. The Times of India, May 18. Retrieved from https://timesofindia.indiatimes.com/city/kolkata/transport-system-gears-up-for-monday-test-as-offices-reopen/articleshow/75795456.cms

Bandyopadhyay K, TNN (2020) Lockdown, rain make Kolkata's air cleanest in recorded history. The Times of India, 29 April. Retrieved from https://timesofindia.indiatimes.com/city/kolkata/lockdown-rain-make-kolkatas-air-cleanest-in-recorded-history/articleshow/75436576.cm

BBC (2020) Amphan: Kolkata devastated as cyclone kills scores in India and Bangladesh. BBC News. Retrieved from https://www.bbc.com/news/world-asia-india-52749935#:~:text=Amphan%3A%20Kolkata%20devastated%20as%20cyclone%20kills%20scores%20in%20India%20and%20Bangladesh,-21%20May&text=The%20eastern%20Indian%20city%20of,with%20ferocious%20wind%20and%20rain

Carbon Brief (2020) EMISSIONS, analysis: coronavirus temporarily reduced China's CO2 emissions by a quarter, 19 February. Available at https://www.carbonbrief.org/

Census of India (2001, 2011) West Bengal, District Census Handbook Kolkata, Village and Town wise Primary Census Abstract (PCA). Series 20 (Pt XII-A & B)

Chen K et al (2020) Air pollution reduction and mortality benefit during the COVID-19 outbreak in China. medRxiv preprint. https://doi.org/10.1101/2020.03.23.20039842

Delhi, Mumbai, Bengaluru and Kolkata attain 95% of clean air targets during lockdown (2020) Climate Trends, 3 July. News Desk. https://citizenmatters.in/lockdown-reduces-air-pollution-in-indian-megacities-delhi-mumbai-kolkata-bengaluru-19370

Economic Outlook (2020) Accessed from http://economicoutlook.cmie.com/

Economy of Kolkata. Wikipedia. https://en.wikipedia.org/wiki/Economy_of_Kolkata

ESA (2020) Sulphur dioxide concentrations drop over India during CovID-19, July. https://www.esa.int/ESA_Multimedia/Images/2020/06/Sulphur_dioxide_concentrations_over_India

Global CO_2 emissions to decline 8 per cent in 2020, surpassing any previous declines: IEA

Global Alliance on Health and Pollution (GAHP) Report (2019) Pollution and health metrics: global, regional and country analysis

Global Energy Review (2020) The Economic Times, 24 May. Accessed from https://energy.economictimes.indiatimes.com/

Health and Family Welfare Department, Government of West Bengal (2020) Data accessed from http://ccu.wbhealth.gov.in:8001/

India suffers most pollution linked deaths in world, study finds (2020) WION News, 19 December. Accessed from https://www.wionews.com/india-news/india-leads-world-in-pollution-linked-deaths-study-269127

IQAir (2019) 'World Air Quality Report', 2019 World Air Quality Report, pp 1–22. Available at https://www.iqair.com/world-most-polluted-cities/world-air-quality-report-2019-en.pdf

IQAir (2020) COVID-19Air Quality Report, pp 1–13. Available at https://www.iqair.com/us/blog/air-quality/report-impact-of-covid-19-on-global-air-quality-earth-day

Javed Z (2020). Kolkata hint of eco turnaround in reverse migration trend. The Times of India. Published and accessed on 13 June. http://timesofindia.indiatimes.com/articleshow/76350299.cms?utm_source=contentofinterest&utm_medium=text&utm_campaign=cppst

Kolkata Municipal Corporation (CMC) Kolkata city map. Accessed from https://www.kmcgov.in/KMCPortal/jsp/UAABoroughWiseMap.jsp

Kota SH, Guo H, Myllyvirta L, Hu J, Sahu SK, Garaga R et al (2018) Year-long simulation of gaseous and particulate air pollutants in India. Atmos Environ 180:244–255. https://doi.org/10.1016/j.atmosenv.2018.03.003

Le Quere C et al (2020) Temporary reduction in daily global CO_2 emissions during the COVID-19 forced confinement. Nat Climate Change 10:647–653. https://www.nature.com/articles/s41558-020-0797-x#citeas

Location of Kolkata Metropolitan City (2020) Created through Google Earth. http://google.com/earth

Lockdown: Jute industry in Bengal allowed to deploy 100 percent workforce (2020) The New Indian Express, 29 May. Accessed from https://www.newindianexpress.com/nation/2020/may/29/lockdown-jute-industry-in-bengal-allowed-to-deploy-100-per-cent-workforce-2149724.html

MoEFC. Ministry of Environment, Forest and Climate Change (2019) In: SNk, Sundaray, DSR, Bharadwaj (eds) National Clean Air Programme, New Delhi

Multi-pollutant emission inventory (2020) Accessed from https://urbanemissions.info/india-apna/kolkata-india/

NAQI (National Air Quality Index) (2020) Central Pollution Control Board (CPCB). https://app.cpcbccr.com/AQI_India/

National Atlas and Thematic Mapping Organization (NATMO) West Bengal, India Base Map. Accesed from http://www.natmo.gov.in/

PHFI and CEH (2017) Air pollution and health in India: a review of the current evidence and opportunities for the future. Public Health Foundation of India and Center for Environmental Health, p 64

Sharma S, Zhang M, Gao AJ, Zhang H, Kota H (2020) Effect of restricted emissions during COVID-19 on air quality in India. Sci Total Environ 728:138878. https://doi.org/10.1016/j.scitotenv.2020.138878

United in Science Report (2020) Accessed from https://trello-attachments.s3.amazonaws.com/5f560af19197118edf74cf93/5f59f8b11a9063544de4bf39/cdb10977949b38128408f5322f9f676d/United_In_Science_2020_8_Sep_FINAL_LowResBetterQuality.pdf

Urbanemission.info retrieved from various associated pages of URL www.urbanemission.info

Wang P, Guo H, Hu J, Kota H, Qi Ying, Zhang H (2019) Responses of $PM_{2.5}$ and O_3 concentrations to changes of meteorology and emissions in China. Sci Total Environ 662:297–306. https://doi.org/10.1016/j.scitotenv.2019.01.227

West Bengal Transport Corporation (WBTC)

WHO (2005) WHO Air quality guidelines for particulate matter, ozone, nitrogen dioxide and sulfur dioxide: global update 2005; summary of risk assessment. https://apps.who.int/iris/bitstream/handle/10665/69477/WHO_SDE_PHE_OEH_06.02_eng.pdf;jsessionid=5F91D15DE7CC85A08337639001B8B5BD?sequence=1

WHO (2016) Mortality and burden of disease from ambient air pollution: situation and trends. GHO. Accessed from https://www.who.int/gho/phe/outdoor_air_pollution/burden_text/en/

WHO (2020) How does COVID-19 spread? Accessed 20 July 2020. https://www.who.int/news-room/q-a-detail

World's Biggest Lockdown May Have Cost Rs 7–8 Lakh Crore to Indian Economy. The Economic Times (2020) Retrieved from www.economictimes.indiatimes.com/news/economy/finance/worlds-biggest-lockdown-may-have-cost-rs-7-8-lakh-crore-to-indianconomy/articleshow/75123004.cms?from=mdr

World Health Organization (2020) WHO Director general's opening remarks at the media briefing on covid-19. Accessed from https://www.who.int/dg/speeches/detail/who-director-general-s-opening-remarks-at-the-media-briefing-on-covid-19—27-july-2020

Xing J, Pleim J, Mathur R, Pouliot G, Hogrefe C, Gan CM, Wei C (2013) Historical gaseous and primary aerosol emissions in the United States from 1990 to 2010. Atmos Chem Phys 13(15):7531–7549. https://acp.copernicus.org/articles/13/7531/2013/acp-13-7531-2013.html

Zhang H, Hu J, Kleeman M, Ying Q (2014) Source apportionment of sulfate and nitrate particulate matter in the eastern United States and effectiveness of emission control programs. Sci Total Environ 490:171–181. https://doi.org/10.1016/j.scitotenv.2014.04.064

Zhipeng P, Han G, Ma X, Su H, Gon W (2020) Response of major air pollutants to COVID-19 lockdowns in China. Sci Total Environ 743:140879. https://doi.org/10.1016/j.scitotenv.2020.140879

Kushagra Rajendra is Associate Professor (Earth and Environmental Science) and Head of School of Earth and Environmental Science, Amity University Haryana. He holds a Doctorate in Water resources from Jawaharlal Nehru University. He is educated and trained in interdisciplinary aspects of Environmental Science. His research area falls in applied aspects of interdisciplinarity of Earth and Environment, including water resource, pollution control, the social, and economic perspective of environmental conservation. He is a VLIR UOS fellow from Belgium and recipient of the President of India Gold Medal. His research publication includes Water Treatment, Geogenic Contamination, Social context of Environmental Conservation, and presently working on Environmental Payment System (EPS) and Planning aspects of Aravalli conservation.

Namita Yadav is a Ph.D. scholar at the School of Earth and Environmental Sciences, Amity University Haryana. She did her Master in Environmental Science and now prospecting her thesis work on Climate Change Induced Variability in Atmospheric Factors like heat waves, Monsoonal precipitation, and others.

Dipanjana Chakraborty is an Assistant Professor of the Department of Geography at Sitalkuchi College, affiliated to CBPB University. She holds a Doctorate in Urban morphology from the University of North Bengal. Her research chiefly focuses on different aspects of urban life, built environment, eco-urban planning, and interdisciplinary application of Social Sciences. She has been awarded a doctoral fellowship from the Indian Council of Social Science Research (ICSSR). Her research publication includes Urban Waste Management, Eco-Tourism, Flood Hazard Mitigation, and Management and Women Empowerment. Presently she is supervising a research project on the Changing Paradigm of Tea Plantation in Jalpaiguri District.

Chapter 6
Questioning the Green Recovery: A Take on Post-COVID Scenario

Anindya Basu and Lopamudra Bakshi Basu

Abstract As the COVID-19 pandemic rampaged the world, many found solace in the improving air and water quality, more fabulous greener cover, and invigorating bio-diversity. However, the question arises whether these positive changes could be maintained in the post-COVID world, with an increase in economic activities alongside the unrestricted mobility of people and goods. The paper tries to gauge the situation beyond this short-lived moment of environmental rejuvenation, which seems quite grave. The entire work has been based on extensive literature review and perception studies. The adequate provision of sanitary landfills and medical incinerators for safe disposal of medical wastes was already in question in developing countries. The high usage of protective gear like personal protective equipment, gloves, masks, face shields, etc. will incite unscientific disposal and burning—leading to soil and air pollution. The quantum of air pollution may get higher due to avoidance of mass and para-transit modes of communication and over-reliance on personal vehicles to continue social distancing. Besides, maintaining self-hygiene at regular intervals, an antidote for corona infectivity, in a water-stressed world where 2.2 billion people do not have access to safe potable water will add to water scarcity. Further, indiscriminate usage of low-quality sanitizer to fight the virus causes serious health hazards and affects the environment detrimentally. For reaping the benefit of the current green environment, sustainable strategies have to be developed to cope with the new normal.

Keywords Environmental rejuvenation · Medical waste · Social distancing · Water scarcity · Self-hygiene · New normal

---

A. Basu (✉)
Department of Geography, Diamond Harbour Women's University, Sarisha, West Bengal, India
e-mail: basu_anindya2004@yahoo.com

L. B. Basu
Department of Geography, Heramba Chandra College, Kolkata, West Bengal, India

M. Mishra and R. B. Singh, *COVID-19 Pandemic Trajectory in the Developing World*, Advances in Geographical and Environmental Sciences,
https://doi.org/10.1007/978-981-33-6440-0_6

6.1 Backdrop

World Health Organization (WHO), on 30 January 2020, declared COVID-19 as a public health emergency of international concern. Globally, as on 7 August 2020, there have been 18,902,735 confirmed cases of COVID-19, including 709,511 deaths, across 213 countries reported to WHO; region-wise, the number of cases is highest in America, followed by Europe and South-east Asia and India registering 2,027,074 confirmed cases as of now (WHO 2020a). To date, when the chapter is being written, no sanctioned medication or vaccine is available to deal with this novel coronavirus, and the infection rate is increasing unabated.

COVID-19 pandemic is an enormous challenge to societies and economies across the world. The immediate priority for governments has been to deal with the health crisis and save lives. In the current setup, preventive measures and a healthy lifestyle with an efficient immune system have been suggested by WHO to fight and stay safe from COVID-19. So, to forefend the infection, countries have strictly regulated the movement of people and transportation, reduced human interactions, imposed strict quarantine, barred large-scale private and public congregations, encouraged social distancing, restricted private and public transportations and economic behaviour (Fernandes 2020; Wang and Su 2020). There is huge pressure on the health care sector, which is now challenged with the mammoth patient influx and limited medical infrastructure. Upgrading medical facilities, enormous testing campaigns, and recalibration of health policies are required to counter this public health crisis (WHO 2020b). As strict containment measures have resulted in a severe drop in economic activities (OECD 2020), another key priority has been to quickly adopt the support policies that minimize the destruction of jobs, incomes, value chains, and production capacity. These interludes will not just be an interim issue, but can also have long-term effects on the various cohorts.

Amid all the daily activities and business that the pandemic has impacted, the transport sector in general and travel behaviour of the people in particular which would likely drive the entire world shift towards the "new normal". The unprecedented situation has impelled people to rethink their choices and behaviour. In this period of widespread lockdowns, i.e. complete cease of movements, the world has witnessed the positive environmental changes with cleaner rivers and clearer skies (Gardiner 2020a). As the popular saying goes, "Every crisis brings along with it an opportunity"—the COVID-19, despite posing specific difficulties, has provided us with the opportunity to let the Earth heal itself by switching to eco-friendly practices for the sake of the public health revival.

6.2 On Gauging the Crisis

Due to the highly transmissible nature of the disease, people have been left with no choice but to stay at home and take several safety measures at a personal level. The situation has been quite challenging from disinfecting every single material fetched for daily use to avoid social interfaces. This extremely tough situation has compelled everyone to think about managing nature and its resources differently and efficiently to cope with this pandemic.

This chapter's main objective is to highlight the probable situations that are supposed to arise beyond this "short-lived moment" of environmental rejuvenation. Therefore, it is vital to reconsider the conventional principle of environmental sustainability. The unwarranted use of plastics, sanitizers, and water might lead to a severe crisis in the entire world. The study identifies the challenges exposed by the pandemic, which needs immediate redressal. New ideas for handling the environment post-COVID-19 have to be considered for sustainable development and safe living. For the purpose of keeping the focus on saving the green environment, safe living of the people for health security, the principle of sustainable development needs to be reinforced adequately by adding new controlling parameters. However, the critical question remains, "What needs to be prioritized or added or modified to the existing principle of protecting the environ?"

This study was conducted to gauge the probable crisis and environmental deterioration, especially in the developing world, that would ensue in the post-COVID period. Most of the research paper is based on secondary data with heavy reliance on available "online" materials. COVID-19, lockdown, environment, pollutants, biomedical waste, water scarcity/crisis, and hand sanitizer were the primary search theme, followed by associated terms. A systematic assessment has been carried out, based on available literature in PubMed, Google Scholar, and Scopus database up to August 2020. More than 69 million indexed documents were found, and the search was refined by specifying the language and the document type. The study's theme, i.e. COVID-19 and environmental impact, was manually checked while scrutinizing the text.

Besides, a qualitative study was made, primarily relying on perceptions of randomly selected residents of several parts from the developing Global South countries to bring out how their daily behaviour trajectories were affected by the pandemic, which had an indirect bearing on their immediate environment.

6.3 Issues to Ponder Over

The interrelationship between COVID-19 and the environment is an evolving research topic. Though the present pandemic situation is showing a positive impact on the environmental condition, it is also pertinent to think about the world's post-COVID scenario. The excessive use of plastics, their disposal, the shift in the choice of the transportation mode—all these are likely to put a combined adverse impact on the environment. The health and hygiene norms that demand using water and sanitizer may pose a problem of water crisis along with health hazard. To highlight the grave situation that is likely to come up in the future, an in-depth literature review was done and based on that, four sections have been identified which need to be addressed. The four clusters are (a) waste generation, which is an important aspect of the current situation. The increased use of personal protective equipment, gloves, masks, face shields, along with sanitizer bottles made of plastics is exerting pressure on the waste management system and increasing environmental concerns. The improper disposal of waste materials will adversely influence the soil, groundwater, and air quality. (b) Air pollution is the other parameter that requires a thorough study. The unscientific burning of wastes will add to the particulate matters present in the atmosphere. (c) Water stress is likely to follow if strict health hygiene has to be maintained. The excessive use of water will reduce the groundwater level and also lead to chemical concentration in soil and (d) chemical exposure, highlighting the impact of sanitizers on health. A comprehensive replication of the study under each cluster has been done.

6.3.1 Waste Generation

Among several impacts of the current pandemic, municipal waste management is an important aspect (Smart Waste Report European Union 2020). This pandemic has completely altered the waste generation dynamics, giving way to a new "throw-away" culture, creating a dilemma regarding management practices among policymakers (Mallapur 2020). Few countries have already developed efficient municipal waste segregation, treatment, and disposal facilities. Japan, for instance, manages to incinerate 74% and recycles 17% of the country's waste (Mollica and Balestieri 2020). Several European countries have also traced on technological innovations to facilitate energy recovery from wastes (Kyriakis et al. 2019; Istrate et al. 2020). On the other hand, the developing nations, lacking with the technologies, need to limit the huge amount of non-beneficial residual waste generated from different sources as an immediate action.

6.3.1.1 Plastic Waste

In modern society, plastics, with their enhanced physicochemical properties, play a pivotal role in various sectors (Geyer et al. 2017). However, the usage of single-use plastics (SUP) has raised environmental concerns because of their low recyclability (UNEP 2020b; Silva et al. 2020). According to Plastics Europe (2019) 359 million metric tons (Mt) of plastic were produced in 2018 while plastic waste generation reached 6.9 Mt. Out of the total plastic-based waste generated, a whopping 42% was inefficiently treated that raises the alarm (Hahladakis et al. 2018; OECD Statistics

2018). Zheng and Suh (2019) highlighted that burning of plastic wastes contributed to 15% of the global greenhouse gas emissions.

The enforcement of total lockdowns and strict physical distancing makes going out for marketing, eating out very infrequent. As an alternative discourse, the reliance on online shopping increases remarkably (Hyun 2020). The home delivery of groceries and food items has surged, generating lots of plastic packaging waste involving mainly polypropylene, polyethylene, terephthalate, polystyrene (Tenenbaum 2020). Moreover, there is an increase in plastic packaging waste from medical industries which are trying hard to meet the demand for essential medical logistics worldwide (WHO 2020c). The concept of Plastic Waste Footprint (PWF) used to capture the environmental footprint of a plastic product throughout its entire life cycle proves how detrimental uncontrolled plastic use can be (Boucher and Billard 2019). The present pandemic further introduces disruptive changes creating "butterfly effects" on post-COVID-19 economies and environment (Klemeš et al. 2020).

Normal recycling of plastics has got reduced now (Kaufman and Chasan 2020), and handling of an increased amount of general plastic waste has become a challenge (Ferronato and Torretta 2019). Few countries like the US, Brazil, and few European nations have rolled back their usual waste disposal and recycling industry in this situation, fearing widespread contamination making the situation more critical. In contrast, nations like the UK, France, Spain, Italy, despite all odds, have continued business in the waste management segment (BIR 2020). The post-COVID-19 slump in oil prices is also hampering the growth of the plastics recycling sector, while the relatively high-priced recycled plastics has always been a concern for creating broader markets for waste recycling (Bell 2016). This pandemic has forced people to take two-steps back regarding rollback policies against single-use plastic bags (Sinclair et al. 2018). It is feared that with these temporary relaxations, the market again will be flooded with unsustainable plastic usage having a long-term influence on consumer behaviours (Tenenbaum 2020). Even WHO has urged to increase the production of plastics by 40% to meet rising global demand (WHO 2020d). However, Climate Action Tracker (2020) has harped that greener products like bioplastics should be incentivized. Long-yearned sustainability may be achieved if the principle of the 4-R Programme—Refuse, Reduce, Reuse, and Recycle is embraced by the citizens (Huang 2016).

6.3.1.2 Biomedical Waste

Infectious wastes are broadly classified into—bio-medical waste, regulated medical waste, and clinical waste (International Solid Waste Association 2020). WHO defines the waste generated by various categories of medical institutes as bio-medical (BMW) and healthcare waste (HCW) (WHO 2015). The BMW includes waste of sharps, infectious, pathological, pharmaceutical, chemical, and radioactive nature with hazardous compositions (WHO 2017).

The experiences from SARS-CoV, Ebola, and MERS-CoV disease outbreaks highlighted the requirement for safe bio-medical waste management for infection prevention and control (Sharma et al. 2020b). Before this crisis, about 2 billion people worldwide lacked access to the waste collection, while 3 billion were out of the ambit of controlled waste disposal facilities (Wilson et al. 2015b). To handle the COVID-19-related infectious wastes, WHO, US Occupational Safety and Health Administration (OSHA), EU, and other agencies have issued a set of new handling guidelines (ACR + 2020). Indian Central Pollution Control Board (CPCB) has also issued a detailed course of action in this regard (CPCB 2020).

Public health advisories and recommendations from several sources tend to generate a high amount of bio-medical wastes, comprising the used and contaminated plastic-based gears like personal protective equipment (PPEs), face masks, gloves, wipes, and other protective equipment (UNEP 2020b). However, most of which, if not appropriately managed after disposal, might pose environmental and health threats (Kampf et al. 2020). Various studies indicate that the coronavirus can survive for quite a few days on different material surfaces, which seeks scientific management of bio-medical wastes as the utmost important (van Doremalen et al. 2020).

Though many governments across the world have existing legislation and regulations in place for proper disposal of infectious medical wastes, the increasing amount of bio-medical waste in the COVID-19 situation has overwhelmed existing medical transport and disposal infrastructure. In developed countries, green and sustainable waste management strategies are in place (Nzediegwu and Chang 2020). The scenario of most of the developing countries is quite different. The developing countries, which are limited by the lack of technologies and economic resources, treat the solid wastes unscientifically—dumping in the open. Landfills are poorly managed, and the waste-pickers are exposed to risks as there is lack of protective gear for them (World Bank 2019; Mol and Caldas 2020). The WHO model estimated that there would be a requirement of 89 million medical each month along with 76 million examination gloves. At the same time, international demand for goggles will clock 1.6 million per month (WHO 2020d). There must be a surge in the volume of single-use plastics due to growing concerns about personal health and hygiene.

The amount of infectious medical waste has been increased remarkably by 600%—from 40 tonnes per day to 240 tonnes per day in Hubei Province in the People's Republic of China (Shi and Zheng 2020). It has been estimated by the Asian Development Bank (ADB) (2020) that in several cities of south-east Asia like Manila, Jakarta, Kuala Lumpur, Bangkok, and Ha Noi, the additional amount of medical waste would vary from 280 metric tonnes to 154 metric tonnes per day depending on population taking per head 3.4 kilogrammes of infectious medical waste as the general norm. To deal with such mammoth infectious wastes, the public authorities of Wuhan deploys mobile incineration facilities, which has increased the capacity by four-fold (Saadat et al. 2020).

The risk of contact transmission from bio-medical wastes is not the only concern. The improper disposal might cause physical injuries through sharps, and adverse environmental effects like soil and groundwater contamination through sludge flow, which are also quite common (Datta et al. 2018). If not managed soundly, uncontrolled incineration causes toxins release in the open-air, leading to the secondary transmission of diseases to humans. However, installing the technology-based, highly automated infrastructure in short notice is difficult. So the short term responses are sensible alternatives. Creating detailed inventory from on-site, mobile, and off-site units, adopting 3S methodology (involving Sorting, Segregation, Storage) may be useful. Implementing the SAICM (Strategic Approach to International Chemicals Management) to study the impacts on wider environment may be solicited to handle the situation. As a stop-gap arrangement, special waste collection buckets can be arranged for collecting disposable bio-medical wastes following the Nigeria Centre for Disease Control (NCDC) guidelines (NCDC 2020). It will allow time to chalk out the strategies for long-term responses. Initiating legislative changes, creating robust systems for waste segregation, collection & management, investing on Sustainability Assessment of Technologies (SAT), opting for Best Available Technology (BAT) for source segregation, primary disposal, and destruction of waste or recovery of materials and carrying out Best Environmental Practices (BEP) to manage the environment soundly, in line with the Stockholm Convention can be carried out (UNEP 2020a) involving social viability alongside environmental and economic feasibility.

For safe disposal, it has been suggested that pandemic-related waste should be double-bagged, "swan neck" tied, and the outside sprayed with a 0.5% chlorine disinfectant solution (ACR + 2020). If the disposal system is under severe stress, then though not preferable, it has been suggested that the double-bagged waste should be stored for 72 h before being disposed of with the general household waste. Safe disposal of greywater or water from washing PPE, surfaces, and floors is also an area of concern (WHO 2020e). Stringent enforcement of air quality controls to limit toxic pollutants' emissions requires increasing the capacity of advanced technologies (White 2020). Encapsulation, sanitary landfills, mobile incinerators, autoclave (steam sterilization) units, gas sterilization, industrial furnaces, thermal inactivation, and micro-wave or radio-wave treatment (Liu et al. 2015) has to be put in place following the guidelines of the WHO, the Basel Convention, and the United Nations Environment Programme (Sharma et al. 2020a). Life cycle assessment (LCA) and related approaches indicate that incineration with waste heat recovery is an alternative through which plastics' chemical energy content can be recovered for useful purposes (Hong et al. 2018). But certain issues like trace emissions of dioxins and furans (Makarichi et al. 2018) are linked with the widespread use of incineration with heat recovery. A new technology, hydrothermal carbonization, is being implemented, coupled with high pressure and temperature autoclaving technique (Shen et al. 2017). When the period of normalcy returns, these ramped-up treatment facilities can be repurposed to treat municipal solid waste. Apart from the disposal, safe transporting of the wastes is also essential, which requires labelling, specific colour coding, sterilization, trained drivers and waste collectors, dedicated routes, waste tracking systems, and emission monitoring (Refer Fig. 6.1). As adopting and implementing of modern technologies is subject to affordability and adaptability United Nations Environment Programme (UNEP) has been proactive in managing the situation and assessing capabilities, and its director Inger Andersen pledged that:

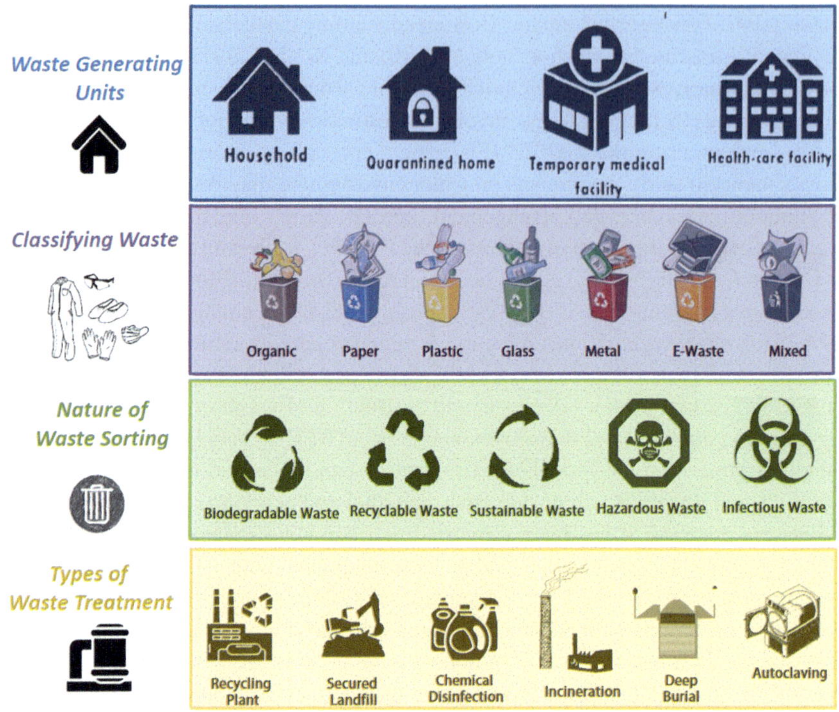

Fig. 6.1 Structural framework of solid waste management during COVID times (*Source* Compiled by authors)

> Our response is to support Member States in addressing immediate challenges from the medical emergency, such as strengthening waste management systems. (UNEP 2020a)

Most urban areas in the developing world lack resources like vacant plots, technical know-how to tackle the increasing amount of waste generated (Hoornweg and Bhada-Tata 2012), and COVID-19 has created a further logistical challenge. In India, there is an infrastructural set up for recycling 60% of the plastic waste (Alpizar et al. 2020), but a substantial increase of infectious plastic waste is leading to uncontrolled landfilling and unscientific burning to avoid virus contagion (Corburn et al. 2020). It is estimated that India, facing the pandemic's extreme wrath, needs 25 lakh units of Personal Protective Equipment (PPE) every day. For every 1,000 Covid-19 tests that are carried out, about 22 kg of plastic waste is generated, which is creating massive pressure on the disposal system (Mishra 2020). Basu (2020) discussed how the Council of Scientific and Industrial Research has planned to utilize the COVID-19 plastic surge by converting the single-use plastics of the PPE kits into standardized plastic pellets and utilizing them in road construction, plastic cover manufacturing, etc.

6.3.2 *Air Pollution*

Fresh air is the best ventilator and can serve as a unique natural remedial measure. The testament to that is not from COVID-19, but from SARS. A relevant case study to support this statement is the outbreak of SARS during 2003 in Vietnam. It was observed that there were deaths and extensive transmission in hospitals having a closed air condition system. On the contrary, in a hospital with spacious rooms, high ceilings, ceiling fans, and large windows kept open for cross-ventilation, there were no cases of transmission (Le Dang Ha et al. 2004).

Air pollution has become a topic of robust debate at all levels mainly because of the heightened anthropogenic activities, e.g. rapid urbanization, higher population growth, increased energy consumption, and vehicular emission and industrial emission (Dadhich et al. 2018; Ghose et al. 2005; Gupta et al. 2008). With this pandemic, nature is recuperating while people stay at home.

6.3.2.1 Vehicular Emission Scenario

Process of unlocking across the world the situation has started to show a change. Now, the fear is that the world faces perils with more traffic, more pollution, and climate change in the future that worsen faster than before (Gardiner 2020a). In April, with worldwide shutdowns, regular global carbon emissions went down by 17% in comparison to last year. As of 11 June, the fresh datasets show that it is only about 5% lower than at the same point in 2019, though usual activities have not yet fully resumed (Gardiner 2020b; National Geographic Channel 2020).

Most of the developing cities have witnessed a rapid rise in the pollution level. Due to densely distributed small-scale industries without any pollution controlling measures, unplanned traffic flow, and congestion, the Indian cities are to withstand higher pollution of ambient air (Gupta et al. 2008). Today, the utmost threat on Earth's planet is the conquest by the tiny particle causing Corona Virus Disease (COVID-19). COVID-19 is transmitted via airborne droplets. Therefore, reducing the effective population density in public meeting places is the basic rationale behind the quarantine strategy favoured by most countries responding to the outbreak. The moral coming out from this discussion is to stay inside with open windows and not to flock so that the air can diffuse the germ. However, it is necessary to check the air quality as poor air quality with GHGs and SPM may give rise to several other health problems, which may intensify the infection.

The cumulative concentration of greenhouse gases (CO_2, CH_4, N_2O, etc.) has led to global warming. Humans have been mainly responsible for this situation as they destroyed nature as per their own whims and desire. As an inevitable consequence, environmental pollution has become a big issue of the present day almost globally. The restricted movement has also led to a decrease in the industrial emission throughout the world and, therefore, a reduction in the use of fossil fuel (Bremer et al. 2019; Coutts et al. 2010). Various studies related to the environment have been conducted, highlighting a drop in the particulate matters and a lower carbon dioxide (Agarwal et al. 2017; Steinle et al. 2013; IRENA 2020). The lockdown phase unveiled a significant decrease in CO_2 levels in all the developing cities of the world. The Indian cities are also showing signs of recovery in the pollution level. Due to pandemic COVID-19, industries have been shut; transport sectors are dwindling due to domestic and international flights' complete shutdown. Tourism activities have stopped, and hence all hotels, tourism units, homestays, and other recreational activities related to tourism have been stationary. The running of trains, buses, and passenger-vessel services have also been ceased. All these have significantly abridged the consumption of fossil fuels; and, subsequent emission of CO_2 has also touched the trough in the existing CO_2 profile of the region. Just a week of lockdown had brought about a difference of nearly 20% in the values of particulate matter (PM_1 and $PM_{2.5}$).

Figure 6.2 shows a drastic fall in the level of nitrogen dioxide during the lockdown phase in India's metro cities. However, the year 2019 shows a high rate of increase of the same air element during the same period (i.e. the month of March). This scenario is quite alarming as it includes high usage of the mass transit system. The post-COVID scenario will, therefore, need a comprehensive management & supervision mechanism to implement. With the start of the unlocking phases, the situation is demanding to become normal. India's road transport system has long been struggling in the city with inefficiencies, such as high congestion levels, limited multimodal integration, inadequate public transit systems, degraded footpaths, and non-existing cycle tracks (TERI 2020). The COVID-19 crisis will craft many new challenges for this sector, especially in urban areas with high travel demand. Increased risks associated with crowded places pooled with social distancing measures in public and shared transport are likely to influence commuters' modal picks. People are most likely to change their mode of transport for work trips post COVID-19. A sharp decline can be assumed in the usage of bus and metro services, and cases of shared mobility might drop as well. Thus, a shift towards private vehicles and intermediate public transport (IPT) such as taxis and autorickshaws, seems inescapable. The percentage share of non-motorized modes of transport may also increase, especially for short-distance trips. Therefore, the problem of congestion and pollution that plague different cities of the country could worsen in the near future.

As cities begin to ease out lockdown restrictions, public transport services will need to play with limited capacity to conform social distancing norms. Even cabs and other shared services will be affected due to similar rules. People may try to shift from public and shared transport services due to higher professed risks, leading to increased use of private travel modes. As Indian cities are struggling to reduce the dependence on private motor vehicles, this crisis may further give way to adverse outcomes created by high motorization rates. Further, the increased demand for contactless services provided by e-commerce platforms indicates a rise in the number of light commercial vehicles and 2-wheelers. This could lead to adverse outcomes for congestion, pollution, and of course, road safety. These passenger and freight mobility sectors' changes may impact the demand for energy and pollution levels in most cities of the country. The effect on greenhouse gas (GHG) emissions will also

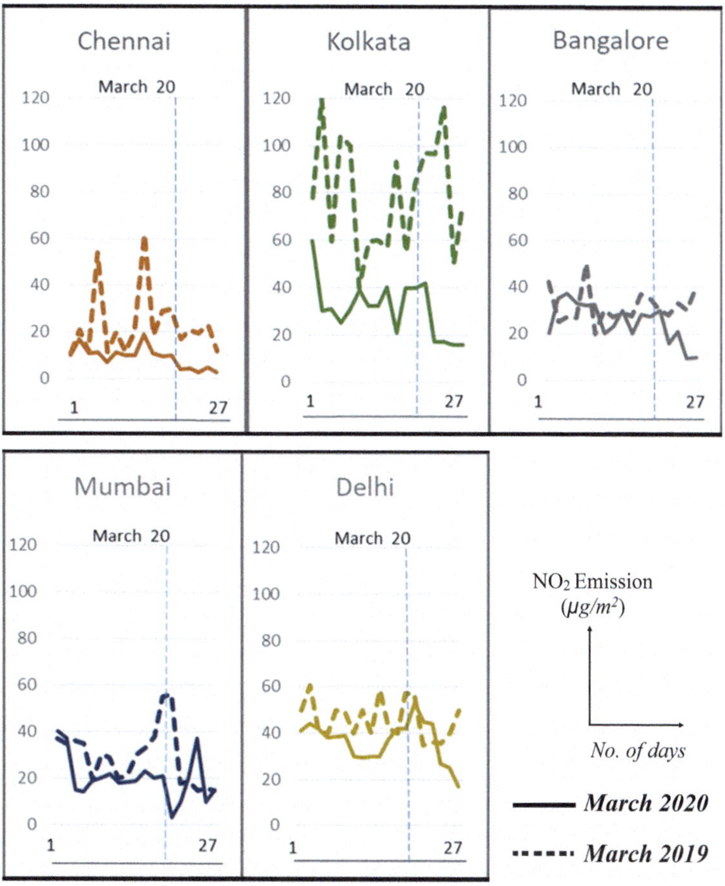

Fig. 6.2 Average daily nitrogen dioxide emission (micrograms/sq metre) in March 2020 in major cities of India (*Data Source* Centre for Research on Clean Air and Energy, Central Pollution Control Board)

depend on the technology used in these vehicles. Hence, the outlining of policies for the implementation of cleaner technologies will be crucial to boosting the sector's sustainability in the long term.

Various past surveys (Nesheli et al. 2017; Bansal and Kockelman 2016; Banerjee et al. 2010) revealed personalized vehicles' preference as the primary mode of travel to work and recreational places. Public transport was well represented, as a fair size of the population availed the metro rail and bus services. The intermediate public transport (IPT) modes, such as private taxis and autorickshaws, also played a crucial role as a feeder service. The high usage of public transport despite the availability of private cars suggests the prevalence of choice users, while a very small section of the people either walk or use a bicycle to the workplaces. Distance between home and workplace might hinder the efforts of cycling to work, but this could also be attributed to the lack of cycling infrastructure in Indian metropolitan cities. The motorized two-wheelers were also representing a fair share. A modal shift seems inevitable post lockdown. The crisis would alter people's choices. A sizeable decrease will be observed in the usage of public transport services. People will prefer to use private cars and two-wheelers. Substitution of metro services with intermediate public transport, such as autorickshaws, taxis, and even preference of shared cabs might be noticed. The decrease will also be witnessed in the use of local trains, mostly by residents in Mumbai. The decrease in the usage of mass transit systems will have an immense impact on the environment. This will not only increase the congestion but also further degrade the air quality of the urban cities of India. With the growing demand for online shopping, the increase in the number of motorized two-wheelers is unavoidable. Constrain of road space will surely seek proper traffic management as the congestion and road accidents are likely to increase. The internationally settled target is to reduce emissions by at least 7.6% every year up to 2050 to keep global warming below 1.5 °C (above pre-industrial levels). It seems distressingly unachievable. According to Simon Evans of climate science website Carbon Brief

> It shows that the challenge of avoiding dangerous climate change and getting to zero emissions is unbelievably hard. (The Guardian 2020)

6.3.2.2 Air Travel Aspect

Unfettered carbon pollution from aviation is now a fast-growing source of greenhouse gas discharges, which is driving global climate change. This problem will only worsen as demand for air travel rises. In 2010, the aviation industry carried 2.4 billion passengers; in 2050, that number is forecast to rise to 16 billion (EPUK 2012). Airport operations are a significant factor in the economy, for tourism, imports, exports, and business. However, these aids must be evaluated against air travel's impact on the quality of life and the environment. Noise and air pollution, both from the aircraft and from airport ground operations, are sources of concern, especially in and around the airport. Noise parameters to delimit the pollution level have been introduced at the designated airports in India. The rules state that busy airports referring to a civil airport with more than 50,000 aircraft movements per year (take-off or a landing)—should not exceed noise levels beyond 75 dB(A) Leq (decibels) during the day time (6 am–10 pm) and 65 dB(A) during the night (10 pm–6 am) (Chatterjee 2018). The combustion of fuel is generally efficient in aircraft engines, and smoke emissions from jets are fairly low. However, there has been a rise in pollutant emissions from aircraft at ground level with aircraft movement. In addition, a large amount of air pollution around airports is also generated by surface traffic (EPUK 2012). The key pollutant of alarm around airports is nitrogen dioxide (NO_2). NO_2 is formed by nitrogen oxide (NOx) emissions from surface traffic, aircraft, and airport operations (CBD 2000). $PM_{2.5}$ is also a concern since particulate emissions from jet exhausts are almost all in this fine fraction. Now in this pandemic situation, the use of private jets is likely to increase, which will add to this pollution level.

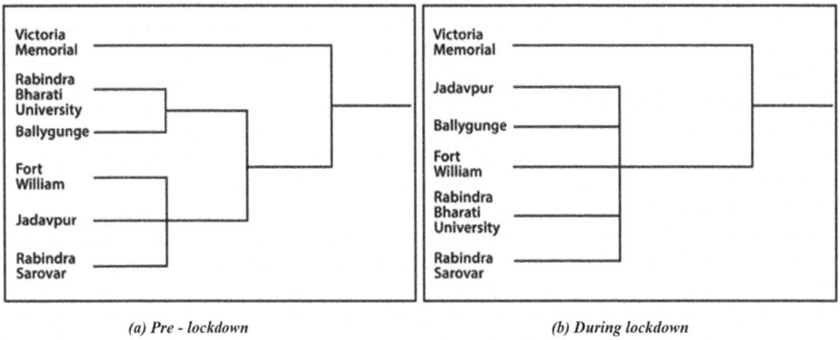

Fig. 6.3 Clustering of monitoring stations based on the concentration of pollutants (*Source* Central Pollution Control Board 2020)

Kolkata, unceremoniously called "the dusty city" (Haque and Singh 2017), is the second most polluted metropolis in India, next only to Delhi, and suffers from the highest pollution levels among eight tropical Asian countries (Mukherjee et al. 1998; Upadhyay et al. 2014; Roy et al. 2015; Financial Express 2020), typically a representative of Global South megalopolis. Apart from emission from the Kolkata-Howrah industrial belt, vehicular emissions from continuously increasing vehicles contribute a large in polluting air, and the number has increased 8 times from 1951 to 2001 (Chowdhury 2015) for the city. The $PM_{2.5}$ and PM_{10} are taken to be the main pollutants and have a persistent concentration above the permissible level in Kolkata (Das et al. 2015).

The cluster analysis (Fig. 6.3) shows a massive change in the concentration of atmospheric pollutants in the city of Kolkata. It clearly distinguishes between the scenario of pre-lockdown and during the lockdown phase. Two distinct clusters can be seen in Fig. 6.3a. One is consisting of Victoria Memorial, RBU, and Ballygunge; the other being Jadavpur, Fort William, and Rabindra Sarovar. The scenario in Fig. 6.3b shows a complete transformation of clustering. The locations show almost equal values of concentration of pollutants. It is unreasonable at this situation to either rejoice or to think it to last forever, as due to the factors discussed above, the situation in the post-COVID time is again going to resume to the previous level.

6.3.3 Water Stress

Water is the foundation of life and livelihood as it is aptly said, "there is no health without water". Initially, it was believed that freshwater is a renewable resource, but given the rate of consumption and residence time of several sources like streams, lakes, groundwater, it is clear that the replenishment rate is too low. The growing population and water demand-consumption have virtually made it to a finite one if one takes into account a smaller time-frame in mind. One of the aspirations of Sustainable

Development Goal was to ensure availability and sustainable management of water and sanitation for all by 2030, and this pandemic has kind of increased both the extent and consequences of the safe water access gap more than ever before (Butler 2020).

Singhal (2020) referred to the report "The Water Gap – Water Aid's State of the World's Water 2018", which stated that the global population without access to clean water was around 844 million while Matto and Singhal (2020) reported that 2.2 billion people live with high water stress. About 4 billion population experience severe water scarcity at least one month in a year (World Water Development Report 2019). The fact that many areas, especially in North and South Africa alongside South and Central Asia, are getting water-stressed is not only due to population rise but also because of an increase in demand and usage. The expected rise was estimated by OECD (2012) as 400% in the industrial sector and 130% in the domestic sphere. In the twentieth century, the population tripled while water usage escalated remarkably to six-fold (FAO 2009). The drivers that caused the increase in water stress before the COVID-19 pandemic were population growth, rising per capita water requirement, global climate change, collapsing wetland systems, rapid urbanization, carbon-intensive technology, demanding high water requirements, and age-old infrastructure (Pegram 2010; Butler 2020). In water-scarce areas, resort to water trucks is taken to provide water access to the deprived ones, but during the pandemics like Ebola in 2014 (TKG 2016) and COVID-19 in recent times hampered the regular movements of the trucks and, in turn, that of the water due to the quarantine, lockdowns, containments (Cooper 2020a).

For the prevention of the spread of coronavirus, there is a need for frequent handwashing, which demands access to clean water. It has been estimated that COVID-19 mitigation measures like more cleaning and disinfecting would lead to at least 5% average increase in water demand at the household level (United Nations ESCWA 2020). Senapati (2020) clarified that for washing hands an extra 4–5 times daily, a family of five members will need 20–40 litres of water more than normal. It was approximated that 1-second-dripping faucet wastes over 5 gallons of water per day and just less than 2,083 gallons per year. It was further roughly estimated that if a tap-water is left running during the twenty-second period of handwash, it leads to a loss of 1.5–2 litres of water, which can cumulatively go up to 15–20 litres of water for an individual daily. A section of experts is afraid that these extra litres of water used for handwashing in a nation that is already struggling with water scarcity may worsen the situation (Matto and Singhal 2020). Frequent washing of hands surely cannot be avoided through prudence and maybe by using taps with sensors, but the negligent wastage can definitely be curbed. This growing demand for water has occurred at a time when the potential for augmenting supply is limited, water tables are declining, and water quality issues have surfaced too. Apart from the depletion of groundwater, too much soap and detergent usage might lead to eutrophication in the surface water bodies.

Dobe (2020) carried out a study in seven India cities and found that the average per capita water consumption in domestic households was about 92 litres per capita per day, whereas 100 litres per capita per day is required to maintain a minimum standard of health and hygiene. It was predicted a few years back that cities like

Chennai in India, Cape Town in South Africa, and Mexico City in Mexico are at risk of running out of the water very soon (Matto and Singhal 2020). In the developing nations, including India, the section of people residing in abject poverty in informal urban settlements, rural interiors lack the adequate quality and quantity of water required in general and more so in this pandemic situation, making them susceptible to infection (International Commission of Jurists 2020).

To cope with this sudden surge in water demand, wastewater treatment has to be implemented at war-footing. Guppy and Anderson (2017) documented that 80% of the wastewater is left untreated. In low-income countries, only 8% of wastewater receives treatment, and in the lower-middle-income countries the treatment hovers around 28% (WWAP 2017). Moreover, through leakage, 30% of abstracted water is lost, and it is predicted that by 2030, there will be 40% gap between water demand and water available. The global water demand is supposed to increase by 50% (WWAP 2012). It is the time when wastewater and rainwater are increasingly considered a largely untapped resource to augment water supply that can ultimately reduce water stress.

Experts led by the UN Special Rapporteur on the Human Rights to Safe Drinking Water and Sanitation indicated that vulnerable persons "need to have continuous access to sufficient and affordable water" conforming to the hygiene measures required to combat COVID-19 (OHCHR 2020). Anim and Ofori-Asenso (2020) feel that strategies have to be implemented to ensure the provision of water for all stakeholders, regardless of any other factors like socio-economic conditions, legal status, through green solutions regarding the conservation of available water. Cooper (2020b) emphasized that the focus has to be on aspects like—sufficient water availability, satisfactory water quality, judicious water resources management, and affordable access to Water, Sanitation, and Hygiene (WASH), to strengthen water security in this troubled time. Butler (2020) added few other parameters like addressing historical gaps in water supply, emphasizing the construction of water utility infrastructure with a higher degree of automation, undertaking communication campaigns to raise awareness of the importance of good hygiene practices and water conservation too. While policymakers recognize that rationalizing the usage and pricing of water is necessary to curb wasteful practices, it remains politically contentious. Hoque and Wichelns (2013) conducted a survey in 60 cities across the continents to suggest effective water tariff policies like metering through price signals, keeping a minimum charge for all so that all the households use water wisely, and charging higher rates for the wealthier consumers who use greater volumes of water.

The effective management of water resources across the globe has been hampered by factors like—water wastage, infrastructural dysfunction, low accountability, and unethical practices. It can be hoped that the COVID-19 will act as a wake-up call to hasten the process of forming of water-smart society (de Melo et al. 2020), which would deliver improved water services during future infectious disease outbreaks.

Surie (2020) pointed out that in 2019 the Indian government formed a new integrated water ministry—"Jal Shakti" (Water Power)—merging the former ministries of Water Resources, River Development, and Ganga Rejuvenation, and Drinking Water and Sanitation to raise public awareness about water conservation. It involves the initiatives like the Jal Shakti Abhiyan (for promoting water conservation in 256 most water-stressed districts in India), Jal Jeevan Mission (for providing piped water connections to 146 million rural households by 2024), and Atal Bhujal (Groundwater) Yojana (for improving groundwater management through community participation in seven Indian states). If these integrated and holistic measures are adequately implemented, then the water crisis, which is being further magnified by COVID-19 induced surplus water usage, could be managed efficiently, assuring access to water for all.

6.3.4 Chemical Exposure

This COVID-19 has also made people aware of health and hygiene. Hand washing is considered as a means to prevent serious diseases and infections. The Global Handwashing Partnership introduced the Global Handwashing Day, and it is being celebrated since 2008 (CDC 2020), on 15 October every year. It is a platform shaped to promote and create new ways to encourage people to wash their hands. It also advocates the fact that handwashing with soap and water is the easiest and best way to do the same. However, going by the ethos of this day, where people always opt for easier alternatives, hand sanitizers to clean hands have become highly popular. Adaptation of effective hand hygiene is vital, where one of the best advice by WHO is to wash or sanitize your hands frequently with soap or >60% alcoholic hand sanitizer, respectively. WHO suggested two alcohols-based formulations for hand hygiene in healthcare to sanitize the hands for reducing the spread of coronavirus (WHO 2020f). These recommendations are based on fast, effective, and broad-spectrum antimicrobial activity combined with easy availability and safety.

Hand sanitizers are commercially obtainable in various types and forms such as antimicrobial soaps, water-based or alcohol-based hand sanitizers. The different types of formulations involve—gel, foam, cream, spray, and wipes (Jones et al. 2000). WHO recommended that alcohol-based hand sanitizers are mainly made up of ethanol, isopropyl alcohols, and hydrogen peroxides in different combinations (Fig. 6.4).

These preparations may become toxic to human health and the environment when misused or overused. These chemicals have identified toxic and hazardous impacts on the environment when released by evaporation. It is recognized that the ingestion of a low concentration of hydrogen peroxide (3% solution) is responsible for minor gastrointestinal tract irritation. In a few cases, it is also responsible for portal vein embolism and mild mucosal irritating and vomiting. Ingestion of isopropyl alcohol accidentally or deliberately leads to severe respiratory or central nervous system depression (Slaughter et al. 2014). The active reagents of alcohol-based hand sanitizers are ethanol or isopropyl alcohol at 60–95% concentration. Langer et al. (2004) reported that alcohol-based solutions with ethanol, 2-propanol, and distilled water and disinfectant with octenidine dihydrochloride and phenoxyethanol are the vital reagents. The popular hand sanitizers, such as Dettol® contain Alcohol Denat,

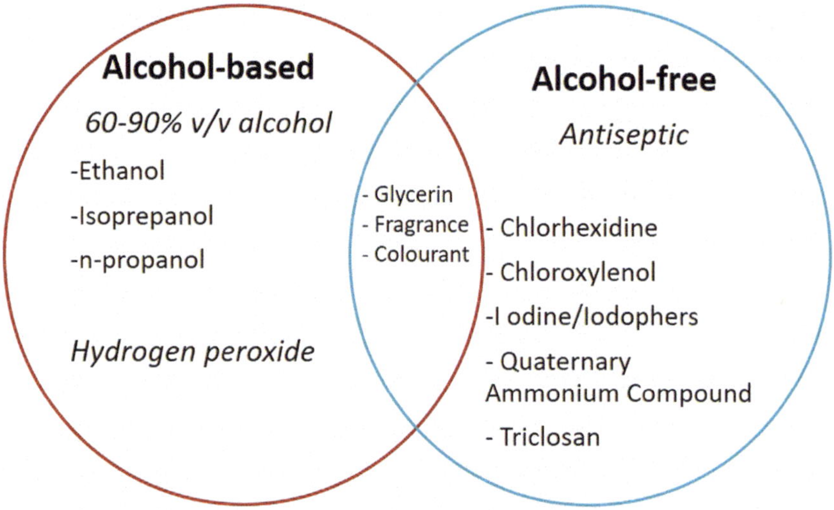

Fig. 6.4 Classifying Hand sanitizers (*Source* Compiled by author)

and Purell® includes ethyl alcohol 70% *v*/v, isopropyl alcohol, and aminomethyl propanol.

This pandemic has fetched a shortage of hand sanitizers all around the world. WHO has directed the local manufacturers regarding the preparation of hand sanitizers. WHO suggested two formulations for lesser volume production; one with ethanol (96%) and the other with isopropyl alcohol (99.8%). Final product concentration suggested by WHO for household or local production is ethanol (80%) *v*/v, hydrogen peroxide (0.125%) v/v and glycerol (1.45%) v/v for formulation A and isopropyl alcohol (75%) v/v, hydrogen peroxide (0.125%) v/v and glycerol (1.45%) v/v for formulation B. (Refer Table 6.1).

As the motto of the Global Handwashing Day is to encourage washing hands with soap and water, what is it about hand sanitizers that many are ignoring? It is for the side-effects they have. Proper cleaning of hands using the liquid sanitizer is appropriate whenever washing hands is always not possible. It can save us from the infection by bacteria and viruses. When outside, riding, playing in a park or shopping, it is not always possible to use soap and water to wash hands. This is when we all have to depend on alcohol-based hand sanitizers. Sanitizer works extremely well for most of the bacteria and viruses, and that is how it keeps us safe from a variety of illnesses. However, sanitizers can affect the micro-biomes of the body, which can have a negative impact on health. These alcohol-based sanitizers can be beneficial for killing the bacteria but, in turn, can wreak havoc in our healthy bacterial community. Another issue, especially in India, is that due to the rising demand for hand sanitizers, often spurious products are flooding the market, and there is no procedure for stringent quality control. So, many physicians have suggested preparing a home-made variant

Table 6.1 Effects of acute and chronic toxicity by active ingredients of hand sanitizers

Active ingredients	Acute harmfulness	Chronic harmfulness	Source
Ethanol	Central nervous system and respiratory depression, Lactic acidosis, Nausea	Cardiac arrhythmia, Acute liver damage, Hypokalaemia, Hypocalcaemia, Hypophosphatemia, Cardiac arrest and death	Wilson et al. (2015a), Vonghia et al. (2008)
Isopropanol	Alike ethanol including central nervous system and respiratory depression, skin and mucous membrane irritation	Ketosis, Osmolal gap ketonemia, Myoglobinuria, Acute renal failure and death	Zaman et al. (2002), Jersey Department of Health (2016)
3% H_2O_2	Vomiting, skin infection, slight gastrointestinal and mucosal irritation	Air embolism, death in few cases	Moon et al. (2006), New Jersey Department of Health (2016)

Source Compiled by authors

that might be of much riskless (TOI 2020), but that too comes with certain conditions (Fig. 6.5).

Medical experts have started to warn against the excessive use of alcohol-based hand sanitizer as a preventive measure against coronavirus. Too much use of sanitizer against the virus is responsible for skin damage and reduce its ability to work as

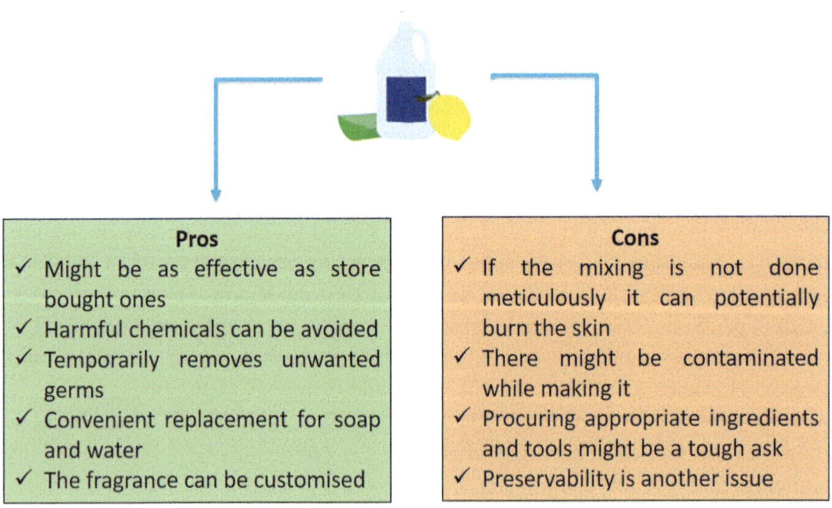

Fig. 6.5 Having one's own mix—Pros and Cons of self-manufacturing hand sanitizer (*Source* Compiled by authors)

a barrier against other harmful viruses (Wang and Su 2020). Sanitizers have been frequently used all over the world as a disinfectant for better hand hygiene. Excessive use of alcohol-based sanitizer increases the skin's permeability and deprives oil and water of the skin, leading to skin roughness and irritation. Dry and damaged skin is a hotbed for many diseases. Repeated exposure of disinfectant, antibiotics or other genotoxic chemicals to microbes tend to mutate through the natural process, making them resistant to survive from repeated use of hand sanitizer (Tachikawa 2020). Research reports have indicated that overuse of sanitizers in some cases may increase the risk of viral outbreaks (Larson 2001). CDC (2020) warns that the excess use of sanitizers may lead to hormonal disruptions and cause bacteria to adapt to its antimicrobial properties, which produce more antibiotic-resistant strains. The use might also weaken the immune system and make people more susceptible to allergies. The scented sanitizers are likely to be loaded with chemicals. Synthetic fragrances contain phthalates, which could alter genital development (Bonner 2017). Animal studies have shown that the compound could change the way hormones work in the body, raising concerns and necessitating further investigation to understand better how they might affect humans (Bonner 2017).

6.4 Instances of Behavioural Trajectories

The regulations levied by the countries from the global south in response to the pandemic have been relatively uniform when viewed from a distance—there is the imposition of quarantine, social distancing, encouraging the usage of protective gear, and sanitizers. However, these mandates have generated varied responses in different countries during the "crisis" and "new normal" (Table 6.2). On the factors like the severity of the virus outbreak, population characteristics, economic standings, and the way people handled the exceptional situation vary across the nations. But one thing that has been common for most of these developing nations is that all are fighting tooth and nail to stop the spread of the pandemic and are not really taking into account the immediate and far-reaching environmental impact of these measures at this point in time. Apart from governmental initiatives, it is also the responsibility of the aware and sensitized citizens to act judiciously in this hour of need so that sustainability is maintained.

Table 6.2 Facing the pandemic head-on: excerpts of ground reality

Countries / Parameters	Respondents' views			
	Respondent-1 Female (40 years) Kuala Lumpur, Malaysia	Respondent-2 Female (34 years) Bandar Seri Begawan, Brunei	Respondent-3 Male (65 years) Singapore	Respondent-4 Male (52 years) Delhi-NCR, India
Waste Generation	• Segregated waste disposal was already in place • Plastic waste generation has increased with compulsory mask usage and imposition of a hefty fine if not followed	• Zero plastic waste initiative has been there for the last two years • The use of plastics, even for the protective gear, is also restricted as far as possible	• Plastic use is very limited, as stringent laws are in place • Even for the protective gear, the stress is on the use of recyclable plastics	• No stringent waste segregation is done • SUP usage is back with a bang in the form of packaging material and protective equipment • Very few are opting for eco-friendly products with higher price tags • Often medical wastes are recycling unauthorizedly and sold in the black market
Air Pollution	• With lesser vehicular movements, during the lockdown, it has gone down	• The level of pollution was low already	• The level of air pollution was not pronounced beforehand too	• Air pollution levels have gone down drastically during the lockdown phas
Water Stress	• Water usage has increased way bit but not drastically	• Water wastage is limited, as one needs to pay tax as per usage	• Water usage has been on the rise, but there has been so widespread fear of scarcity	• Due to hygiene issues, water use has gone up and is leading to an impending water crisis
Chemical Exposure	• All are using sanitizers in both public and private spheres, so there is high demand in spite of the spike in price	• Though initially sanitizer market did not pick up, gradually, it gained ground and was even sold at twice the market rate	• Initially, sanitizers were not used; reliance was on soap and water. But from the month of May, sanitizer sales have picked up	• Sanitizer usage is very high, and there is no strict quality control measure

Source Primary Survey, 2020

6.5 Ways Ahead

In this unprecedented crisis of COVID-19, the protection of lives and livelihoods has become the core of government decisions and actions at every level, with a specific focus on the health care sector (Prata et al. 2020). Various guidelines have been issued to control the spread of this highly communicable disease, which has emphasized social distancing, frequent hand washing, and practicing proper respiratory etiquette by wearing masks. Due to lesser mobility and limited economic activities the world has witnessed an ecological rejuvenation, which is going to be transitory if the disease control measures are not adequately chalked out. The COVID-19 crisis should not be solved at the expense of inviting a longer-term environmental crisis. The stakeholders like policymakers, scientists, and common people need to try to figure out innovative, holistic solutions in handling the existing challenges of the current crisis. For designing efficient management systems that would be viable in pandemic and post-pandemic world, holistic public-private initiatives have been encouraged. These efforts can be made as part of the broad disaster management planning in the new normal conditions to better comply with pollution standards and to ensure sustainability.

References

ACR + (Association of Cities and Regions for Sustainable Resource Management) (2020) Municipal waste management and COVID 19. Available at https://www.acrplus.org/en/municipal-waste-managementcovid-19. Accessed 26 Apr 2020

ADB (Asian Development Bank) (2020) Managing infectious medical waste during the COVID-19 pandemic. Available at https://www.adb.org/sites/default/files/publication/578771/managingmedical-waste-covid19.pdf. Accessed 4 July 2020

Agarwal S, Banerjee K, Pal N, Mallik K, Bal G, Pramanick P, Mitra A (2017) Carbon sequestration by mangrove vegetations: a case study from Mahanadi mangrove wetland. J Environ Sci Comput Sci Eng Technol 7(1):16–29

Alpizar F, Carlsson F, Lanza G, Carney B, Daniels RC, Jaime M, Ho T, Nie Z, Salazar C, Tibesigwa B, Wahdera S (2020) A framework for selecting and designing policies to reduce marine plastics pollution in developing countries. Environ Sci Policy 109:25–65

Anim DO, Ofori-Asenso R (2020) Water scarcity and COVID-19 in sub-Saharan Africa. J Infect, Letter to the Editor

Banerjee I, Walker J, Deakin E, Kanafani A (2010) Vehicle choice in 22 India: household choice among motorized vehicle segments. Available at http://metrostudies.berkeley.edu/pubs/reports/Walker_Working_IpsitaWCTR.pdf. Accessed 4 Aug 2020

Bansal P, Kockelman KM (2016). Indian vehicle ownership: insights from 25 literature review, expert interviews, and state-level model. Under review for publication in 26 Transportation Research Record. Available at http://www.caee.utexas.edu/prof/kockelman/public_html/TRB16IndianVehOwn.pdf. Accessed 4 Aug 2020

Basu M (2020). Today's PPE kits could be tomorrow's roads, fuel—CSIR's plan to tackle Covid plastic surge. Available at https://theprint.in/health/todays-ppe-kits-could-be-tomorrows-roads-fuel-csirs-plan-to-tackle-covid-plastic-surge/478760/. Accessed 12 Aug 2020

Bell B (2016) How do low oil prices affect recycling. WM Med. Room Available at https://mediaroom.wm.com/how-do-low-oil-prices-affect-recycling/. Accessed 26 July 2020

BIR (Bureau of International Recycling) (2020) COVID-19: Update by BIR member national associations—the world moves tentatively into reopening phase. Available at https://bir.org/news-press/news/item/covid-19-update-by-bir-member-nationalassociations-the-world-moves-tentatively-into-reopening-phase. Accessed 9 July 2020

Bonner L (2017) CDC report calls attention to hand sanitizer risk in children. Pharma Today. Available at https://www.pharmacytoday.org/article/S1042-0991(17)30602-3/pdf. Accessed 10 July 2020

Boucher J, Billard G (2019) The challenges of measuring plastic pollution. Field Actions Science Reports, SI19 2019:68–75. IUCN, Review of Plastic Footprint Methodologies. Available at http://portals.iucn.org/library/sites/library/files/documents/2019-027-En.pdf. Accessed 10 July 2020

Bremer S, Schneider P, Glavovic B (2019) Climate change and amplified representations of natural hazards in institutional cultures. Oxford Research Encyclopedia of Natural Hazard Science. Accessed 4 July 2020

Butler G (2020) The impact of COVID-19 on the water and sanitation sector. International Finance Corporation, World Bank Group. Available at https://www.ifc.org/wps/wcm/connect/126b1a18-23d9-46f3-beb7047c20885bf6/The+Impact+of+COVID_Water%26Sanitation_final_web.pdf?MOD=AJPERES&CVID=ncaG-hA. Accessed 1 July 2020

CBD (Centre for Biological Diversity) (2000) Airplane emissions. Available at https://www.biologicaldiversity.org/programs/climate_law_institute/transportation_and_global_warming/airplane_emissions/. Accessed 19 July 2020

CDC (Centre for Disease Control and Prevention) (2020) Handwashing: clean hands save lives, 17 August. Available at https://www.cdc.gov/handwashing/hand-sanitizer-use.html. Accessed 19 Aug 2020

Chatterjee B (Hindustan Times) (2018) Govt sets noise standards for airports across India. Available at https://www.hindustantimes.com/india-news/govt-sets-noise-standards-for-airports-across-india/storyulukaD7YUwFf26DqwqMnBJ.html#:~:text=The%20rules%20state%20that%20busy%20airports%20%E2%80%94%20a%20civil%20airport%20with,night%20(10pm%20to%206am). Accessed 14 July 2020

Chowdhury IR (2015) Scenario of vehicular emissions and its effect on human health in Kolkata city. Int J Humanities Soc Sci Invent 4(5):1–9

Climate Action Tracker (2020) A government roadmap for addressing the climate and post COVID-19 economic crises. CAT (Climate Action Tracker). Available at https://climateactiontracker.org/publications/addressing-theclimate-and-post-covid-19-economic-crises/. Accessed 19 July 2020

Cooper R (2020a) Water for the urban poor and Covid-19. K4D helpdesk report 826. Institute of Development Studies, Brighton, UK

Cooper R (2020b) Water security beyond Covid-19. K4D helpdesk report 803. Institute of Development Studies, Brighton, UK

Corburn J, Vlahov D, Mberu B, Riley L, Caiaffa WT, Rashid SF, Ko A, Patel S, Jukur S, Martínez-Herrera E, Jayasinghe S, Jayasinghe S (2020) Slum health: arresting COVID-19 and improving well-being in urban informal settlements. J Urban Health:1–10

Coutts A, Beringer J, Tapper N (2010) Changing urban climate and CO_2 emissions: implications for the development of policies for sustainable cities. Urban Policy Res 28(1):27–47

CPCB (2020) Guidelines for handling, treatment and disposal of waste generated during treatment/diagnosis/ quarantine of COVID-19 patients. Central Pollution Control Board, India. Available at http://www.cpcb.nic.in/. Accessed 17 July 2020

Dadhich AP, Goyal R, Dadhich PN (2018) Assessment of spatio-temporal variations in air quality of Jaipur city, Rajasthan, India. Egypt J Remote Sens Space Sci 21(2):173–181

Das R, Khezri B, Srivastava B, Datta S, Sikdar PK, Webster RD, Wang X (2015) Trace element composition of PM2.5 and PM10 from Kolkata—a heavily polluted Indian metropolis. Atmos Pollut Res 6(5):742–750

Datta P, Mohi GK, Chander J (2018) Biomedical waste management in India: critical appraisal. J Lab Phys 10(1):6–14

de Melo MC, Santos AS, Vieira JMP (2020) The new centrality of water and sanitation post Covid19. Revista Augustus 25(51):294–315
Dobe M (2020) Safe water and COVID-19—where the twain meet. Available at https://www.downtoearth.org.in/blog/water/safe-water-and-covid-19-where-the-twain-meet-70048. Accessed 1 July 2020
EPUK (Environmental Protection UK) (2012) Aviation Pollution. Available at https://www.environmental-protection.org.uk/policy-areas/air-quality/air-pollution-and-transport/aviation-pollution/. Accessed 11 July 2020
FAO (Food and Agriculture Organization of the United Nations) (2009) How to feed the world in 2050. Available at http://www.fao.org/fileadmin/templates/wsfs/docs/expert_paper/How_to_Feed_the_World_in_2050.pdf. Accessed 1 July 2020
Fernandes N (2020) Economic effects of coronavirus outbreak (COVID-19) on the world economy. SSRN Electron J:3557504
Ferronato N, Torretta V (2019) Waste mismanagement in developing countries: a review of global issues. Int J Environ Res Public Health 16(6):1060
Financial Express (2020) Kolkata most polluted metro; India has highest PAH level, says Tokyo University study. Accessed 11 July 2020
Gardiner B (2020a) Pollution made COVID-19 worse. now, lockdowns are clearing the air. National Geographic. Available at https://www.nationalgeographic.com/science/2020/04/pollution-made-the-pandemic-worse-but-lockdowns-clean-the-sky/. Accessed 15 July 2020
Gardiner B (2020b) Why COVID-19 will end up harming the environment. National Geographic Channel. Available at https://www.nationalgeographic.com/science/2020/06/why-covid-19-will-end-up-harming-the-environment/. Accessed 1 Aug 2020
Geyer R, Jambeck JR, Law KL (2017) Producción, uso y destino de todos los plásticos jamás fabricados (Production, use, and fate of all plastics ever made). Sci Adv 3(7):1207–1221
Ghose MK, Paul R, Banerjee SK (2005) Assessment of the impact on human health of exposure to urban air pollutants: an Indian case study. Int J Environ Stud 62(2):201–214
Guppy L, Anderson K (2017) Global water crisis: the facts. United Nations University Institute for Water, Environment and Health, United Nations University. Available at https://inweh.unu.edu/wp-content/uploads/2017/11/Global-Water-Crisis-The-Facts.pdf. Accessed 1 July 2020
Gupta AK, Karar K, Ayoob S, John K (2008) Spatio-temporal characteristics of gaseous and particulate pollutants in an urban region of Kolkata, India. Atmos Res 87(2):103–115
Hahladakis JN, Velis CA, Weber R, Iacovidou E, Purnell P (2018) An overview of chemical additives present in plastics: migration, release, fate and environmental impact during their use, disposal and recycling. J Hazard Mater 344:179–199. Available at https://doi.org/10.1016/j.jhazmat.2017. https://doi.org/10.014. Accessed 17 July 2020
Haque MS, Singh RB (2017) Air pollution and human health in Kolkata, India: a case study. Climate 5(4):1–16
Hong J, Zhan S, Yu Z, Hong J, Qi C (2018) Life-cycle environmental and economic assessment of medical waste treatment. J Clean Prod 174:65–73
Hoornweg D, Bhada-Tata P (2012) What a waste: a global review of solid waste management. Urban development series; knowledge papers No. 15. World Bank, Washington, DC. Available at https://openknowledge.worldbank.org/handle/10986/17388. Accessed 15 July 2020
Hoque SF, Wichelns D (2013) State-of-the-art review: designing urban water tariffs to recover costs and promote wise use. Int J Water Resour Dev 29(3):472–491
Huang H (2016) Media use, environmental beliefs, self-efficacy, and pro-environmental behaviour. J Bus Res 69(6):2206–2212
Hyun MC (2020) Korea sees steep rise in online shopping during COVID-19 pandemic. ZD Net. Available at https://www.zdnet.com/article/korea-sees-steep-rise-inonline-shopping-during-covid-19-pandemic. Accessed 19 July 2020
International Commission of Jurists (2020) COVID-19 pandemic in India: the right to water a briefing paper. Available at https://www.icj.org/wp-content/uploads/2020/06/India-Right-to-Water-COVID-19-Briefing-Paper-2020-ENG.pdf. Accessed 1 July 2020

International Solid Waste Association (2020) Waste management during the covid-19 pandemic. ISWA's recommendations. Available at https://www.iswa.org/iswa/covid-19/. Accessed 12 July 2020
IRENA (International Renewable Energy Agency) (2020) Post-COVID recovery: an agenda for resilience, development and equality. Available at https://www.irena.org/publications. Accessed 5 July 2020
Istrate I-R, Iribarren D, Galvez-Martos J-L, Dufour J (2020) Review of life-cycle environmental consequences of waste-to-energy solutions on the municipal solid waste management system. Resour Conserv Recycl 157:104778
Jones RD, Jampani HB, Newman JL, Lee AS. (2000). A review of effectiveness and safety in health care settings. Am J Infect Control 28(2):184–196
Kampf G, Todt D, Pfaender S, Steinmann E (2020) Persistence of corona viruses on inanimate surfaces and their inactivation with biocidal agents. J Hosp Infect 104(3):246–251
Kaufman L, Chasan E (2020) Cities wonder whether recycling counts as essential during the virus. Bloomberg Green. Available at https://www.bloomberg.com/news/articles/2020-03-27/cities-wonder-whether-recycling-counts-as-essential-during-the-virus. Accessed 26 Apr 2020
Klemeš JJ, Van Fan Y, Tan RR, Jiang P (2020) Minimising the present and future plastic waste, energy and environmental footprints related to COVID-19. Renew Sustain Energy Rev 127:109883
Kyriakis E, Psomopoulos C, Kalkanis K (2019) Investigating the correlation of purchase power parity with the adopted waste management method in EU28. Soc Sci 8(5):162
Langer S, Salakdeh MS, Goertz O, Steinau HU, Steinstraesser L, Homann HH (2004) The impact of topical antiseptics on skin microcirculation. Eur J Med Res 9(9):449–454
Larson E (2001) Hygiene of the skin: when is clean too clean? Emerg Infect Dis 7(2):225
Le Dang Ha SAB, Hien NQ, Maloney SA, Le Quynh Mai KCL, Anh BH, Reynolds MG, Montgomery JM, Comer JA, Horby P, Plant AJ (2004) Lack of SARS transmission among public hospital workers, Vietnam. Emerg Infect Dis 10(2):265–268
Liu HC, You JX, Lu C, Chen YZ (2015) Evaluating healthcare waste treatment technologies using a hybrid multi-criteria decision-making model. Renew Sustain Energy Rev 41:932–942
Makarichi L, Jutidamrongphan W, Techato KA (2018) The evolution of waste-to-energy incineration: a review. Renew Sustain Energy Rev 91:812–821
Mallapur C (2020) Sanitation workers at risk from discarded medical waste related to COVID-19. IndiaSpend. Available at https://www.indiaspend.com/sanitation-workersat-risk-from-discarded-medical-waste-related-tocovid-19/. Accessed 26 July 2020
Matto M, Singhal S (2020) COVID-19: the need is to decentralise how we manage wastewater. Available at https://www.downtoearth.org.in/blog/water/covid-19-the-need-is-to-decentralise-how-we-manage-wastewater-70991. Accessed 1 July 2020
Mishra L (2020) Experts stress on proper disposal of PPEs. Available at https://www.thehindu.com/news/national/experts-stress-on-proper-disposal-of-plastic-ppes/article31475746.ece. Accessed 14 May 2020
Mol MPG, Caldas S (2020) Can the human coronavirus epidemic also spread through solid waste? Waste Manage Res 38(5):485–486
Mollica GJG, Balestieri JAP (2020) Is it worth generating energy with garbage? defining a carbon tax to encourage waste-to-energy cycles. Appl Therm Eng 173:115195
Moon JM, Chun BJ, Min YI (2006) Hemorrhagic gastritis and gas emboli after ingesting 3% hydrogen peroxide. Int J Emerg Med 30(4):403–406
Mukherjee A, Mukherjee G, Banerjee U, Muichopadryay SP (1998) Occupational exposure of the traffic, personnel of Calcutta to lead and carbon monoxide. Pollut Res 17:359–362
National Geographic Channel (2020) Why COVID-19 will end up harming the environment. Available at https://www.nationalgeographic.com/science/2020/06/why-covid-19-will-end-up-harming-the-environment/. Accessed 1 August 2020
Nesheli MM, Ceder A, Brissaud R (2017) Public transport service-quality elements based on real-time operational tactics. Transportation 44(5):957–975

New Jersey Department of Health (2016) Hazardous substance fact sheet (isopropyl alcohol). Available at https://www.nj.gov/health/eoh/rtkweb/documents/fs/1076. Accessed 14 July 2020
Nigeria Centre for Disease Control (2020) Coronavirus (COVID-19) highlights. Available at https://covid19.ncdc.gov.ng. Accessed 14 May 2020
Nzediegwu C, Chang SX (2020) Improper solid waste management increases potential for COVID-19 spread in developing countries. Resour Conserv Recycl 161:104947
OECD (Canada and the Organisation for Economic Co-operation and Development) (2012) Environmental Outlook to 2050: the consequences of inaction. Available at http://www.oecd.org/env/indicators-modelling-outlooks/oecd-environmental-outlook-1999155x.html. Accessed 1 July 2020
OECD (Canada and the Organisation for Economic Co-operation and Development) (2020) OECD policy responses to coronavirus (COVID-19)—cities policy responses. Organisation for Economic Co-operation and Development. Available at http://www.oecd.org/coronavirus/policy-responses/cities-policyresponses-fd1053ff/. Accessed 19 July 2020
OECD Statistics (2018) Organization for economic co-operation and development. Available at https://stats.oecd.org/Index.aspx?DataSetCode=WSECTOR. Accessed 14 May 2020
OHCHR (Office of the High Commissioner Human Rights) (2020) COVID-19 will not be stopped without providing safe water to people living in vulnerability-UN experts. Available at https://www.ohchr.org/en/NewsEvents/Pages/DisplayNews.aspx?NewsID=25738&LangID=E. Accessed 1 July 2020
Pegram G (2010) Briefing Global water scarcity: risks and challenges for business. Lloyd's and WWF, 32
Plastics Europe (2019) Plastics—the facts 2019. An analysis of European plastics production, demand and waste data. Available at https://www.plasticseurope.org/application/files/9715/7129/9584/FINAL_web_version_Plastics_the_facts2019_14102019.pdf. Accessed 1 July 2020
Prata JC, Patrício Silva AL, Walker TR, Duarte AC, Rocha Santos T (2020) COVID-19 pandemic repercussions on the use and management of plastics. Environ Sci Technol 54(13):7760–7765
Roy S Sen, Singh RB, Kumar M (2015) An analysis of local spatial temperature patterns in the Delhi Metropolitan Area. Phys Geogr 32(2):114–138
Saadat S, Rawtani D, Hussain CM (2020) Environmental perspective of COVID-19. Sci Total Environ, 138870
Senapati MR (2020) We should prepare for severe water shortage in a post COVID-19 world. Outlook. Available at https://www.outlookindia.com/website/story/opinion-we-should-prepare-for-severe-water-shortage-in-a-post-covid-19-world/352792. Accessed 1 July 2020
Sharma HB, Sarmah AK, Dubey B (2020a) Hydrothermal carbonization of renewable waste biomass for solid biofuel production: a discussion on process mechanism, the influence of process parameters, environmental performance and fuel properties of hydrochar. Renew Sustain Energy Rev, 123, 109761
Sharma HB, Vanapalli KR, Cheela VS, Ranjan VP, Jaglan AK, Dubey B, Goel S, Bhattacharya J (2020b) Challenges, opportunities, and innovations for effective solid waste management during and post COVID-19 pandemic. Resour Conserv Recycl, 105052
Shen Y, Yu S, Ge S, Chen X, Ge X, Chen M (2017) Hydrothermal carbonization of medical wastes and lignocellulosic biomass for solid fuel production from lab-scale to pilot-scale. Energy 118:312–323
Shi J, Zheng W (2020) Coronavirus: China struggling to deal with mountains of medical waste created by epidemic. Available at https://www.scmp.com/news/china/society/article/3065049/coronaviruschina-struggling-deal-mountain-medical-waste-created. Accessed 1 July 2020
Silva ALP, Prata JC, Walker TR, Campos D, Duarte AC, Soares AM, Barcelò D, Rocha-Santos T (2020) Rethinking and optimising plastic waste management under COVID-19 pandemic: policy solutions based on redesign and reduction of single-use plastics and personal protective equipment. Sci Total Environ:140565
Sinclair R, Fahnestock L, Feliz A, Patel J, Perry C (2018) The spread of a norovirus surrogate via reusable grocery bags in a grocery supermarket. J Environ Health 80(10):8–14

Singhal S (2020) Every drop counts in these times of COVID-19. Available at https://www.downtoearth.org.in/blog/water/covid-19-the-need-is-to-decentralise-how-we-manage-wastewater-70991. Accessed 1 July 2020

Slaughter RJ, Mason RW, Beasley DMG, Vale JA, Schep LJ (2014) Isopropanol poisoning. Clin Toxicol 52(5):470–478

Smart Waste Report European Union (2020) COVID-19 and municipal waste management. https://www.interregeurope.eu/smartwaste/news/news-article/8127/covid-19-andmunicipal-waste-management/. Accessed 7 May 2020

Steinle S, Reis S, Sabel CE (2013) Quantifying human exposure to air pollution–moving from static monitoring to spatio-temporally resolved personal exposure assessment. Sci Total Environ 443:184–193

Surie MD (2020) Will Covid-19 force India to face up to its water crisis? Available at https://www.thethirdpole.net/2020/04/14/will-covid-19-force-india-to-face-up-to-its-water-crisis/. Accessed 14 July 2020

Tachikawa T (2020) Overreliance on hand sanitizers may increase risk of coronavirus infection. Kyodo News. Available at https://english.kyodonews.net/news/2020/02/690b9e1c9b86-focus-overreliance-on-hand-sanitizers-may-increase-risk-of-virus-infection.html. Accessed 20 July 2020

Tenenbaum L (2020). The amount of plastic waste is surging because of the coronavirus pandemic. Forbes. Available at https://www.forbes.com/sites/lauratenenbaum/2020/04/25/plastic-waste-during-the-time-of-covid-19/#7c4e661f7e48. Accessed 17 July 2020

TERI (The Energy and Resources Institute) (2020) Impact of COVID-19 on urban mobility in India: evidence from a perception study. Available at https://www.teriin.org/sites/default/files/2020-05/behavioural-effects-covid19_0.pdf. Accessed 10 July 2020

The Guardian (2020) After the Covid-19 crisis, will we get a greener world? Available at https://www.theguardian.com/environment/2020/may/17/after-the-covid-19-crisis-will-we-get-a-greener-world. Accessed 10 July 2020

TKG (The Khabna Group) (2016) CARE international DEC Ebola emergency response project: final evaluation report. New York. Available at https://www.alnap.org/help-library/care-international-dec-ebola-emergency-response-project. Accessed 14 July 2020

TOI (Times of India) (2020) 7 situations in which you should not use a hand sanitizer. Available at https://timesofindia.indiatimes.com/life-style/health-fitness/health-news/7-situations-in-which-you-should-not-use-a-hand-sanitizer/photostory/76006211.cms. Accessed 15 July 2020

UNEP (United Nations Environment Programme) (2020a) Covid-19 waste management factsheet. Available at https://www.unenvironment.org/resources/factsheet/covid-19-waste-management-factsheets. Accessed 1 July 2020

UNEP (United Nations Environment Programme) (2020b) Waste management and essential public service in the fight to beat COVID-19. United Nations Environment Programme and the Basel Convention. Available at https://www.unenvironment.org/news-and-stories/press-release/waste-management-essential-public-service-fight-beat-covid-19. Accessed 17 July 2020

United Nations ESCWA (2020) The impact of COVID-19 on the water-scarce Arab region. Available at https://www.unescwa.org/sites/www.unescwa.org/files/20-00150_covid-19_water-scarcity-en.pdf. Accessed 17 July 2020

Upadhyay A, Kanchan K, Goyal P, Yerramilli A, Gorai AK (2014) Development of a fuzzy pattern recognition model for air quality assessment of Howrah City. Aerosol Air Quali Res 14(6):1639–1652

van Doremalen N, Bushmaker T, Morris DH, Holbrook MG, Gamble A, Williamson BN, Lloyd-Smith JO (2020) Aerosol and surface stability of SARS-CoV- 2 as compared with SARS-CoV-1. N Engl J Med 382(16):1564–1567

Vonghia L, Leggio L, Ferrulli A, Bertini M, Gasbarrini G, Addolorato G (2008) Acute alcohol intoxication. Eur J Internal Med 8:561–567

Wang Q, Su M (2020) A preliminary assessment of the impact of COVID-19 on environment—a case study of China. Sci Total Environ 728:138915

White AO (2020) Plastic pollution: why chemical recycling could provide a solution. Available at https://theconversation.com/plastic-pollution-whychemical-recycling-could-provide-a-solution-129917. Accessed 1 July 2020
Wilson DC, Rodic L, Modak P, Soos R, Carpintero A, Velis K, Simonett O (2015b) Global waste management outlook. United Nations Environment Programme. Available at https://www.unenvironment.org/resources/report/global-wastemanagement-outlook. Accessed 20 July 2020
Wilson ME, Guru PK, Park JG (2015a) Recurrent lactic acidosis secondary to hand sanitizer ingestion. Indian J Nephrol 25(1):57–59
WHO (2015) Safe management of wastes from health-care activities. The World Health Organization. https://apps.who.int/iris/bitstream/handle/10665/85349/9789241548564_eng.pdf;jsessionid=404F93479E9BF429DE9880E2F24A3590? sequence=1. Accessed 26 July 2020
WHO (2017) Report on health-care waste management (HCWM) status in Countries of the South-East Asia Region (SEA-EH-593). The World Health Organization. https://apps.who.int/iris/handle/10665/258761. Accessed 27 July 2020
WHO (World Health Organisation) (2020a) Coronavirus disease (COVID-19), data last updated: 2020/8/7. Available at https://covid19.who.int/. Accessed 7 August 2020
WHO (World Health Organisation) (2020b) Situation Report—2020 Available at https://www.who.int/docs/default-source/coronaviruse/situation-reports/20200807-covid-19-sitrep-200.pdf?sfvrsn=2799bc0f_2. Accessed 1 July 2020
WHO (World Health Organisation) (2020c) COVID-19 strategy up date. The World Health Organization. Available at https://www.who.int/emergencies/diseases/novel-coronavirus-2019. Accessed 1 July 2020
WHO (World Health Organisation) (2020d) Shortage of personal protective equipment endangering health workers worldwide. The World Health Organization. Available at https://www.who.int/news-room/detail/03-03-2020-shortage-of-personalprotective-equipment-endangering-healthworkers-worldwide. Accessed 1 July 2020
WHO (World Health Organisation) (2020e) Sanitation, hygiene and waste management for the COVID-19 virus. World Heal Organ:1–11. WHO/2019-nCoV/IPC_WASH/2020.3. Available at https://www.who.int/publications/i/item/water-sanitation-hygiene-and-waste-management-for-covid-19. Accessed 3 July 2020
WHO (World Health Organisation) (2020f) WHO guidelines on hand hygiene in health care: first global patient safety challenge clean care is safer care. Available at https://apps.who.int/iris/bitstream/handle/10665/44102/9789241597906_eng.pdf;jsessionid=8357C66599885B2DD3B99F346656EF52?sequence=1. Accessed 10 Aug 2020
World Bank (2019) World development indicators: solid waste management. Accessed 23 Apr 2020. Available at https://www.worldbank.org/en/topic/urbandevelopment/brief/solid-waste-management. Accessed 1 July 2020
WWAP (World Water Assessment Programme) (2012) The United Nations world water development report 4: managing water under uncertainty and risk. UNESCO, Paris
WWAP (World Water Assessment Programme) (2017) The United Nations world water development report 2017: wastewater the untapped resource. United Nations Educational, Scientific and Cultural Organization (UNESCO), Paris. Available at http://unesdoc.unesco.org/images/0024/002471/247153e.pdf. Accessed 1 July 2020
Zaman F, Pervez A, Abreo K (2002) Isopropyl alcohol intoxication: a diagnostic challenge. Am J Kidney Dis 40(3):1–4
Zheng J, Suh S (2019) Strategies to reduce the global carbon footprint of plastics. Nat Clim Change 9:374–378

Anindya Basu is working as an Assistant Professor in the Department of Geography, Diamond Harbour Women's University. She is a rank-holder in Bachelor's and Master's degree programme in Geography from the University of Calcutta, and she was a recipient of the UGC-Junior Research Fellowship. Dr. Basu was awarded a Ph.D. from the University of Calcutta. She has also received a Post Graduate Diploma in RS-GIS from Jadavpur University. She has a keen interest in socio-political geography, urban-environmental issues, and tourism geography. Basu presented several papers in national and international conferences and seminars, apart from publishing several research papers and book chapters.

Lopamudra Bakshi Basu is working as an Assistant Professor in the Department of Geography, Heramba Chandra College. She has been a gold medallist in Master's degree programme in Geography from the University of Calcutta. She was awarded a Ph.D. from the University of Calcutta. Her research interest is concerned with the various aspects of social geography, urban-environmental issues, urban planning, and transport geography. She has presented several papers in both national and international conferences and seminars. She has also published research papers and book chapters.

Chapter 7
Restructuring State, Society, and Human Development: Projecting Post-COVID Pandemic Equations

Siddhartha Sankar Manna

Abstract The COVID-19 epidemic has not merely affected financial thinking on the path of GDP growth but also the pathway of human development. It has impacted one of the necessary foundations of human progress and development, i.e., a healthy and extensive way of life. The Coronavirus, an epidemic virus, outbroken from China, has been spreading all over the place of the world. The transformation of this virus from China to Europe and other parts of the world forced the people to continue in a social separation that affected social integration. The mode of human relations, interdependence, and interaction have been changed due to the emergence of epidemic conditions in social phenomena. In this paper, I highlight three significant aspects resulting from the COVID-19: (i) *Society as a form of integration*; (ii) *Politics as power*; (iii) *Economy as material gain*. The Coronavirus has affected the country's monetary growth relentlessly, as the lockdown due to infection is producing momentous trouble across several sectors—manufacturing, oil, economics, and many other fields. Numerous estimations are available on the capital loss and the world may face a monetary crisis, and the crisis of human development. If not accurately tackled through political strategy, the social disintegration produced by the COVID-19 disease may develop the discrimination, isolation, inequity, and unemployment in an extensive period. So, the economic strategy may take a policy of inclusiveness, and collective social protection schemes, which may strongly function in safeguarding the right to work and life of the people in society. Subsequently, they perform as automatic stabilizers in decreasing the occurrence of poverty. The economic system in the post-COVID-19 must be settled by the state's political arrangements, which would protect people's basic income security. So, it can enhance people's capability to control trouble and overcome the traumatization.

Keywords Impact · Civil society · COVID-19 · Society · Economy · Human development · Power · Politics

---

S. S. Manna (✉)
Department of Political Science, University of Gour Banga, Malda, West Bengal, India
e-mail: manna.siddhartha@yahoo.in

M. Mishra and R. B. Singh, *COVID-19 Pandemic Trajectory in the Developing World*, Advances in Geographical and Environmental Sciences,
https://doi.org/10.1007/978-981-33-6440-0_7

7.1 Introduction

Human civilization has been suffering from an unprecedented worldwide health crisis. This crisis has been killing the people, increasing human distress, and culminating in people's lives. The COVID-19 epidemic has not merely affected financial thinking on the path of GDP growth but also the pathway of human development. It has impacted one of the basic foundations of the notion of human progress as well as development, i.e., a healthy and extensive way of life. The Coronavirus, an epidemic virus, outbroken from China, has been spreading all over the place of the world. The hostile virus, migrating to Europe and other parts of the world, forced people to maintain social separation, causing a strong negative effect on social integration. The mode of human relations, interdependence, and interaction has been changed due to the emergence of the epidemic condition, disrupting almost all social phenomena.

In this paper, I highlight three significant aspects resulting from the COVID-19: (i) *Society as a form of integration*; (ii) *Politics as power*; (iii) *Economy as material gain* (Fig. 7.1). The Coronavirus has affected India's monetary growth relentlessly, as the lockdown due to infection is producing momentous trouble across several sectors, as well as manufacturing, oil, economics, and many other fields.

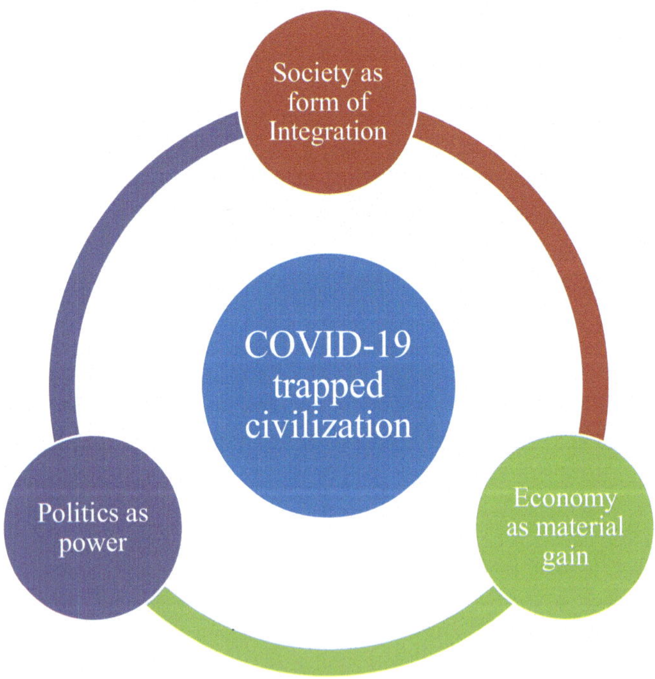

Fig. 7.1 The interrelationship between the three aspects which this chapter deals with

7.2 Society as a Form of Integration

As states start to disburden the COVID-19 epidemic lockdown procedures, demands to not return to normalcy are intensifying. The contagion has revealed the general imperfections and foregrounded the extensive discrimination in the worldwide Neo-Liberal structural society. If not accurately tackled through political strategy, the social disintegration produced by the COVID-19 disease may as well develop the discrimination, isolation, inequity, and unemployment in an extensive period. So, the economic strategy may take the policy of inclusiveness and collective social protection schemes. They might strongly perform their functions in safeguarding the workers, labors, and unorganized workers. Subsequently, they perform as automatic stabilizers in decreasing the occurrence of poverty.

It is important to state that billions of people have been suffering from the effect of the worldwide disease of COVID-19. Its eruption affects all sections of the world population and is mainly harmful to members of those social groups in the most helpless conditions, continues to affect the different sections of populations, as well as people existing in poverty conditions, aged persons, people with disabilities, young men, and aboriginal peoples. There are three indispensable roles of civil society to combat the COVID-19. Three "*ins*" and "*outs*" wherefore the civil society has been prerequisite to COVID-19 resuming (Rick 2020):

- The COVID-19 epidemic has uncovered the broad-spectrum of imperfections and drastic disparity;
- Civil society has a serious task to play in the universal contagion and monetary or commercial revival;
- The civil society is responsible for providing and assisting the people and can play its role as a campaigner, a supervisor, and a reliable institute during the crisis.

It is the state's responsibility to guarantee the right to food, livelihood security, and healthcare of every citizen. In this context, the government should discharge its role to reconstruct the society, based on the values of the *right to equality*, the *right to freedom*, and *integrity*, which consolidate the solidarity of the society. The epidemic has also uncovered the embedded disparities: extensive discrimination and inequalities, operational variations, antique prejudices, insecurity of the healthcare, and further basic needs, vulnerability of labors, weak or insufficient health arrangements, and growing domestic violence or aggression.[1] On the other hand, the socioeconomic effects have been suffered much more by women who prepare the majority of the necessary workforce and have been more likely to function the unpaid household-works and maintenance effort (Human Rights & Democracy Network 2020). Furthermore, a significant number of downgraded and deprived residents in the civil society have been most severely affected not only by the COVID-19 contagions but also by

[1]The Universal Declaration of Human Rights (UDHR), adopted in 1948, was itself a response to half a century of turmoil: two world wars, a pandemic and economic depression. It drew on previous charters and aimed to provide a broad framework covering all aspects of our lives.

the international trade and economic crisis. Also, the disease has exposed the essential cost of under-paid situations in the healthcare and food manufacturing sections (Beardmore and Gibbons 2020).

The extensive spread of COVID-19 has been unparalleled. In this regard, Sarah Beardmore and Michael Gibbons's opinion seems to be appropriate and significant. They (Beardmore and Gibbons 2020) rightly pointed out that:

> This public health emergency will likely result in a global economic crisis – both of which will hit the poorest and most marginalized communities the hardest. These twin crises expose the widening inequality around the world and threaten to exacerbate the gap between rich and poor, sending us back decades in our progress towards a more equal world.

The capacity to diminish these effects trusts profoundly on the capability of civil society to carry on its function and provide a voice to social groups most likely to be left behind in the community emergency outcome. In spite of that, the crisis will also obstruct civil society's capability to react and act in response. The blocking of civil places, controls on movement, and growing authoritarian strategies in several states have made the situation for support and responsibility an extremely hard one to act in.

7.2.1 Relations of Civil Society, Government, and People

In this way, the civil society will also face deteriorating resources, as the contributions that permit them to perform in support of the community interest desiccate as a consequence of the extensive monetary crisis. The contagion could breakdown the roles of civil society in several states. Although this might be one of the greatest problematic milieus for civil society, it has also been a serious moment where its effort has been more unimportant. So, the civil society has an immediate role to play in monitoring the effect of academic-institution closings on the underprivileged, particularly girls, and to support for operative responses to the contagion. Again, Sarah Beardmore and Michael Gibbons stated that (Beardmore and Gibbons 2020):

> Governments struggling to respond and recover will need active open feedback loops helping them understand in real-time how decisions made are impacting their citizens. Civil society can facilitate engagement with affected communities, a critical part of ensuring contextually relevant responses.

The civil networks will make the revival of the academic arrangements. These civil networks would function dynamically with social groups to ensure the educational rights of underprivileged and deprived people in society. The educational arrangements must be strengthened to assist and assimilate the most marginalized sections (Beardmore and Gibbons 2020)—

- To articulate the receptive strategies of responsibility to counter the critical issues in the developing policy backgrounds;

- To develop the clarity, participation, and responsibility of education segment plan negotiation;
- To upsurge the accessibility, sources, and various forms of information to articulate and have an effect on pertinent strategy resolutions;
- To mobilize the national citizens, who have been entitled to enjoy the rights in the state as well society, to perform an effective function in monitoring the execution of education plans and monetary strategies and use this report to hold duty conveyors at all stages responsible.

It is essential to state that there are three types of power (Dey 2020): (i) *Position as the capacity to govern*, (ii) *People as the source of the authoritative-power*, and (iii) *Money-economy as the base of superstructure.* So, the government gets a hold of authoritative-power from its position. It either acquires or captures this position in non-democratic states, or has been assumed this position through the electoral processes. Moreover, the world-trade acquires its power from the money-economy, which determines the superstructures. The third type of power has been the people who validate the power to govern and undermine the sovereign power, which we may well connect with civil society.[2]

The emergence of the crisis necessitates the civil society to create a system of the network that cuts across the sovereign state. It is essential to find out the various organizations and volunteer groups that provide relief in every region, district, the zone of the block, and rural community. The civil society will be reorganized with the various units which would be responsible for their total zone identifying the total population in the zone, the relief desired, the lack of administrative relief provided by the government, the challenges of the zone, and so on. In civil society, these units should have to enable in the particular region for the assistance to each other like providing material or constructing supply connections. It is the most significant to state that the system of the network should have a voice at the national level. The people are entitled to enjoy the opportunities assisted by the government even if they have not the ration cards, or else they are the active labor-force under the scheme of MGNREGA.

The civil society can establish an apolitical system that provides social opportunities. It may include: bringing rations, food-distribution to the people, primary health services, and caregiving. Moreover, it could point out to the government how citizens can be used to engage in these functions. It will not only assist societies affected by the epidemic, but the apparatus of doing so could help others in turn. The civil society must include and promote the constructive and best practices in a state to be reproduced in others (Dey 2020). In this regard, it is essential to state that civil society has the opportunity to focus on the despondency of migrant workers who got the opportunities to attain their job and way of living conditions before the outbreak. The labor-force has been the backbone of all economies. So, the state and

[2]Details see in Dey (2020)—We will need to map the different organizations and groups providing relief in every district, block, and down to every village. We can do this because we have volunteers and workers—from field staff of nonprofits to government school teachers.

society must take care of their life and security. The state and society both would create social opportunities that assist in empowering themselves in the way of life.

The civil society network might be a network of accessibility during the crisis. The government has not been to reach in various zones; the agents of government may use the networks of civil society to reach there. Citizens need to use their responsibility and mechanisms of transparency to supervise the administrative efforts and ensure the welfares of people that state capitals or resources have been adequately allocated and distributed in society. Furthermore, each citizen needs to think about the method in which they can assist the society for the wellbeing. In this regard, the comment made by the Secretary-General of the United Nations (23 March 2020) seems to be pertinent. He said that[3]:

> We must come to the aid of the ultra-vulnerable – millions upon millions of people who are least able to protect themselves. This is a matter of basic human solidarity. It is also crucial for combating the virus. This is the moment to step up for the vulnerable. (United Nations Department of Economics and Social Affairs 2020)

So, we, the people, have to utilize the Public Distribution Systems for the social phenomenon.

The government arrangement has been functioning well in various parts of the state, but it does not cover its functions in all corners equally. The government cannot widely perform its own functions all over the state equally unless civil society works in society. Indeed, civil society cannot perform the extensive role of the government in the wake of the emergence of the crisis in the social phenomenon. However, while civil society takes charge or responsibility of an area/region, the benefits of the state are supposed to reach everyone equally. Moreover, the state could discharge its duties of welfare and could guarantee the social and economic securities of its people with the help of local governments in the rural and urban areas.

The function of the state with local government reflects the government's response toward the people in the society. In this regard, the local government, i.e., panchayats and local self-governments, also have a very significant role to play in this attempt. If the state is to reach out to the people, it must use the civil society as an apolitical apparatus. This tendency can be called the transition from the function of the political society to the function of the apolitical society.

And, finally, let us not disregard or ignore the democracy at this time of crisis—human rights, like "*the right to expression*," "*the right to challenge*," "*the right to argue*" for the reason that now, the one thing a large number of underprivileged people have, is a voice (Dey 2020). In this regard, it is essential to state that we have to intensify the voice of vulnerable people to ensure support.

---

[3]The UN Department of Economics and Social Affairs (UNDESA) is a pioneer of sustainable development and the home of the Sustainable Development Goals (SDGs), where each goal finds it space and where all stakeholders can do their part to leave no one behind. UN DESA through the Division for Inclusive Social Development (DISD), monitors national and global socioeconomic trends, identifies emerging issues, and assesses their implications for social policy at the national and international levels.

7.3 Politics as Power

The politics have been largely applied to the behavior of civil society, but it is perceived in all interactions and relations of the human group. It has been composed of social relations involving authoritative-power, the rule of political sections, and the systems used to form and put on the social strategy. Politics have been the process by which people in society have the opportunity to take participate in social decision-making. In this regard, the term is largely applied to the functions within civil governments. However, politics has been considered as an extensive range of human relations, including corporate, educational, and religious organizations. It is composed of social relationships in relation to the power and authority, the rule of political components, and the systems.

Andrew Heywood (Heywood, 2004: 52), the renowned Writer and Director of studies at Croydon College, London, identified three distinct perceptions of politics:

- "Politics has long been associated with the formal institution of government and the activities which take place therein"
- "Politics is commonly linked to public life and public activities, in contrast to what is thought of as private or personal."
- "Politics has been related to the distribution of power, wealth, and resources, something that takes place within all institutions and at every level of social existence."

There are several instances of power strategies that have been relatively common and active every day. Some of these strategies consist of harrying, association, grouchy, decrying, demanding, snatching, avoiding, absurdity, stimulating, controlling, bargaining, socializing, and appealing.

The "Bio-power"[4] occurred with the transformation of power creations in Western civilizations beginning in the seventeenth century, but the furthermost extreme transformation befallen in the nineteenth century (Arnason 2012). This transformation is comprised of several systems of power, of several modes of managing the way of life. In this context, Michel Foucault's interpretation of Bio-power seems to be relevant. He (ibid.) defined:

> (bio-power as a) "set of mechanisms through which the basic biological features of the human species became the object of a political strategy, of a general strategy of power."

Michel Foucault, an Eminent French Philosopher, demanded that this new-fangled power over the natural life, which has been called *Bio-power*, developed in two systems, which he termed (i) *Anatomo-politics of the human body*; (ii) *biopolitics of the population*. Foucault's understanding of power and discipline in his work

---

[4]The concepts of "biopower" and "biopolitics" are perhaps the most elusive, and arguably the most compelling (given the attention they have subsequently received), concepts of Michel Foucault's *oeuvre*. Within his published work, these concepts featured only in the last chapter of the slim first volume of *History of Sexuality* (*The Will to Knowledge: History of Sexuality Volume I* 1976).

Discipline and Punish (1975) had been a consideration of anatomo-politics of the human body, but concentrating obviously on the disciplinary arrangement, on jails, and penalty. Even now, there he had pointed to the extensive application of the disciplinary systems, for example, in academic institutes, infirmaries, army camps, etc. In this regard, Foucault depicted a mainly interesting instance of the transformation of the warrior during the beginning of the seventeenth century to the end of the eighteenth century. He argued that during the beginning of the seventeenth century, the soldier was a number of persons—not for a reason that of his training but for his physical power, bravery, and elegance. One might perhaps recognize soldier equipment by such appearances. However, on the other hand, in the end of the eighteenth century, the soldier was one who was made through training and discipline (Oleinik 2020). He argued that practically someone might be formed into a soldier with new techniques and skills of disciplining the human-figure.

According to Foucault, parallel arenas of biopower have developed to surround patients, school children, nursery-children, along with workforces. He pointed out that the wide-ranging networks of power and knowledge function to form their bodies functional and submissive. This does not occur through any planned collusion of the government of technologically advanced societies. However, it is, to some extent, the combined outcome of numerous strategies and policies at several points in the social order throughout industrial-development and up to this period. Michel Foucault described the second form of power. He pointed out that this form of power, the biopolitics of the population, occurred for the period of the eighteenth century. According to him (Arnason 2012):

> it was concerned not with the human body but with the human species or human populations. Managing populations means managing reproduction, births and deaths, behavior, and health and sanitation.

The *biopolitics of the population* comprises of several types of systems to interfere and regulate the members of the society. These interferences require the collection of a huge number of information about populations, data analysis, and, ultimately, the creation of knowledge. In the eighteenth century, this creation of knowledge controlled accurately the origin of human-ecology.

In this regard, the notion of biopower is pertinent at this moment for various subjects in applied ethics, furthermost perceptibly for subjects connected to the cases considered by Foucault's *Punishment and Sexuality*. G. Arnason's analysis of biopower seems to be pertinent here. He revealed that (ibid.):

> The concept is also highly relevant for a great variety of other topics, for example reproduction, transformation of the body (from genetic and pharmacological enhancement to body-building, cosmetic-surgery, and body art), public health, medical practice, biomedical-sciences, race, and disability, to name just a few. The management of bodies and populations, and the sciences involved in that management, are so ubiquitous, that most topics in applied ethics may seem to invite an analysis of the mechanisms and micro-relations of power and knowledge.

By the way, the power/knowledge consideration does not assist us to a great extent if we have been largely interested in considering jeopardies, troubles, and benefits.

Subsequently, the society has been an important mechanism for power relations. Therefore, we cannot abscond and avoid the power relations without leaving society behind, but we can develop and expand our position in the power relations, protect ourselves against their influence, and expose new prospects for action and existence.

In "*Discipline and Punish: The Birth of the Prison*" and "*History of Sexuality*," another text, where Foucault recognized a modern form of power, an innovative understanding of power relations in society, and the notion of bio-power, which he assumes "*as its object as life itself*" (Arnason 2012). The biopower works by founding its objects of knowledge, to which it then retorts with the progress of new knowledge, technologies, and techniques of government. In this regard, Arnason perfectly underlined the apparatuses of modern power. According to him (ibid.):

> Far from being 'neutral' and power-free, the biological, medical, and social sciences are seen as parts of the apparatus of modern power. Through their scientific discourses, analytic categories, and intellectual practices, the natural and human sciences help to construct 'fertility,' 'the body,' 'population,' and 'the social,' in the process subjecting people to new, distinctly modern forms of power that affect how they constitute themselves as subjects. Given the sciences' close relationships with bureaucratic states in most settings, scientific knowledges of reproduction have often served the political objectives of regimes in promoting such goals as socioeconomic modernity, eugenic, purity, and national power.

In Foucault's interpretation, modern-power has been exemplified in dialogues and functional activities that have not been focused in any organization, such as the state or medical institution, but extensively isolated through the social structure. Even this concise analysis of some Foucauldian notions made it clear that Foucault's concepts unsettle the accepted conditions of dilemmas and debates about the reproduction. On the other hand, they focus on the language analysis, an important focus of study, set out the power, problematize the categories in which reproduction has been comprehended, and metamorphose the sciences, extensive understanding to be beyond the sphere of politics, into objectives of political analysis.

It is important to state that Foucauldian effort on the population focuses on the function of common people, sociology, social activity, social sanitization, urban arrangement, and other factors in constituting improper, i.e., very high or very low, fecundity or reproduction as a problem of society. So far, the Foucauldian conception of "population" has been a stimulating and significant paradigm that is positive to generate further such analysis and academic interpretations. Although a COVID-19 vaccine has been quiet in the producing, the notion of bio-power might assist make a better understanding of how we perceive the state handle and tackle the ongoing epidemic and wide-ranging contagion in the society.

7.4 Economy as Material Gain

Numerous estimations are nowadays available on the capital loss, and the world may face the monetary crisis. If not accurately tackled through political strategy, the social disintegration produced by the COVID-19 disease may as well develop the discrimination, isolation, inequity, and unemployment in an extensive period of time. So, the economic strategy may take a policy of inclusiveness, and collective social protection schemes, which may strongly function in safeguarding the workers and labors. Subsequently, they perform as automatic stabilizers in decreasing the occurrence of poverty. The economic system in the post-COVID-19 must be settled by the political arrangements in the state, which would provide the protection of the basic income security of People. So, it can enhance people's capability to control trouble and overcome traumatizes.

The COVID-19 epidemic challenges all sections of the population in the society and has been particularly harmful to people of those social groups who are in the maximum vulnerable conditions. It is continuously affecting the extensive segment of people, as well as people who are living in dearth conditions, older-persons, handicap-people, youth, and aboriginal persons. In this context, early indication suggests that the health and economic effects of the virus have been being tolerated disproportionately by the marginalized and underprivileged persons. For instance, displaced people, for the reason that they may not be able to take shelter in a secure place, are extremely unprotected to the peril of the contagion (United Nations Department… 2020). In the epidemic situation, the people without access to running water, migrants, asylum-seeker, or displaced persons as well stand to suffer excessively both from the epidemic and its outcome—whether owing to restricted movement, fewer employment prospects and opportunities developed racial intolerance, etc.

Indeed, this Covid-19 has plunged the global economy at significant risk and threat. It affects the economic practicalities of the global business. Observers have recognized this occurrence as a result of hyper-globalization or opening of de-globalization. However, the global system is going to face a downturn and world-wide damages. This damage and downturn might go beyond the World Wars I and II combined. Simultaneously, the deteriorating world price of unpolished oil has added additional concerns and disquiets. Amidst the troubled time, when the noble coronavirus was spreading in the different parts of the world, United Nations Secretary-General (on 23 March 2020) accentuated the Humanitarian Response Plan. He told that (United Nations Department… 2020):

> We must come to the aid of the ultra-vulnerable – millions upon millions of people who are least able to protect themselves. This is a matter of basic human solidarity. It is also crucial for combating the virus. This is the moment to step up for the vulnerable.

The COVID-19 is at presently spreading beyond urban to rural zones in various parts of the world. In low-income states, countryside health arrangements are being burdened, and lockdowns and further constrictions have been reducing the income and earnings of the people. By the way, the Governments are countering to the monetary disorder with an arrangement of social safeguard plans. In this regard, it is important to state that safeguarding the excellent governance and setting-up of services in rural parts is serious for livelihood security and progress—and therefore, central to COVID-19 strategy reactions. So far, academicians and experts have concentrated frequently on the complications of governance that COVID-19 pretends in urban zones, given larger exposure high-risks for contagion there.

On the contrary rural parts face a different type of epidemic challenges and threats deserving extraordinary consideration (Kosec and Ragasa 2020). These include the followings:

- The logistic-assistance and communication obstructs obfuscate the facilities of services, as well as vital epidemic related health-cares services and agrarian amenities and other supports;
- Rural parts have been comparatively underprivileged, and particularly dependent on government facilities like these; they are not well linked with the central government, and in this situation, the COVID-19 is probable to more undermine linkages, actually weakening the receptiveness of decision-makers to rural demands and requirements;
- Several workers are coming back to their rural sections—perhaps increasing infection, damaging the local workforce in marketplaces, or stimulating conflicts;
- The food system of the state and society is essentially dependent on rural zones, where the ranchers produce the maximum food-grains; agriculturalists need access to marketplaces not only for their productions, but also for essential inputs and facilities.

A small number of recent COVID-19 plans, however, emphasize on carrying on the production of agriculture and cultivation. Furthermore, in several developing states, Ministers of agriculture and cultivation have not been appropriately present in nationwide and regional COVID-19 Response Committees.

In this regard, Countering to COVID-19 and guaranteeing that high-quality facilities arrive at the rural poor claims a series of activities by regimes, donors, and systems of government. They should render high-quality information to preserve countryside inhabitants informed of necessary public health information about the infection and its spread, strategy retorts, and accessible facilities; rouse rural ventures and production of food-grains to alleviate disturbances to food supply manacles and rural livings, and galvanize citizen supervising of the administrative regime to promote reciprocal and mutual interaction between governments and rural people. Subsequently, the rural parts of the state, with an excessively high share of the marginal people, are minimum prepared to put up with the expenses of following such guidance, particularly within a serious monetary downfall. In various low-income states, concerns of the COVID-19 contagion have kept people from looking for essential health treatment, and disinformation about transmission has even abbreviated solid food and meat consumption—perhaps presenting a lost opportunity for developing the quality food and nutrition in village areas.

To make sure that rural inhabitants can receive and have faith in the high-quality reports and information, the governments and development practitioners must function with organizations and effective-institutions that people entrust. It is observed that the reliable and faithful information and communications campaigns have been arranged to reject the misinformation in city zones. Parallel policies should be addressed in rural zones. For instance, information propagation in several African states has embroiled employing rural leaders, conventional healers, personalities of religion, and young men to confirm that public wellbeing information reaches and resonates with the people (Kosec and Ragasa 2020). In addition to this, the low-tech solutions, e.g., the *Taking Books*, have been assisting in interconnecting culturally suitable ideas and reporting on COVID-19 in rural parts and establish civic response networks.

7.5 Rouse Rural Initiatives and Production of Food Grains

At the same time, as it unsettles the process of the food supply, the COVID-19 instantaneously represents income-generating chances for rural people. Several states are inflicting export constraints to protect national food supplies, which could reduce food accessibility and increase prices in low-income states that import a large number of their foods and essential products. Then it is also a chance to make grow the homegrown food products, as well as farmstead horticultural, to improve the food and nutrition security, which helps to generate the income of the marginal people like migrant labors, unemployed persons, and other marginal people. In this context, the agrarian and essential food trades, as well as related projects, must be held-open. The agrarian products should be permitted to free transport to take away the supply-side constraints and controls. Accordingly, some Loan plans and provisional relinquishments on excise charges and trade taxes can assist the traders of agricultural crops and service wage-earners. The provisional input parcels, money transfer, or lend-campaigns must be executed without deferring to assist marginal producers, manufacturers or cultivators, and businessmen to deal with the distractions and keep on in commercial occupations. Manufacturers and workforces will correspondingly need the protective equipment, the free test of COVID-19, and better-quality of hygiene and cleanliness.

On the other hand, the seed distribution and agrarian development are more necessary than continue to promote and accelerate domestic food production and crops produced from homelands in the period of crisis. The *Information and communications technologies (ICT)* can transmit accurate news and smooth the progress of imbursements and logistics—but then are often inadequately accessible and available in rural ranges. Subsidized information programs and training plans on their usage might help and support in accessing the opportunities regarding state assistance. In rural parts of the state, which have been often outside of the limelight of mass-media, and where healthcare employees might be struggling with a small number of equipment or facing a smaller amount of test than their civic parts, citizen involvement is enormously significant as new health, and social safeguard responses have been rolled-out.

The COVID-19-related administration and regime are needed to trace the infection and contagions in society and retorting to people. Subsequently, the particular goods and health facilities are much required e.g., public handwashing places, individual protective apparatus, test-facilities, or sanatorium services, and infirmary facilities for people. It is also essential to know where people are receiving effectual treatment or hospitality vs. being dismissed. In this regard, ICT could facilitate these objectives in the course of lockdowns, isolations, and social distancing actions. By way of ICT, rural inhabitants can specify what is needed or not. A robust public unions like *Rapid Action Teams* and associations should also be organized and reinforced for active and inclusive planning, project, and supervising of government's arrangements and schemes for development.

7.6 COVID-19: Reconsidering Human Development

The extensive dispersion of COVID-19 is not only the international health crisis but also a systemic crisis of human development, demonstrating our relations with the ecology and environment, which has been now affecting the aspects of the socioeconomic development in unparalleled systems. Strategies to decrease susceptibilities and develop capabilities to confront the health issues are essential to empower persons and the social order to better conditions and get well from these traumatizes (Kovacevic and Jahic 2020). The COVID-19 epidemic has been releasing the crises of human development and obstructing the possible capabilities of individual in society. On certain aspects of human development, situations currently are equivalent to stages of deficiency last perceived during the time of 1980s. However, the wide-ranging problem is exterminating all productive aspects of human development. The extensive global lockdowns indicate various people have to depend on the internet facilities to perform the functional activities, carry on their studies or academic training, and make the interaction with others in society. The digital share has turned out to be more important than ever, as a large number of people in the world still do not have the opportunity to access reliable broadband internet connections (ibid.).

The Human Development is based on three principal elements (UNDP 2020):

- *Education* (with effective out-of-school charges—implying, accounting for the incapability to retrieve the net connection—is expected to decrease or discontinue to the particular levels of original rate of the mid-1980s in primary schooling);
- *Health* (exactly causing a demise toll over 300,000 and obliquely leading possibly to an extra 6,000 child demises per-day from avoidable reasons over the subsequent 6 months);
- (iii) *Income (per Capita)* (with the enormous reduction in monetary motion since the Excessive Depression).

These are the visible outcome, including raising the gender-based violence; it will yet be totally recorded. Owing to slow the amplification of COVID-19 viruses, several countries stopped practically all nonessential functions, like sealed borders of the state and inflicted changing levels of testing, tracking the communication, solitary-confinement, and social isolation. The social lockdown imposed by the state has dramatically reduced the business and marketing sales, industrial production, and condensed the employment opportunities and job security (Caparini 2020).

The developed economies have tried to diminish several socioeconomic impacts resulted from the prolonged lockdowns by actions and strategies like the extension of tax submission, augmenting personnel, funding wage-earners salaries, protecting mislaid pay-cheques, and offering further jobless or redundant insurable coverage. These actions and policies would be funded by deriving from their Central Banks to generate monetary-space. On the other hand, underdeveloped states have very fewer means to accomplish the effect and deal with not only a monetary crisis but it is also a social crisis which develops the social disintegration and provides the wrong impression in the social orders what the United N Development Programme (UNDP) entitles a "*Systematic Human Development Crisis*" (Caparini 2020). In these contexts, shocks have a major effect on human development, with previous crunches indicating two different outlines (UNDP 2020):

Firstly, the shocks have longstanding outcomes on human development and can be passed to the upcoming generations. Even after the end of contagion or monetary development, the effects of the shock can bring about long-term devastation,[5] and, *secondly*, the corollaries are inequitably allotted, with impressionable groups arbitrarily exaggerated.

These arrangements highlight the significance of an equity lens. As a crisis reveals, an effective approach recognizing its impacts and transmission apparatuses can apprise timely and reasonable effort. In this regard, it is necessary to differentiate between the short-term and longer-term effects of epidemics—and foremost, shocks as a whole. Depicting from an extensive and intense historical study, Walter Scheidel, a distinguished Historian of Austria, exposed that major shocks like wars and epidemics can decrease the income disparity but that the effect depends on the response of the strategies and arrangements. According to him:

> When pandemics result in high mortality, the relative returns to labour increase compared with the returns to capital because workers demand higher compensation—in part because there is lower labour supply due to mortality and in part because they are afraid to be infected and demand more to show up for work. (UNDP 2020)

7.7 Arrangement of States to Counter to COVID-19

In the phase of the wide-spreading of COVID-19, each society, group of people, and individuals are vulnerable to opposing actions. Nonetheless, when contagions like the COVID-19 epidemic occur, the capability of people to respond has been expressively lower and inequitably disseminated. Table 7.1 shows a sequence of statistics for human development sets, areas, and 189 states that present the echelon of preparation to respond and tackle the effects of COVID-19 infection, as well as a state's degree of human development, the capacity of the healthcare arrangement, and internet access

[5] The pandemic was superimposed on unresolved tensions between people and technology, between people and the planet, between the haves and the have-nots. These tensions were already shaping a new generation of inequalities—pertaining to enhanced capabilities, the new necessities of the twenty-first century, as defined in the 2019 Human Development Report.

Table 7.1 Recommended non-pharmaceutical interventions

Area	Intervention	Enabler
Home	• Isolate sick people; • Quarantine household members of sick people;	• Access to computers and to the internet, particularly broadband; • Accessible digital services (for ordering goods, food, and entertainment); • Safe space (violence-free); • Balanced care work; • Continuous flow of income—drawn from existing assets, government transfers or the ability to work remotely for "nonessential" occupations;
School	• Close schools and childcare facilities; • Reduce children's social contacts outside school;	
Workplace	• Hold conference calls instead of face-to-face meetings; • Modify work schedules and have employees work from home;	
Community	• Cancel or postpone large public gatherings; • Increase distance between people;	

Source UNDP

to the people. In this way, it is found that the level of human development and its variation, simultaneously with the facilities of healthcare arrangement, can reveal the preparation of states to counter effectually and competently to a health emergency.

Let's say, a very developed state has on usual 55 hospital beds, above 30 medical doctors, and 81 nurses per 10,000 people, put side by side to 7 hospital beds, 2.5 doctors, and 6 nurses in an underdeveloped state (UNDP 2020). It is an analysis of 'economic growth before people's model of development'. So, the current pattern of analyzing human development has not been properly implemented because there has been no emphasis on the widespread pandemic prevention mechanisms in the hospital and public health when calculating the performances of states. With an emphasis on life expectancy, HDI has so far concentrated on decreasing the effect and transmission of local ailments without paying equivalent attention to the pandemics.

As systemic upsurges in the number of public health crises (SARS, MERS, H1N1, Avian virus) leading to pandemics over the years have augmented, it has become significant to calculate the wide-ranging pandemic management dimensions of states to avert development in life expectancy.

In the background of the continuing COVID-19 infection, the study of the information in Table 7.1 for viruses and mortalities in states by means of their HDI positions reveals this inadequacy. It has been found that though states like Germany, Sweden, the USA, Italy, and the UK position very high on the HDI and have the uppermost life expectancies in the world, they are surprisingly the most affected. As of 31 May 2020, these states outdid the diagrams of coronavirus demises worldwide per million of the people.[6] An epigrammatic study of this bizarre phenomenon indicates that if healthcare services have been well arranged in a habitual period of

[6]See the datasets on COVID-19, provided by the World Health Organization (WHO), which is available under the URL: https://covid19.who.int/table [Accessed on 31 August 2020].

time, all the improvements made are got rid of in a particular case of the epidemic (Fig. 7.2).

On the other hand, both South Korea and New Zealand states have performed a significant role in increasing a high life expectancy of 83.50 years and 82.80 years correspondingly, and they protected high HDI for the people in society. Their ranks are very high in the contest of the HDI. The systems and tactics executed by these states in maintaining a little number of contagions and demises have been being addressed as an ideal type for replication of the various states of the globe. However, another important factor in the study is the extraordinary performance of emerging states like India, Vietnam, and South Africa. Accordingly, these middle HDI positioned states have been able to maintain average life expectancies because of high occurrences of extensive viruses, and they are capable of confronting the COVID-19 contagions and reduce the mortality rates.

Consequently, what could be the possible decisions for the offbeat tendency viewed in this time? *Firstly*, success in administering widespread health crises in the normal periods need not be reformed in times of an epidemic; *Second*, the diseases have few traits owing to the emphasis on increasing the quality of living, but it can also have a flip sideways as well. States began to improve their health systems or arrangements for granted to provide the health facilities to the people; *Third*, taking away of policies of privatization by the state could provide the essential services and fulfill the basic needs like food, shelter, and healthcare amenities at the time of infections. In this regard, the state could minimize the excessive privatization and develop the quality of life; *Lastly*, effective government's intercession and interference can have a constructive outcome on managing healthcare crises such as contagion or epidemic in society.

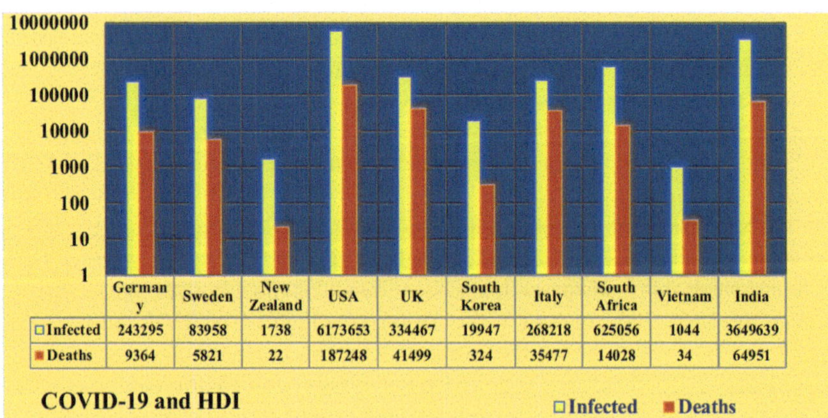

Fig. 7.2 COVID-19 and HDI in a single frame (*Data source* World Health Organization)

7.8 Reconsidering Human Development: Epidemic Preparation

After studying the effect of epidemics on the HDI, understood with the high mortality in states taking high life expectancies, the need for calculating pandemic management turn out to be essential. A new indicator termed '*pandemic preparedness*' is the need of the hour. This new indicator has three fundamental standards of consideration, which are as follows (ibid.): (i) Strategy response by the government/state; (ii) Determining health structure capability to manage epidemics; (iii) Mobility and Utilization of resources.

7.8.1 Strategy Response by the Government/State

The response of a government to any crisis is significant. A stable and comprehensible strategy arranged according to expert suggestion and using a scientific method can have an enormous effect in taking care of public health crises like COVID-19. The *policy response* indicator pursues to study the time taken by the state and government to construct a comprehensible response to manage the epidemic in human society. In the lack of pharmaceutical measures, it also tries to focus on a number of non-pharmaceutical courses of actions and policies, i.e., lockdowns, home isolation, quarantine, social distancing at the initial stages of health crises like COVID-19. But this is not a long-lasting solution. The pharmaceutical solutions, which can protect the human society and civilization, are needed to get rid of this crisis.

Figure 7.3 has been curated using the doubling period of founded COVID-19 cases (three-day duration). Using the five-day average times when the cases extended 100

	Germany	Sweden	New Zealand	USA	UK	South Korea	Italy	South Africa	Vietnam	India
Infected	243295	83958	1738	6173653	334467	19947	268218	625056	1044	3649639
Death	9364	5821	22	187248	41499	324	35477	14028	34	64951

Fig. 7.3 Ratio of doubling time of confirmed COVID cases (3-day period) (*Data source* Observer Research Foundation)

or so, and two days before and after that day—the time was counted twice.[7] So, therefore, the same type of average was calculated again using a similar system accurately one month later when 100 or more cases were attained. The one month limit was the period of time when governments started to introduce measures like lockdowns and social isolation, and home confinement.

Lastly, the "Ratio of Doubling Time" mentioned in Fig. 7.3 was estimated in the first limited number of cases after using strict procedures to reduce the harmful infection in society (Mittal and Mishra 2020):

$$Ratio = \frac{Average\, doubling\, time\, after\, the\, precautions}{Average\, doubling\, time\, for\, the\, first\, few\, cases}$$

It shows the efficiency of non-pharmaceutical interferences by governments, like lockdowns, home confinement, quarantine, and social distancing. The higher the ratio, the better it is. It is observed from Fig. 7.2 that the positive effect of operative government response to the harmful virus has been successful in states like South Korea, New Zealand, and Vietnam (Mittal and Mishra 2020). Furthermore, successes are similarly observed in other developing states like India and South Africa. These states may not have the strong health structures and healthcare facilities, but they have restricted the rapid growth of infection in the community and have been able to assist states in terrible necessity by way of exports of medicines, serious pharmaceuticals, PPEs, etc. These fruitful effects have not only been effective in reducing and controlling the functional growth of the virus in the initial stage but have curtailed in the wave of the second stage of the virus or prevented it on the whole. In this regard, all these achievements could be projected to an effective and systematic governmental response, which is necessary to represent in the scope of HDI.

7.8.2 Determining Health Structure Capability to Manage Epidemics

The important factor that enables confronting epidemics is having a strong healthcare structure and pharmaceutical facilities that can resist the anxieties and provide prerequisites to enhance the remedy in using the pharmaceutical vaccines, which prevent the infections. In this context, the COVID-19 has appeared as an incomparable challenge for the medicinal alliance and pharmaceutical nuts and bolts.

In the present epidemic, states which have been providing the advanced healthcare facilities and developed infrastructural services they are found lacking in terms of drugs, medicines, testing facilities, and critical medicinal apparatuses such as ventilators, oxygen-mask, PPEs breathing apparatus, and masks. In a few states such

---

[7] See the web-resource on "COVID-19: Rethinking Human Development" by the Observer Research Foundation, which is available under the URL: https://www.orfonline.org/expert-speak/covid-19-rethinking-human-development-66366/ [Accessed on 31 August 2020].

as the United States, Italy, and France, the condition has deteriorated and has been getting worse by the day. The small cases of widespread infections because of an upgraded standard of living have formed a healthcare structural arrangement which is lacking in managing the virus Covid-19 as an epidemic crisis.

The planned index seeks to measure these severe insufficiencies in the health-care structural facilities and services of states so as to make them for any awaited epidemic occurrence. The essential values will be measured would consist of:—*first*, the existing capacity of infirmaries, public health center, nursing hospices, etc., to manage an outburst in incidents (10,000 or more); *second*, studying the inadequacies in the standard technical response to an epidemic like testing facilities, procurement, and manufacture of vital medicinal products such as PPEs, surgical-mask, and necessary hand-gloves etc.; *lastly* and furthermost significantly, guaranteeing that there has been a minimum disruption of the established healthcare facilities and services while fighting against the unprecedented epidemic crises.

7.8.3 Mobility and Utilization of Resources

The epidemics disrupt numerous features of daily lives as well as access to necessary goods and public services like food, medicines, pharmaceutical facilities, and health facilities. The states are forced to establish precise apparatuses to defend people in society. Deployment of resources and capital turns into a dynamic responsibility in such periods. In this context, it is, therefore, important to understand what mobilization of resources and capitals means in the course of an epidemic and why it is vital to assess a state's preparation in terms of its capability to maintain an impeccable arrangement.

The deployment of resources at the time of an epidemic implies reducing the effect of shock, and confronting deficiency of farmhouse requirements, food scarcities, lack of necessary medical kit effectually through out-of-the-box resolutions. For instance, India transmuted its unaccustomed railway sections, guest houses, private hospital, or nursing homes, and boarding houses into quarantine accommodations to safeguard that there is no lack in a long period. The notion is similarly vital to measure for the reason that it decreases the twofold burden and guarantees that all resources have been used to confront the infection and wide-ranging disease.

The planned indicator will logically consider the scope to which a number of vital supply manacles of food and medicinal goods are resistant to commotions instigated by such catastrophes. It will similarly assess the time taken by states to remedy the commotions in the supply chain and the innovative systems employed to work so. Lastly, the indicator would similarly ascertain to what scope a state is self-sufficient in essential products such as food, medical commodities and kit, pharmaceutical facilities, and healthcare arrangement.

7.9 The Human Development to Respond to Virus COVID-19

The strategy response to COVID-19 has to balance between the public health primacies with monetary and social accomplishments, adjusting the short-term courses of action to alleviate the widespread of the infection and their longstanding outcomes. Therefore, the healthcare and economic responses to the COVID-19 need to be formed and systematically arranged to protect and develop abilities during the spread to infection in society and afterward the emergency: The health response to develop the extensive healthy lives, the monetary response to adjust a well-standardized *downtime* with the fortification of the standards of living (Atkeson 2020).

From the perspective of the people's capabilities and the health response, it is important to state that the inequities in human development demonstrate a scarcity of proficiencies for a large section of the population in the world. In emergencies, these inequities must be curtailed by the political arrangements in a concise period. Consequently, the priority must be decreasing these disparities by enhancing the abilities of those who were previously deteriorating behind before the infection and wide-ranging syndrome. A tactic in accordance with this strategy devolves on the accessibility of resources. Lacking investments, insurance schemes, or access to open markets, which is based on capitalism, the national and global public region has to intervene and accelerate transfers to overcome momentary shocks. This unit explains how abilities and their allocations matter to the healthcare and monetary responses to the emergency. The support for essential abilities is crucial to cover the indirect harmful impacts of COVID-19 on public sphere. Improved abilities like easy access to technology, networking facilities, and standard healthcare facilities have been not a luxury, but they provide the basic necessity of people and function a key role in managing the critical emergency in both accommodation and extenuation.

7.9.1 People's Abilities and the Health Response to COVID-19

People's abilities perform an essential function in response to the COVID-19 emergency. Non-pharmaceutical interferences are related to enablers that form the interference less expensive or facilitate its achievement (Table 7.1). All the interferences act for a system of social isolation that distresses peoples' capability to interrelate with others in the work-place, academic institution, shopping, refreshment, and social life.[8]

[8]See UNDP report entitle "COVID-19 And Human Development: Assessing the Crisis, Envisioning the Recovery," published by the Human Development Report Office based on PAHO, available under the URL: http://hdr.undp.org/sites/default/files/covid-19_and_human_development_0.pdf [Accessed on 28 August 2020].

The enabled persons could decrease the human development damages in accompanying with COVID-19 constraints in numerous scopes, beginning alternative abilities: opportunities in education, access to income-generating actions, facilities of goods and services, and opportunities in social life and recreation. They equally improve the likelihood of the interferences' success and diminish their human development detriments. On the other hand, without these enablers, there has been the jeopardy of a disastrous choice between non-pharmaceutical interferences at a mitigating human cost and deficiency of non-pharmaceutical intercession use.

In this regard, various enablers have been interrelated to the enhanced abilities—the new requirements of the twenty-first century—which are inequitably allocated to the entire population in the world. The gaps and disparities have been widening in the context of the COVID-19. These developed abilities can diminish the effect of the interruption to overcome the healthcare emergency caused by Corona infections. Thus, in underdeveloped states, the non-pharmaceutical interferences will tax people's wellbeing more and, therefore, can also be less effectual. Creating improved abilities—even in these serious periods—would decrease such inequalities. So, the emphasis on improved abilities does not indicate that the work on basic abilities is worked on. As per the Report of Human Development 2019, 785 million people do not have the facility to access the fundamental sources of pure drinking water, and about 3 billion persons do not have the facilities of handwashing with cleanser and soaps in their domestic functions. In the absence of addressing basic abilities could even reverse the transformation in response to the COVID-19 emergency.

In this situation, Public schooling means, in part, to be an equivalent to the amount that it can pause the intergenerational transition of discrimination. Quality education, irrespective of the educational situations of the parent, means to provide the services of equal opportunity to everybody. By disrupting education, in the epidemic situation, millions of children are taking away from the opportunities of education and creating the disadvantages among the intergeneration conversions. In several states, academic organizations and institutions of higher education, Colleges, Universities have learned their courses through the online mode. As per the Human Development Report of 2019, access to technological facilities is not equivalent and has been insufficient in all states. The digital capabilities like mobile phones, network subscription digital amenities have been enhanced but not sufficiently and satisfactorily provided between states and within the states. The epidemic interactions and the discrimination in improved capabilities mean that various states lack the opportunities for online education and to move their lessons and schoolwork online. If things keep on, the states left behind will also lack this choice in the future (discrepancy). The digital gap is accountable for excessive dissemination in effective school dropout in 2020. The following table signifies the index of Human Development:

The human development should be enhanced with two different conditions: (a) Directly Enhancing Human Capabilities; (b) Creating Conditions for Human Development. These two types of capabilities must be arranged by state policies. If not accurately tackled through political strategy, the social disintegration produced by the COVID-19 disease may as well develop the discrimination, isolation, inequity, and unemployment in an extensive period. So, the economic strategy may take a policy of inclusiveness, and collective social protection schemes, which may strongly function in safeguarding the workers and labors. Subsequently, they perform as automatic stabilizers in decreasing the occurrence of poverty.

The economic system in the post-COVID-19 must be established by the political arrangements in the state, which would provide the protection of the basic income security of People. In this regard, Marni Evans's revealing observation seems to be relevant. Evans (Evans 2020) argues that environmental sustainability has been concerned by means of whether environmental resources will be preserved and well-maintained for the upcoming generation. At what time preparing economic decisions, it is necessary to emphasize not only the existent day but also on the implications for future generations as well as human civilization. Therefore, the protection and safeguard of the environmental have been considered as the third pillar, which should be the primary concern for the protection of human civilization. The protection of the ecological arrangements, in terms of the quality of air, restoration of resources, and emphasis on the many aspects of the ecosystem that stress on the environment need to be maintained and safeguarded.

The Strategies to safeguard the environment of the earth does not depreciate to a fact where the coming generations should face water deficiencies, dangerous weather actions, and extreme temperature. People in society have a function to perform, but so do organizations that contribute to the cause on a larger scale (Evans 2020). The systems in which we can altogether live much more sustainably can take numerous forms of the arrangement, for example (ibid.):

- To restructure livelihood conditions in the procedure of *eco-villages*, *eco-municipalities*, and sustainable towns;
- To reconstruct the financial fields like green foundation, sustainable agronomy, e.g., sustainable planning;
- To improve innovative technologies and apparatuses, green machinery technologies, renewable power and resources, etc.;
- To make adjustments in individual lives that preserve the natural resources in the ecosystem.

Figure 7.4 shows features of human development that are directly enhancing the human capabilities considered as the foundational portion of human development; and features that are more relevant. They help to make the conditions in which men have the opportunities to be their best selves. Three foundations for human development (Education, Health; and Per Capita Income) have been necessary to enhance human capabilities. On the other hand, many other significant features may well be arranged to enhance the conditions for human development such as political participation, rights, and security, equality and justice, environmental sustainability. When necessary conditions for human development increase, the creating conditions automatically enlarged. The necessary conditions make guarantees for creating conditions and creating conditions assist in enhancing the necessary conditions. When one rises, the other increases, and when one reduces, the other declines. There has been intimate relationship between the basic human capabilities and creating conditions of human development.

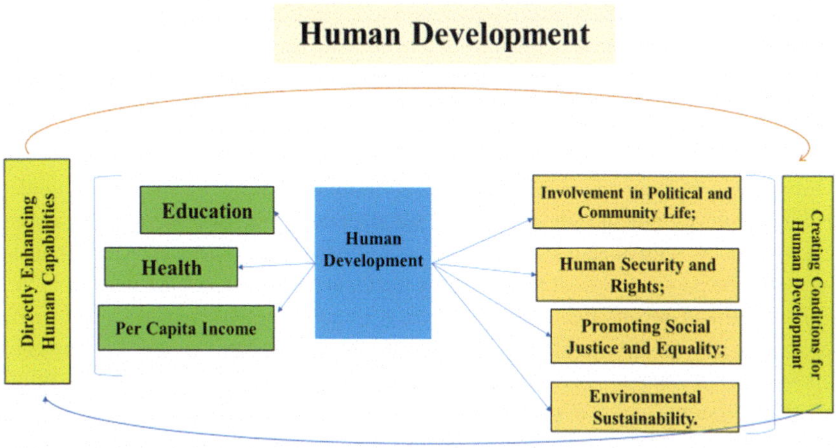

Fig. 7.4 Features of human development that are directly enhancing the human capabilities considered as the foundational portion of human development

7.9.2 Transmuting the World

As history teaches, the effect and responses to epidemics have the potential to restructure the state, society, and human conditions for the development as well as the world for future generations. As the aftermaths of the emergency revealed as well as the impacts of responses amid excessive uncertainty—enunciating a hallucination can donate to set up plans for consequences associated with the objectives of the 2030 Program for the Sustainable Development Aims. The unprecedented plans between March and April 2020 disrupted the normal and regular functions of the socioeconomic life. At present, the concern is affecting rapidly toward the economic components—correctly thus, given the intense reduction in productivity and its social entities. This tendency will reinforce as states and groups of people go through the epidemic waves to discover themselves under economic stress. In this framework, it is necessary to safeguard the lens system of human development, to concentrate on human beings.

This statement focuses three factors for a concept for the strategy response:

- Examine the response through the lens of equality. States, society, and groups of people before now in improved abilities will be most impacted, and placing them more behind will have longstanding effects in progressing the human development;
- Emphasis on the longstanding capabilities of people. It could comprise the ostensible adjustments between public health and economic functions (a means to the end of growing abilities) but would correspondingly assist build resilience for upcoming crises and dangers;

- Abide by a reasonable multidimensional tactic. In view of the fact that the crisis has various interrelated dimensions like health and socioeconomic factors. Therefore, a systemic approach is important. The United Nations has already represented a guideline along with these outlines in its original structure for essential socioeconomic retorts.

Equality is the concept of fairness for each individual, between male and female. Sustainable development is the notion that we, the people, have the right to livelihood security that can withstand our existences and have access to a more even distribution of goods. The participation of people is the necessary condition of the development. In this regard, it means that further effective social welfare programs need to be arranged by the government for the wellbeing of the people in society. Therefore, empowerment means a process by which liberty of the individuals is granted to influence the decisions, and it improves the capabilities of people to directly participate in the development processes that have an effect on their existence. The security proposes people for the development opportunities spontaneously and securely with self-confidence that they will not evaporate abruptly in the future (United Nations Development Programme… 2020).

The United Nations articulates that the COVID-19 has been considered as much more than a medicinal crisis. The epidemic has been retreating human development, which is rotating the regulator back to the 1980s. The UN declares that we should begin by protecting health facilities and social security. Then states must as well confront disparity and shape societal consistency. The UN concentrates on the five main concern for the socioeconomic development from the infection are (The UN's 5-point framework… 2020):

(1) To develop and protect the healthcare arrangements and facilities;
(2) To build up social security;
(3) To safeguard the jobs, small and medium-sized industries and safeguard the workers in unorganized divisions;
(4) To make the macro-economic strategies work for all in the world-society;
(5) To promote peace and harmony, accountability and good governance, and trust to develop social unity and integrity.

In this regard, the state and governments have been accelerating strategies to support the healthcare systems and improve the deteriorating economy. The plans to be executed have the potential to have an effect on the shape of scientific innovation, which mobilizes the new technology, mixing the power between renewables and fossil fuels, and the distribution of resources. There will be numerous inevitable functional choices like saving the productive sections, advocating various types of innovations with several impacts on employment opportunities, reshaping social facilities, and infrastructural investment. And so far, at the various stages, people around the globe have replied jointly. The acceptance of social distancing policy, which in various cases began before formal strategies came in force—was not probably wholly implemented and could depend on the voluntary teamwork of billions of people.

Consequently, there must be transparency and responsibility in stages of human development. The people have the ability to lead their lives in harmony with the environment. It is important to state that there have been five primary concern and priorities for allocating the financial resources that would improve both the welfare effect and ecological goals (UNDP 2020):

- Investment in the hygienic physical infrastructure;
- Emphasis on the skill-recovery structure;
- Put stress on the investment in education and training to combat the immediate joblessness from the COVID-19 pandemic and structural redundancy for the environmental change;
- Investing in the natural wealth for ecological resilience and redevelopment; and
- Adequate investment in Research & Development.

Learning from the COVID-19 emergency, in addition to the development in morals and priorities, will have an effect on our opinions of what human development appearances like as we move toward the middle of the twenty-first century and of what we are willing to do there. The world is off-target to prevent the environmental change, and it is corroborating a new generation of discriminations in human development, and even anticipating to have 800 million people existing in tremendous deficiency and scarcity by 2030 (COVID-19 And Human ...). Subsequently, the digital economy and society turn out to be the only technique to maintain economic functions and public interactions in society. More usage of tele-education and telemedicine could develop access to these services if investments are arranged in decreasing disparities in improved abilities.

More essentially, the emergency is absolutely a reminder that civilization is unlikely to be healthy in a sickening world. We disregard the barriers of nature in our threat. But the emergency showed the capacity of humans to work together to confront the common global challenges. Absolutely, the response was blemished, traumatized, and illogical, but practically far and wide, billions of people reformed their behavioral manners to face a shared peril. It makes it copiously perfect that we have within our reach to address other common threats—from weather variation and biological death of species to increasing disparities and discriminations in improved abilities.

7.10 Conclusion

It is concluded that this systemic emergency knockouts a world that has been taking care of the unsettled tensions: between individuals and technology, between man and the world, and between the haves and the have-nots—these have been creating a new generation of disparities. But the response to the crisis is an opportunity to reimagine how those tensions are addressed. The epidemic produced by COVID-19 has been an enormous systemic traumatize involving both public health devastation and an extreme economic emergency. The twofold emergency rests bare cleavages and organizational vulnerabilities both in individual states and in the extremely interrelated worldwide structure. This is a serious challenge to governments and multifaceted organizations to form and execute an operative and comprehensible response.

It intensifies the pre-existing dynamics of disparity and frangibility both within countries and in the global organization. The Human Development Index is a significant indicator that measures a state on its socioeconomic aspects. The coronavirus, therefore, gave a wake-up call to reconsider the HDI in terms of epidemic preparation for the future. The outcomes determine that developments in education, healthcare organizations, and income have been jeopardized by the coronavirus epidemic that generated the greatest retrenchment in financial activities.

References

Arnason G (2012) Biopower (Foucault), encyclopedia of applied ethics, 2nd edn. https://www.sciencedirect.com/topics/medicine-and-dentistry/biopower. Accessed 30 July 2020

Atkeson A (2020) What will be the economic impact of COVID-19 in the US? rough estimates of disease scenarios. Working Paper 26867. National Bureau of Economic Research, Cambridge, MA

Beardmore S, Gibbons M (2020) Civil society has a crucial role to play in response to the coronavirus pandemic, Published on 28 June 2020. https://www.globalpartnership.org/blog/civil-society-has-crucial-role-play-response-coronavirus-pandemic. Accessed 28 June 2020

Caparini M Dr (2020) Coronavirus shocks to human development and sustaining peace, Published on 7 May 2020. https://www.sipri.org/commentary/blog/2020/coronavirus-shocks-human-development-and-sustaining-peace. Accessed 22 Aug 2020

Dey N (2020) The role of civil society in times of crisis, Published on 7 May 2020. https://idronline.org/the-role-of-civil-society-in-times-of-crisis/. Accessed 27 June 2020

Evans M (2020) What is environmental sustainability? definition & examples of environmental sustainability, Published on 8 July 2020. https://www.thebalancesmb.com/what-is-sustainability-3157876 Accessed 29 Aug 2020

Heywood A (2004) Political theory: an introduction. Palgrave Macmillan, New York

Human Rights & Democracy Network (2020) The role of civil society organisations. https://hrdn.eu/covid-19/the-role-of-civil-society-organisations-during-and-in-a-post-covid-19-world/. Accessed 28 June 2020

Kosec K, Ragasa, C (2020) How to ensure effective government responses as COVID-19 spreads to rural areas, Published on 22 June 2020. https://www.ifpri.org/blog/how-ensure-effective-government-responses-covid-19-spreads-rural-areas. Accessed 20 Aug 2020

Kovacevic M, Jahic A (2020) COVID-19 And human development exploring global preparedness and vulnerability, Published on 29 April 2020. http://hdr.undp.org/sites/default/files/covid-19_and_human_development.pdf. Accessed 21 Aug 2020

Mittal P, Mishra, S (2020) COVID-19: rethinking human development. Observer Research Foundation. https://www.orfonline.org/expert-speak/covid-19-rethinking-human-development-66366/. Accessed 31 Aug 2020

Oleinik A (2020) The politics behind how governments control coronavirus data, US News. https://www.usnews.com/news/best-countries/articles/2020-06-05/the-politics-behind-how-governments-control-coronavirus-data. Accessed 3 Sept 2020

Rick J (2020) 3 reasons why civil society is essential to COVID-19 recovery. World Economic Forum, Published on 29 May 2020. https://www.weforum.org/agenda/2020/05/why-civil-society-is-essential-to-covid-19-pandemic-recovery/. Accessed 28 June 2020

The UN's 5-point framework for a socioeconomic recovery (2020) ORIUS, Published on 18 June 2020. https://qrius.com/the-uns-5-point-framework-for-a-socio-economic-recovery/. Accessed 4 Sept 2020

UNDP (2020) COVID-19 and human development: assessing the crisis, envisioning the recovery. Human Development Perspectives. http://hdr.undp.org/sites/default/files/covid-19_and_human_development_0.pdf. Accessed 22 Aug 2020

United Nations Department of Economics and Social Affairs (UNDESA) (2020) The social impact of COVID-19, Published on 6 April 2020. https://www.un.org/development/desa/dspd/2020/04/social-impact-of-covid-19/. Accessed 17 July 2020

United Nations Development Programme (2020) What is human development? https://web.archive.org/web/20120601112934/http://www.undp.org.bz/human-development/what-is-human-development/. Accessed 29 Aug 2020

Dr. Siddhartha Sankar Manna is an Assistant Professor in the Department of Political Science at the University of Gour Banga. The focal theme of his research is the analysis of international relations, political theory, democratic politics, and human development. He published more than twenty papers in reputed national and international journals.

Chapter 8
The *New Normal* of the Education System: Issues of Rights and Sustainability in Pandemic Trapped India

Srashta Chowdhury and Sushma Rohatgi

Abstract Education is the guiding light to development, climbing up the social ladder, and having perspectives about how the world can change for good. With time people on earth have felt the need for education, emphasizing education as a right and considering it a Sustainable Development Goal. This paper explores how the COVID-19 pandemic has come as a blow to the education system and sector; the rights, Sustainable goals are being jeopardized due to the sudden closure of the educational institutions and shift in teaching methods from paper to digital. All three education sectors, primary, secondary, and tertiary, are severely affected by the pandemic. While the government had started stressing on the quality of education, the access once again has been put into question. There are a revisiting and framing of new initiatives by India's government to take up digital measures of education for the students. While the future of education and students in India still remain unclear with examinations getting canceled and new academic calendars are being prepared and relooked at, the concern is about the changes that shall come in the education post-pandemic to build better.

Keywords Pandemic · Human rights · Sustainability · Education · Government schemes · Digital learning

8.1 Introduction

The world is vigorously affected by the pandemic outbreak of COVID-19. It has led to disruption in human activities and the basic needs of the population in the world. The traces and spread of the pandemic have first been noticed in Wuhan, Central China's Hubei Province, late in 2019, post which the whole world came under the grasp of the virus. The human-centric socio-economic environments have been witnessing almost complete standstill around the world, leading to a crisis for

S. Chowdhury (✉) · S. Rohatgi
Department of Geography, University of North Bengal, Darjiling, West Bengal, India
e-mail: srashtachowdhury@gmail.com

M. Mishra and R. B. Singh, *COVID-19 Pandemic Trajectory in the Developing World*, Advances in Geographical and Environmental Sciences,
https://doi.org/10.1007/978-981-33-6440-0_8

the human resource. The education sector has been hard hit by the pandemic as most of the countries have decided to keep the educational institutes shut to avoid the disease being spread among young children and adults.

The shreds of evidence from history make it clear that the closure of educational institutions is often considered as the first non-pharmaceutical intervention for the implementation in a pandemic, as students are effective 'agents' in spreading the malicious virus. Timely closure of schools and the cancellation of public gatherings was significantly associated with reduced mortality related to influenza epidemics during the 1918 influenza epidemic in the United States (Chen et al. 2011). More than 1300 public, charter, and private schools in 240 communities across the United States were closed during the *spring wave*[1] of the 2009 (H1N1) pandemic (Navarro et al. 2016).

United Nations Educational, Scientific, and Cultural Organisation (UNESCO 2020a) estimated that 107 countries have temporarily closed their educational institutions nationwide, impacting over 861.7 million children and youth, i.e., 90% of the world's student population.[2] The Government in India, like most of the governments worldwide, has decided on temporary shutdown of the Educational Institutions as a precautionary attempt to prevent the spread of Covid-19.

However, the closure of educational institutes as a part of safety measures severely compromises the ethics of social equality. The article by Cauchemez et al. (2009) points out that school closure raises a range of ethical and social issues, particularly since families from underprivileged backgrounds are likely to be disproportionately affected by this intervention. The pace of development for certain groups of people has been low, and the pandemic acts as the catalyst to magnify the pre-existing constraints of development.

Prior to the onset of COVID-19, youth (aged 15–24) were already three times more likely to be unemployed compared to adults, while 126 million young workers were in extreme and moderate poverty worldwide (UNDESA 2020; International Labour Organization 2020). The governments have deployed measures for learning to continue through digital/online platforms, like television and radio: a most far-reaching experiment in the history of the education system (Giannini et al. 2020). The problem is, however, farfetched in terms of access to these virtual lessons. The young population, being kept away from their spaces of learning, can make the step and advances toward the *Agenda 2030*[3] take a back seat with stagnation in achieving a sustainable future. This chapter takes the discussion into three sections. The first section focuses on the relationship between education, sustainability, and rights. The second section brings forth the Indian context on ways the pandemic has changed the

---

[1]The United States experienced its first wave of 2009 H1N1 pandemic activity in the spring of 2009, followed by a second wave of 2009 H1N1 activity in the fall. The first wave of the H1N1 was known as the 'spring wave'.

[2]Mentioned in *UNESCO's Initiative Against COVID-19* (https://en.unesco.org/sites/default/files/no_teacher_or_student_should_be_left_behind.pdf).

[3]The 2030 Agenda announces a "plan of action for the people, planet and prosperity." It highlights strengthening "universal peace in larger freedom," and recognizes eradicating poverty in all its forms and dimensions.

country's education system, and the third deals with the issues of rights, sustainability, and their violation during the pandemic and highlights the measures that shall be the deciders of a *new normal*[4] after the pandemic ends.

8.2 Rights, Sustainable Development, and Education

Rights are fundamental to civilization entitled to all beings. Humans acquire their right by virtue of birth irrespective of their biological, national, social, cultural, and any other identities. Human Rights are norms or descriptions of certain moral human behavior void of discrimination, Universal, and Egalitarian in nature. The rights enshrined in the Universal Declaration of Human Rights (UDHR) are all interdependent for their successful conferment, as is mentioned in both the treaties—the International Covenant on Civil and Political Rights (ICCPR) and International Covenant of Economic, Social and Cultural Rights (ICESCR), 1966:

> The ideal of free human beings enjoying civil and political freedom and freedom from fear and want can only be achieved if conditions are created whereby everyone may enjoy his civil and political rights, as well as his economic and cultural rights.

The UDHR was formulated in 1948. Whereas '*Our Common Future*' of the Brundtland Commission in 1987 set the cornerstone of sustainable development. Framed as principles for human development, taking into account the sustenance and balance between natural resources and man for the future, the concept of sustainable development was established as:

> (the) development that meets the needs of the present without compromising that ability of the future generation to meet their own needs. (World Commission on Environment and Development 1987)

With the advancement of time, human civilization has progressed in technological know-how and innovation. Human rights violation remains a prevalent issue in most of the developing country, and there is an ever-increasing gap between the rich and the poor. A large proportion of the world's population remains illiterate, and children still live in poverty, challenging one of the key concepts of sustainable development, i.e., '*providing essential needs to the world's poor through overriding priority*' (Shaker 2015). The main target of sustainable development would be attaining equity between the present and the future generation that is to be regarded with the basic needs for a quality life. It refers to the pathways and processes that help achieve sustainability, which is the long-term process that establishes a sustainable global world. However, the documented existence of the rights and the views on a better future remains unfamiliar among a huge section of the world population, which can be overcome by education.

---

[4] A *new normal* is a state, following a crisis, to which an economic or social system settles. It differs from the situation that prevailed prior to the start of the crisis. The term has been used frequently in literature relating to the financial crisis of 2007–2008, the aftermath of the 2008–2012 global recession, and now, during the COVID-19 pandemic.

8.2.1 Human Rights and Sustainable Development Goal

Human exists on the earth by virtue of their interlinking to one another, as well as with their physical, social, and cultural surroundings. The very existence and implementation of human rights create conditions essential for sustainable development. Development can be a key to change in human existence and welfare when the rights are protected and helps modulate the social, economic, cultural, political, and civic well-being of the population and individual. The process of development, however, requires the free and active participation of all individuals. Nobel laureate Professor Amartya Sen (1999) defines development as freedom, which is both the primary objective of development and the principal means of development. In order to achieve this 'freedom,' the human world requires to be free from all kinds of discrimination that is guaranteed by rights granted. As humans are the harbingers of change, the Right to Development forms their basis of linkage between rights and development. Article 1 of the Declaration of Right to Development states:

> The right to development is an inalienable human right by virtue of which every human person and all peoples are entitled to participate in, contribute to, and enjoy economic, social, cultural and political development, in which all human rights and fundamental freedoms can be fully realized. (Declaration on the Right to Development 1986)

There has been a constant emphasis on the recognition of human rights essential for achieving sustainable development.

The Millennium Development Goals (MDGs)[5] have acted as a stand-in for certain economic and social rights, but overlook other human rights linkages. Despite significant improvements in increasing primary school enrolment in some regions, the Millennium Development Goal of achieving universal primary level education by 2015 could not be achieved. The 2030 Agenda for Sustainable Development was adopted in 2015, provides a plan of a peaceful and prosperous world. The 17 Sustainable Development Goals (SDGs') recognize that ending poverty and other deprivations must go hand-in-hand with strategies that improve health and education, reduce inequality, and spur economic growth (United Nations 2015). The SDGs laid down in Agenda 2030 are inclusive, comprehensive, and 'unequivocally anchored' (Office of the High Commissioner for Human Rights 2015) in human rights marking a paradigm shift toward a more balanced sustainable development aiming at equality and non-discrimination. Like the universal nature of human rights, Agenda 2030 is also universal, applicable to all without any form of discrimination. Although the SDGs themselves are not framed explicitly in the language of human rights, most targets explicitly reflect the content of corresponding human rights standards. The SDGs address availability, accessibility, affordability, and quality of education, health, water, and other services related to those rights.

[5]At the beginning of the new millennium, world leaders gathered at the United Nations to shape a broad vision to fight poverty in its many dimensions. That vision, which was translated into eight Millennium Development Goals (MDGs), has remained the overarching development framework for the world for the past 15 years (2000–2015).

8.2.1.1 Importance of Right to Education (RTE)

While the world captured by the virus, understanding and implementing human rights is the most vital and crucial for the upliftment of human life; as UN Secretary-General António Guterres in his brief on COVID-19 and Human Rights (United Nations 2020a) says,

> The virus threatens everyone. Human rights uplift everyone. By respecting human rights in this time of crisis, we will build more effective and inclusive solutions for the emergency of today and the recovery for tomorrow.

Education is the pathway to an aware and a liberated world, leading to social change and a much-developed world. Education as a human right has made its place in Article 26 of the UDHR, 1948, which states that:

> Everyone has the right to education. Education shall be free, at least in the elementary and fundamental stages. Elementary education shall be compulsory. Technical and professional education shall be made generally available, and higher education shall be equally accessible to all on the basis of merit. (Universal Declaration of Human Rights 1948)

As defined by *General Comment No. 13* of the United Nations Committee on Economic, Social and Cultural Rights (UNCESCR),[6]

> education is both a human right in itself and an indispensable means of realizing other human rights. As an empowerment right, education is the primary vehicle by which economically and socially marginalized adults and children can lift themselves out of poverty and obtain the means to participate fully in their communities. The importance of education is not just practical: a well-educated, enlightened and active mind, able to wander freely and widely, is one of the joys and rewards of human existence. (UN Economic and Social Council 1999)

On the other hand, the International Covenant on Economic, Social, and Cultural Rights (ICESCR) identifies the '*Right to Education*' under article 13 and states that:

> The States Parties to the present Covenant recognize the right of everyone to education. They agree that education shall be directed to the full development of the human personality and the sense of its dignity, and shall strengthen the respect for human rights and fundamental freedoms. They further agree that education shall enable all persons to participate effectively in a free society, promote understanding, tolerance and friendship among all nations and all racial, ethnic or religious groups, and further the activities of the United Nations for the maintenance of peace. (International Covenant on Economic, Social and Cultural Rights 1966)

The treaty also finds mention of all forms of education, be it primary, secondary, and higher education, to be accessible to all. Human rights and education during the global crisis can be aligned with one another. While the rights and the rule of law exist for the protection of mankind, education could be the most important and key component to defend the rights that are the very subject, finds mention in an article published by UNESCO:

> Education is a key ally in defending rights and protecting the rule of law and is an essential life skill in crises. (Giannini and Brandolino 2020)

---

[6]UNCESCR is the body in charge of monitoring the implementation of the International Covenant on Economic, Social and Cultural Rights in the States which are party to it.

8.2.1.2 Education as a Sustainable Development Goal

Education becomes the key to achieve all other Sustainable Development Goals. There is believed to exist a positive relationship between education, socio-economic, and environmental protection. Education is both a means and an end, having a domino effect of triggering the intertwining realization and enjoyment of other human rights, which is a crucial and differential factor for sustainable human development. Learning and development are at the very core of the human experience and existence, both at the individual and collective levels. Evidently, the process of learning and the process of development are two interdependent and interrelated processes and concepts (Savić 2018). Education must be both inclusive and non-discriminatory to avoid the exclusion of some groups from mainstream society-ideally one that tolerates diversity. The lack of education is in itself a dimension to poverty.

Education does influence the growth and reduction of poverty levels in developing countries. Study shows that an increase in the '*years of schooling among adult*' (15 years old and over) by two years throughout 1965–2010, helped to lift nearly 60 million people out of poverty (UIS and GEM 2017). The proposed Sustainable Development Goal 4 reads, '*Ensure inclusive and equitable quality education and promote life-long learning opportunities for all.*' Like inclusive quality education, human development in a sustainable manner is also a human right. Discrimination in human society is still prevalent in the global world that leads to a structured inequality and a segregated society. Education, however, is a fundamental right that can ensure sustainability for the present and the future generation as the impact is long term with a positive impact on the other factors crucial for human development. Education can prevent transmission of poverty between generations, the utmost level of sustainability. Education, being the forerunner of development, inclusiveness, and quality are not sufficient enough. Availability, accessibility, adaptability, and acceptability also form the basis of acquiring education.

In 2019, the UN General Assembly recognized Education for Sustainable Development (ESD) as a model for rethinking the learning to achieve the Sustainable Development Goals. ESD reassesses what we learn, where we learn, and how we learn. It develops the knowledge, skills, values, and attitudes that enable learners to make informed decisions and actions on global problems such as the climate crisis. It empowers learners of all ages to change the way they think and work toward a sustainable future (Giannini 2020). UNESCO is to come up with an initiative called '*the future of education*,' which aims to rethink education and reshape the future. The initiative is catalyzing a global debate on how knowledge, education, and learning need to be reimagined in a world of increasing complexity, uncertainty, and precarity. The commission declares knowledge as the most significant renewable resource of humankind to respond to challenges. On this basis, it is sensible to rethink education regarding how inclusive the world can be in building on education for the betterment of humankind and the planet. UNESCO, in its work 'From Green Economies to Green Societies' has pinpointed the vital role education plays in bringing about change in societies:

> Education is a fundamental lever of change contributing to poverty eradication, sustainable development, equity, and inclusiveness. It is also a means of realizing broader social, economic, political, and cultural benefits. It empowers all people of all ages with the knowledge, skills, and confidence they need to shape a better future. (UNESCO 2011)

In green societies, education needs to be grounded on the values of peace, non-discrimination, equality, justice, nonviolence, tolerance, and respect for human dignity. Quality education must be manifested on a human rights-based approach so that human rights are implemented throughout the whole education system and in all learning environments (UNESCO 2011).

8.2.2 SDG 4 and Related Human Rights Issues During a Pandemic

Guaranteeing human rights for everyone poses a challenge for every country around the world to a differing degree. The public health crisis is fast becoming an economic and social crisis, and a protection and human rights crisis rolled into one. The COVID-19 crisis has exacerbated the vulnerability of the least protected in society. It is highlighting deep economic and social inequalities and inadequate health and social protection systems that require urgent attention as part of the public health response. Women and men, children, youth and older persons, refugees and migrants, the poor, people with disabilities, persons in detention, minorities, LGBTI people, among others, are all being affected differently (United Nations 2020a). As mentioned in Table 8.1, the violation of rights leads to a disruption in achieving sustainability, and the SDGs' are grounded on human rights. The pandemic has put to trial the strength of rights and sustainable goals.[7]

António Guterres, *UN Secretary General,* in the High-Level Political Platform (HLPF) 2020, outlines an overarching theme of '*reducing inequality, by making economies more sustainable and just*' which shall be the key strategy to reduce global poverty. Progress toward reduction has slowed in recent years, and it is projected that in 2020 alone, the pandemic could lead to up to 49 million people falling into poverty (UN News 2020). At the end of 2019, millions of children and young people were still out of school, and more than half of those in school was not meeting minimum proficiency standards in reading and numeracy. The closure of schools to slow the spread of COVID-19 readily carries an adverse impact on learning outcomes and the social and behavioral development of children and young people. It has affected more than 90% of the world's student population, 1.5 billion children, and young people. Although remote learning is provided to many students, children and young people in vulnerable and disadvantaged communities, such as those living in remote areas, extreme poverty, fragile states, and refugee camps, do not have the same access to that. The digital divide will widen existing gaps in inequality concerning education (UNECOSOC 2020).

---

[7] See the website of the United Nations: https://www.un.org/sustainabledevelopment/education/.

Table 8.1 SDG 4 and human rights violations during the pandemic

Sustainable development goal 4[a]	Related human rights that will be violated due to the pandemic[b]
Ensure inclusive and equitable quality education and promote life-long learning opportunities for all • **Targets of SDG 4** Targets include universal access to free, quality pre-primary, primary and secondary education; improving vocational skills; equal access to education; expanding education facilities, scholarships, and training of teachers. • **Threatened Targets of SDG 4 due to the pandemic** Target 4.1 Target 4. 2 Target 4. 3 Target 4. 4 Target 4.5 Target 4.6 Target 4.7 Target 4.A	• **Right to education** (UDHR art. 26; ICESCR art. 13), particularly in relation to children [Convention on Child Right (CRC) arts. 28, 29]; persons with disabilities [CRC art. 23(3), Convention in the Rights of Persons with Disabilities (CRPD) art. 24]; and indigenous peoples [United Nations Declaration on the Rights of Indigenous People (UNDRIP) art. 14] • **Equal rights of women and girls in the field of education** [Convention on the Elimination of all forms of Discrimination against Women (CEDAW) art. 10] • **Right to work, including technical, and vocational training** [ICESCR art. 6] • **International cooperation** [UDHR art. 28; Declaration on the Right to Development (DRtD) arts. 3-4], particularly in relation to children [CRC arts. 23(4), 28(3)], persons with disabilities [CRPD art. 32], and indigenous peoples [UNDRIP art. 39]

Data Source [a]United Nations Sustainable Development Goals
[b]Various Human Rights conventions and declarations

The right to education could also be termed as an empowering right which has been put to a position by the pandemic to be violated. The pandemic magnifies the structured inequalities existing in society. The United Nations Development Programme (UNDP) estimates the '*effective out-of-school rate*' due to school closure. The percentage of primary-school-age children is adjusted to reflect those without internet access. The report indicates that 60% of children are not getting an education, leading to lower the global educational attainment level which is not seen since the 1980s (UNDP 2020). According to RTE's '*State obligations and Responsibilities*', the states are obliged to '*take positive measures that enable and assist individuals and communities to enjoy the right to education*' and provide the right '*when an individual or group is unable to realize the right themselves by the means at their disposal*' in general.

However, the pandemic has changed the ideas with the virtual intrusions in education, remaining the only means the obligation of the state is curbed as it cannot provide each student with the digital toolkit required for education. The pandemic and the temporary closure of the educational institutes are likely to widen the already existing disparities in the society, limiting the long run to sustainability. According to UNESCO (2020c) data, 100 countries have not yet announced a date for schools to reopen; 65 have plans for partial or full reopening; while 32 will end the academic year online. For 890 million students, the school calendar has never been so undefined.

Keeping the young away from educational institutes could aggravate the educational inequalities, jeopardize the attainment of Sustainable Development Goal 4, and violate the right to education. To summarize, the role of the SDGs is greatly heightened by the pandemic as a trajectory to shape the future through education for sustainable development (ESD). Where schools and institutions of higher learning are at the receiving end as 'victims' of the viral attack, unable to cope up with the disruptions that they are facing. It goes back to the issue of cleanliness, hygiene, and sanitation—both mental and physical, an age-old precept that has fallen behind in the virtual world. In some places, the IT facilities are better kept than the toilets! Undoubtedly, the ethos of education must be revisited (Razak 2020). School closures in over 180 countries have laid bare inequalities in education, deficiencies in remote learning, the cost of the digital divide, as well as the important role schools play in student health and wellbeing (Giannini 2020). The pandemic has affected students' learning efficacy on a global scale never encountered before.

8.3 The Indian Scenario

India, the second-largest country in the world by population (preceded by the People's Republic of China),[8] has been put to a state of inactivity due to the current pandemic since the end of March. The first case was reported during the end of January in Kerala,[9] and by March, the numbers had surged. By 24 March 2020, the government of India had ordered a complete nationwide lockdown, and the country went to a standstill. Due to the sudden lockdown put to effect without proper policy measures and repeated prior warnings, there have been violations of rights. Though the country has taken to partial lockdowns and unlocking since mid-May in certain sectors of economic involvement, the education sector in India is still continuing from the 'in home' and 'from home' space. In India, with a huge demographic dividend, the future human resource, which is directly or indirectly associated with the education sector, has been pushed to a state of complete mayhem. The future of reopening the educational institutes, adhering to the norm of social distancing, still remains uncertain. The novel nature of the virus makes the return of the education sector a dubious affair and a matter of concern with a maximum of the young population staying away from education. The youth population (15–24 years)[10] in India is nearly

---

[8] According to the World Population Prospects: The 2012 Revisions (https://population.un.org/wpp/Publications/Files/WPP2012_HIGHLIGHTS.pdf).

[9] As has been reported by the World Health Organization (https://www.who.int/india/news/feature-stories/detail/responding-to-covid-19---learnings-from-kerala).

[10] The UN Secretariat uses the terms youth and young people interchangeable to mean age 15–24 with the understanding that member states and other entities use different definitions (https://www.un.org/esa/socdev/documents/youth/fact-sheets/youth-definition.pdf). According to World Health Organisation 'Young people' comprise the age group 10–24 years (https://www.who.int/southeastasia/health-topics/adolescent-health#:~:text=WHO%20defines%20'Adolescents'%20as%20individuals,age%20range%2010%2D24%20years).

Table 8.2 Rural–urban distribution of educational institutes in India

Location	Primary schools	Upper primary schools	Secondary schools	Higher secondary schools	Total
Rural	756336	136334	26877	5970	1247492
Urban	83903	11245	7056	5467	220184

Source U-DISE Flash Statistics 2016–2017

one fifth or 19.1% of the total population (Census of India 2011a), and, by the end of 2020, it is expected to be 34.43% of the total population.[11] The high proportion of youth determines the potential growth of the nation, which is possible with proper exposure to education, training, and generation of employment for the youth in all sectors and geographies. Despite the recognized importance of the youth, the youth population remains underrepresented and impassive in the decision-making process of the country.

The holistic vision of National Youth Policy (NYP 2014) is '*to empower the youth of the country to achieve their full potential, and through them enable India to find its rightful place in the community of nations.*' The pandemic has affected the whole nexus of the population associated with education, threatening the basic human rights and plugging the country's target in achieving sustainability. The sudden shift on the actual prototype of face to face education has put parents, management bodies, and the stakeholders to take necessary steps to deal with the crumbling system and define 'new normal' in the education system. The country has more than 1.4 million schools, with over 227 million students enrolled and more than 36,000 higher education institutes. India has one of the largest higher education systems in the world (Jha and Shenoy 2016).

The discussion relating to the rural–urban representation is rational during the present situation, mostly in discussing the impact it has on education and the related population. The Census of India states that despite the decline in the rural population, 68.84% of people still live in villages in India, keeping only 31.16% in the urban areas, out of which 17.4% live in the urban slums (Census of India 2011b). The disparity in numbers also exists in the literacy rate, with 85% of the urban population being literate and only 68.9% in the rural. The total number of educational institutions are, however, more in the rural area, as shown in Table 8.2. There are 180 million students who are enrolled in rural schools at all levels, and only 71 million students in urban schools (U-DISE 2016). The rural schools have six million teacher representation, whereas two million teachers are working in urban schools. In the case of higher education, there is no clear distinction with Gross Enrolment Ratio (GER) as low as 26.3% (AISHE 2019) and the mobility of the participants in higher education the rural–urban factor remains confined to the location of the Higher Educational Institutes (HEI).

[11] The figures have been mentioned in Youth of India, 2017 a MoSPI Report, p 13 (http://mospi.nic.in/sites/default/files/publication_reports/Youth_in_India-2017.pdf).

The alteration in the mode of imparting education also requires consideration of digital access in the country. The rural internet density is 25.3, while the urban internet density is 97.9 (Parsheera 2019). The digital penetration in India remains segregated by various economic, societal, and geographic factors surmounting to digital exclusion. Whereas the 'Digital India' drive identifies universal access to wireless handset connectivity as one of its main pillars.

8.3.1 COVID-19 and the Indian Education System

India's impetus to digital empowerment, using information and technology, is in the process of a full swing with the partial engagement of the education sector. The education system in India has been acquired by a physical space known as a pathshala, *gurukula*, school, college, or university with actual classroom space to be shared with the peer and teacher. Due to the pandemic Indian education system has faced a complete shift to synchronous and asynchronous modes of learning. According to a report by the UNESCO, UIS, and GEM, there were 49.8 million students already out of school in their Upper Secondary age in 2015 or the latest year; 2.9 million out of school population in the primary age; and 11.1 million in the lower secondary age in India (UIS and GEM 2017).

Nearly six months of complete disconnection with the educational spaces is not the only threat. The prioritization of curbing the spread of the virus among the young population caused the board examinations to terminate, regular classes to cease, the college examinations to being held online, and many entrance exams for professional institutes to put on hold. The sudden shift in the mode of taking lessons has put the students in a dilemma, involving the parents, teachers, and the management in the haul. When considering a fundamental change in the belief system and education of any country, it becomes obvious that this change requires a great deal of work in various fields. This might include areas such as teacher training, curriculum development, and of course, material development, among others (Mohammadnia and Moghadam 2019). The National Education Policy (NEP) of India, 2020 lays emphasis on '*Early Childhood Care and Education as the foundation of learning.*' This means engaging students from 3 to 6 years of age, which shall be a huge challenge if the schools do not reopen. The invisible divide of public–private modes of education and the privileged-unprivileged divide has again become visible, which has been discussed at length in the later sections. The private-run schools are conducting online classes while most of the students in government schools are being left behind from the privileged peers. The emphasis on High-Quality Research and Liberal Higher Education laid down in NEP, 2020 could also be pushed back in time because of the sudden closure of the higher educational institutions.

8.3.1.1 Government Initiatives in Education During the Pandemic

The shift to remote learning has allowed the Indian government to promote and implement the digital programs that were to be a part of the education system in India in times to come, the initial foundations of which had been laid prior to the pandemic situation. As mentioned earlier about the youth of India, they are to reap the demographic dividend of the country. There were already the e-learning platforms and the Massive Open Online Courses (MOOCs) that had been launched by the government. Table 8.3 shows the existing initiative in pre-pandemic times. It allows flexible and personalized learning at the speed of the learner, and one can continuously augment and expand content through digital means. The rapid increase in internet penetration and various government initiatives such as the Digital India campaign has created a conducive environment for moving toward digital education (Ministry of Human Resource Development 2020). Despite the already existing Information and Communication Technology (ICT) initiative, the COVID-19 has posed as a threat and challenge to the existing ideas and plans for achieving goals in education. There is a need to promote critical thinking, creative and communication skills, along with experiential and joyful learning for the students focussing

Table 8.3 Digital initiatives by the government in India

Nature of the programs	Programs
Audio–video content	• Study Webs of Active–Learning for Young Aspiring Minds (SWAYAM) • SWAYAMPRABHA • Gyan Darshan • Gyan Vani (105.6 FM Radio)
Digital Content: Access journals and e-books	• National Digital Library (NDL) • e-Pathshala • e-PG Pathshala • e-SodhSindhu • Sodhganga • e-Gyan Kosh
Hands-on-Learning	• Free/Libre and Open Source Software for Education (FOSSEE) • e-Yantra • Virtual Labs • Spoken Tutorials
Progress Tracking	• Sodh Suddhi (PDS): Plagiarism Detection Software • Indian Research Information Network System (IRINS)
Digital Infrastructure for Teachers	• DIKSHA

Source Compiled by the author from mhrd.gov.in

on learning outcomes. The curriculum must be rooted in the Indian ethos and integrated with global skill requirements. Therefore, it has been decided to prepare a new National Curriculum and Pedagogical Framework for school education, teacher education, and early childhood stage to prepare students and future teachers as per global benchmarks.

The Finance Minister of India has laid stress on investing in human capital for the prosperity and productivity of the country, which is definitely a positive investment toward a better future. On this note, the HRD ministry has called for '*One nation, one digital platform*' and '*One Class One Channel*' for assuring quality education to every corner of the nation. The ministry is of the view that the measure shall usher in a new paradigm in the creation of a new India. The immediate set of the initiative is:

- A comprehensive initiative called PM e-VIDYA will be launched, which unifies all efforts related to digital/online/on-air education. This will enable multi-mode access to education, and includes: DIKSHA (one nation-one digital platform), which will now become the nation's digital infrastructure for providing quality e-content in school education for all the states/UTs; TV (one class-one channel) where one dedicated channel per grade for each of the classes 1–12 will provide access to quality educational material: SWAYAM online courses in MOOCS format for school and higher education; IITPAL for IITJEE/NEET preparation; Air through Community radio and CBSE Shiksha Vani podcast; and study material for the differently abled developed on Digitally Accessible Information System (DAISY) and in sign language on NIOS website/YouTube. This will benefit nearly 25 crore school-going children across the country.
- In this time of the global pandemic, it is vital that we provide psychosocial support to students, teachers, and families for mental health and emotional wellbeing. The Manodarpan initiative is being launched to provide such support through a website, a toll-free helpline, national directory of counselors, interactive chat platform, etc. This initiative will benefit all school-going children in the country, along with their parents, teachers, and the community of stakeholders in school education.
- Government is expanding e-learning in higher education—by liberalizing open, distance, and online education regulatory framework. Top 100 universities will start online courses. Also, the online component in conventional Universities and Open Distance Learning (ODL) programs will also be raised from the present 20 to 40%. This will provide enhanced learning opportunities to nearly 3.7 crore students across different colleges and Universities.

In order to make e-learning more constructive, NCERT and Rotary India digitally signed MoU for e-learning content telecast for class 1–12 overall NCERT TV channels under Vidya Daan 2.0 Rotary International would provide the e-content in the Hindi language to NCERT for classes I to XII for all subjects. Along with this, Rotary International will provide material for special needs children as well as contribute its entirety to the Adult Literacy Mission. They will also provide the Teacher Training (including professional development) to reach our students through radio and TV, where there is no internet or mobile connectivity available, and this MoU is a big step in that direction (Press Information Bureau 2020). E-learning resources are being prepared in eight regional languages. The study material for the differently abled is being developed on Digitally Accessible Information System (DAISY) and in sign language along with radio podcasts, especially for the visually and hearing-impaired students. Online and digital education is a trend being rapidly embraced for high-quality education on a large scale.

8.3.1.2 Digital Access and Divide in Education

The policymakers believe, despite setbacks and contrary evidence, that information technology (IT) will provide solutions for most of India's development problems. This approach of 'technological fix' to an extraordinary humanitarian crisis, including the provision of education and health services through digital platforms, is likely to widen the inequality in accessing those services from all corners equally. The citizens in the 'offline' category will suffer the most in the changing and challenging times (Ahamed and Siddiqui 2020). The shift of base from the classroom teachings to online classroom deliveries put to question the pedagogies present in the present-day education system. Online lectures create a cognitive dissonance between teachers and students as course content is delivered through a medium where students often see only presentation slides with the lecturer's camera switched off, and teachers see only an impersonal student name-tag without an interacting face. This '*faceless*' classroom not only has the potential to exhaust the participants but more worryingly, it threatens to weaken the bond that students share with their teachers. This dilutes the incentive for both professors and students to connect with each other at a human level.[12]

Regarding the online education market, higher education contributes 59.7% of the market size, school education 38.1%, pre-school segment 1.6%, and technology and multi-media the remaining 0.6%. Many colleges do not have sufficient hardware or software to give students complete learning experience (Jha and Shenoy 2016). The massive disruption of activities due to the spread of COVID-19 has turned many schools to the remote learning centre to maintain the teaching–learning process amidst these troubled days. There is a lack of digital infrastructure for both the teachers and students and a disparity in internet connection and access to devices; while all students might have smartphones, not all of them own desktops or laptops which are more suited for educational purposes.[13] The online classes made available are not in at par with the kind that would be viable for the children with special needs. Despite the government announcement of making PM e-Vidya available to the

[12]Indian national daily newspaper *The Statesman* has published an article on different pedagogy during the COVID crisis, circulated on 13 July 2020 (https://www.thestatesman.com/opinion/a-different-pedagogy-1502908267.html).

[13]Indian national daily newspaper, The Statesman has published an articles on the digital divide in education (https://www.thestatesman.com/opinion/digital-divide-indias-education-1502899472.html).

visual and hearing impaired, there is no public information that addresses how digital education, including online classroom teaching, will generally be made accessible.[14]

There are many kinds of schools in the country: some run by the government (central, state, local government bodies) and some by private sectors, i.e., individuals, trusts, and societies. The number of schools managed by central or state governments is almost double the schools managed by other bodies taken together (British Council 2014). In India, the digital gap starts right from the school, where two parallel worlds are taking the shape—government schools and private schools. The government schools receive substantial funds, although most of them are functioning in a miserable condition (Thakur 2014). In 2016–2017, the government schools shared 71.72% and private schools 19.78% by count in India. In the rural areas, the government schools were dominant (78.42%), with only 14.4% private schools. The scene was, however, the reverse in urban areas: 48.66% private schools and 35.32% government schools (U-DISE 2016). In the purview of the rural–urban divide during the pandemic, it is the fact that the rural students will likely be left behind from their urban counterparts in attaining the education. During pre-pandemic times, only 4.09% of the government schools had an internet connection, where 25.07% of the private schools enjoyed the same (U-DISE 2016). The consideration in the present times is the students who have digital access at home would stay tuned, keeping the rest deprived. Only a handful of private schools could adopt online teaching methods. Their low-income private and government school counterparts, on the other hand, have completely shut down for not having access to amicable e-learning solutions.[15]

The aspect of the digital divide is mostly the structural inequality of privileges. Most of the poor students do not have access to smartphones, and even if their 'poor' economy does afford the device, the 'poor' net connectivity arises as a constraint for most of the remote rural hamlets. The situation is often rather complicated in a multi-lingual country like India. The study materials are often not available in vernacular languages. The latest data available from the National Sample Survey (2017–2018) gives a real insight into the ownership of computing devices[16] in India. Only 10.7% of household have computers, and 23.8% of household internet facilities in the country. The rural figures are at a dismal low, with only 4.4% of households with computers and 14.9% with internet facilities. Even in the urban areas, 76.6% of households have no computer, and 58% of households have no internet facilities (Ministry of Statistics and Programme Implementation 2019).

In reality, most schools in urban areas are providing online classes, while the majority of rural schools do not. Very young children are not able to learn through online processes as they can neither handle computers nor mobile phones, and there are major health issues that young learners are supposed to be exposed. A section

[14]An electronic media article in July 2020 pointed out on the leaving behind of the children with special needs, in education during COVID-19 (https://theprint.in/opinion/indias-online-classrooms-are-outdated-for-disabled-kids-covid-just-made-it-worse/463438/).

[15]Mentioned in an article in Et Government.com, on 16 April 2020 (https://bit.ly/2GXjGUv).

[16]Computing devices include devices such as desktop computer, laptop computer, notebook, netbook, palmtop, tablet (or similar handheld devices) except smart phones (NSS, 75th Round, 2017–2018).

of parents in Mumbai had urged school authorities to discontinue online classes for nursery students, citing health reasons.[17] In many households, there is no computer, and in many, children are not allowed smartphones as well (Tahseen 2020).

The government, in the guidelines for digital education during the pandemic, has classified household into six categories (Ministry of Human Resource Development 2020):

- Households with computer/laptop/smartphone and a 4G internet connection as well as television set with DTH/Cable Tv connection
- Households with smartphone with 4G access
- Households with smartphone with limited (3G/2G) access or no internet
- Television set with DTH/cable connection
- Households with radio set or a basic mobile phone with FM
- Households with no digital device.

In order to ensure the access of the students in such cases, the digital reach of education has been categorized into three groups: *online* (where computer and smartphone with internet connectivity is available), *partially online* (computer and smartphone available but no regular internet) and *offline mode* (television and radio).[18]

According to the Telecom Statistics of India (2019), there are 227.01 million internet subscribers in Rural India and 409.72 million in the urban, with a total user of 636.73 million of the 1.21 billion Indian population. The internet density (i.e., internet subscriber per 100 inhabitants) is only 25.36 in the rural areas compared to the 97.94 in urban India, with a total internet density in the country being 48.48. Reportedly, the 99.93% electrification has been achieved by the country in the year 2019,[19] and about 835 million individuals in 197 million homes have access to television (Broadcast India 2018). There are 1161.71 million mobile subscribers in the country (Government of India 2019). Despite all these positives, the Indian Digital World comes with a connectivity problem with an average broadband speed of 34.07 Mbps. Despite all the numeric data, it is the irony that the use of mobile phones, television, and the internet in India is mostly for leisure and entertainment. According to the report by MICA (2019), Centre for Media and Entertainment Studies, following the availability of internet through mobile phones at a massive scale and low rates, the internet dark zones, especially the semi-urban and rural areas, have got access to content. Many of these users are first-generation Over the Top (OTT) content consumers, attracted to regional content. For India, the highest consumed (OTT) platform is YouTube, followed by Amazon Prime and three Indian platforms Hotstar, Jio Cinema, and Voot (MICA 2019).

---

[17] The regional electronic media, Mumbai Mirror, reported on the concerns raised by parents regarding online classes, on 27 June 2020 (https://mumbaimirror.indiatimes.com/coronavirus/news/malad-parents-raise-concerns-over-online-classes-for-nursery-students/articleshow/76658182.cms?utm_source=Articleshow&utm_medium=Organic&utm_campaign=Related_Stories).

[18] The guidelines have been prepared by the NCERT taking into consideration suggestion by various states and UT (https://www.mhrd.gov.in/sites/upload_files/mhrd/files/pragyata-guidelines_0.pdf).

[19] See the website, https://saubhagya.gov.in/.

The change of mode, and moreover, which is sudden, adds a list of duties to be taken up by the institute authorities to make their teaching staff adapt immediately with ICT, digital media, or social networks, leaving little time for their acquisition or improvement of skills. In order to involve the students in remote learning, the government has issued guidelines to conduct surveys regarding the various ICT facilities available at home. They have an additional duty of being in constant communication with the parents (Ministry of Human Resource Development 2020). The World Bank (2020) recognizes that the transition to online learning at scale is a challenging and highly complex undertaking for education systems. Even in the best of circumstances, highly motivated learners, especially those with previous experience in online learning, are the most likely to take the most advantage of online learning opportunities.[20]

8.3.1.3 Effects on the Population Associated with the Education System

Indian education system has been dealing with the problems of access, equity, and quality, for a long time, which has worsened due to the ongoing situation. The key challenges remain in the form of access and quality, funds, and the disparity in the literacy level (Lall 2005). The Indian government however, come up with numerous measures (Sect. 8.3.1.1), especially during the pandemic. The key principle behind the reforms is to ensure access and equity for students.[21] In addition to the existing challenges, lack of trained teachers, inadequate learning materials, makeshift classes, and poor sanitation facilities make learning difficult for many children[22] for long.

UNESCO (2020a)[23] estimates that about 32 crores of students are affected in India, including those in schools and colleges. The hard-hit population is the students in the primary classes (Table 8.4). The argument of the advantages of digital learning arises because the government considers the digital mode of learning as '*flexible and personalized learning at the speed of the learner*'[24] and that '*the most transformational impact of Technology often happens in the lives of the poor.*[25]' With 68.64% of the population living in rural areas (Census of India 2011b), the debate on learners' easy access and means to attain regular digital classes at their own pace

[20]Remote learning and the COVID-19 Outbreak (http://documents1.worldbank.org/curated/en/266811584657843186/pdf/Rapid-Response-Briefing-Note-Remote-Learning-and-COVID-19-Outbreak.pdf).

[21]The Union Minister of Human Resource Development has mentioned in an article in a national print media, *Hindustan Times*, 18 May 2020 (https://www.hindustantimes.com/analysis/ensuring-access-and-equity-in-education-writes-ramesh-pokhriyal-nishank/story-KSqyOXK2zkUVpl9krJO4dP.html).

[22]https://www.unicef.org/education.

[23]See the website, https://en.unesco.org/covid19/educationresponse.

[24]PRAGYATA: Guidelines for Digital Education, MHRD, Government of India, p 1.

[25]The Prime Minister of India writes in his article '*Life in the era of COVID-19*', 19 April 2020 (https://www.narendramodi.in/life-in-the-era-of-covid-19-549324).

Table 8.4 Learners affected in India

Institution type	Affected learners		
	Female	Male	Total
Pre-primary	4557249	5447169	10004418
Primary	72877621	70349806	143227427
Secondary	63983677	69160694	133144371
Tertiary	16739686	17597908	34337594
India	158158233	162555577	320713810

Source UNESCO as on 2 July 2020

of learning has been addressed in Sect. 8.3.1.2. With the advent of cheaper smartphones and a reduction in data plans, the 4G connection is the most preferred choice of internet connectivity (Internet and Mobile Association of India 2019). India has the second-largest digital user base after China,[26] with 433 million active Internet users at present who are 12 years and above and 71 million active Internet users between 5 and 11 years (Internet and Mobile Association of India 2020). However, in practicality, it is injudicious to compare the entertainment activities with the formal participation of the students in an online class. According to the Global Education Census Report (Cambridge 2018), students in India are least likely to use smartphones in lessons (16% versus the 42% global average). The students in India are prone to using blackboards and white chalk for their lessons (67%). Also, Young Indians spend 4 h on average watching television per week and nearly 28 h on mobile, of which 45% of the spent time is dedicated to entertainment (KMPG India 2017; Meeker 2017). A deliberate effort is needed to help students cope up with the loss and also make sure that no student is deprived of obtaining their courses, lessons, and personality development.

8.3.2 *Effect on Students and Parents*

'*School is the second home of the children*' where the students spend maximum hours of their day other than their own residences. Children and youths in schools learn a good deal and acquire expanded social capacities more than those not in school, even when background factors are controlled (see, e.g., Holsinger 1974; Plant 1965; Meyer 1977). According to the India Education Report (2005), students attend school for about 6 h during the day, with each classroom session spanning 30–40 min and a 30-minute lunch break on an average. Some schools operate for a longer duration (about 8 h), and these schools provide two break periods. This practice has, however, changed due to the closure of the educational institutes. The experience of the government school authorities in Madhya Pradesh is worthy of

[26]https://economictimes.indiatimes.com/tech/internet/india-has-second-highest-number-of-internet-users-after-china-report/articleshow/71311705.cms?from=mdr.

being mentioned as Yadav (2020) reports. The authorities could succeed to reach to just 30% of secondary and senior secondary students through regular home-based learning programs. The program 'Humara Ghar, Humara Vidyalaya' was initially planned to reach 22 lakh students from classes 9–12, but only 20% could be connected through TV and 10% through WhatsApp.

The institutes are spaces of free-thinking and learning of students by mingling with their peers and teachers, as the dominant view has it that the educational institutes prepare individuals to socialize (Meyer 1977). The '*new normal*' of synchronous and asynchronous learning lays bare the development of students as a social being of being confined to home spaces with a very intimate (Debrock et al. 2020) form of learning. Students staying away from the teachers and peers face psychological problems along with academic lows; 55% of students still find the lack of social interactions troubling. They learn better with fellow students, and for 45% of students, this could lead to underperformance (Roesch 2020). Other than the academic know-how, the educational spaces also play an essential role in the physical and emotional growth of the students. In the case of online learning, the students are exposed only to the subject curriculum. The extracurricular activities[27] that students are engaged in the educational spaces are invaluable learning experiences impacting the physical and emotional wellbeing of the students (Cambridge 2018). The activities increase leadership and teamwork abilities in students, including higher grades, educational achievement, and self-confidence (Saqib et al. 2018). There are serious health implications that come with the over usage of technology by the young, mostly harms the enterprise skills (Lilliard and Peterson 2011) in students. Halupa (2016) says enterprise skills are key skills children need in life, such as motivation, initiative, creativity, individuality, the ability to get along with others, and strategic thinking. The use of technology is one of the contributing factors to overweight and obesity (Kautiainen et al. 2005; Cawley 2010; Halupa 2016), which is a major problem among the youth in the technology-driven world.

In Kerala, the situation of an increase in the number of suicides among children and adolescents is a worrying concern. The primary reasons as cited are varied, like being confined to the home and not being able to share their problems with peers or teachers. Some are being scolded by the parents for using a mobile phone or because there is a lack of facilities to attend online classes (Express New Service 2020).

The students involved with higher education are worst hit by the sudden closure of the institutes with the stranded syllabi and initiation of a revolution in the form of haphazard online modes of education (Kumar 2020; John 2020). According to the All India Survey of Higher Education (AISHE) 2019 report, India has 51649 higher educational institutes,[28] with 37.4 million students enrolled. The population effected by the closure are mostly in the 18–24 years (young adults) age cohort of India

[27]*Extracurricular activities* kinds of activities and they do not fall into the sphere of normal curriculum and instructional method. The activities can be within the institutes and outside the institutes.

[28]The universities, colleges and standalone universities comprise the higher educational institutes. There are 993 universities, 39931 colleges and 10725 Stand Alone Universities in India (AISHE 2019).

enrolled in the institutes in the country as well as those enrolled abroad. The pandemic has seen the return of many students abroad back home, and the persistence of the pandemic may see a decline in the demand for higher education internationally (Jena 2020). In the national level institutes, most of the external assessments have been canceled, and some have taken the online mode of internal assessment. Delhi University (DU) authorities are mulling over conducting end-semester exams online.[29] Most of these decisions are taken without seeking any input from the students though they are the primary stakeholders of any institution. The higher educational institutes like the schools have students who come from various walks of life, from different socio-economic sections of the society, so two-way problems are to be dealt with. One is the access to digital commodities for inclusiveness in learning, and the other is the social exclusion that may come with the closure of the institutes. According to a World Bank Report (2020), students from better-off households are likely to have access to digital resources and to use them to maintain some continuity with academic work during the closures, disadvantaged students (first-generation college-goers, women, tribal youth, economically disadvantaged) will likely suffer deeper learning losses. The effect of falling back shall imply the social mobility[30] of the youth in the long run. Higher education becomes a mandatory stage in the trajectories of upward mobility virtually as education makes it possible for students of a lower class to acquire higher professional positions and to move up in the class structure (Burlutskaia 2014). The structural differences are profound across gender lines, as the disruption leads to increasing inequality in terms of access and participation. In India, female students (especially in higher education) have serious issues of pre-existing technological literacy, socio-cultural norms, and gender bias that lead to additional gender-based digital exclusion (Manazir et al. 2020).

Moreover, the '*new normal*' of staying away from institutional spaces makes the parents solely responsible for the education of the children, especially the school goers. The closure of schools has made every home a school and every parent a teacher.[31] In an article by Livemint (2020) the writers mention as to how the children in nuclear families are not used to spending 24 h at home with parents, without school and friends. Working parents have difficult times switching between work and family in an undivided home space. In urban India, the constant juggle of the parents between work from home and taking care of young children causes stress (Quartz India 2020). Daniel (2020) points out due to the hiatus caused by the pandemic many parents and guardians may be deeply anxious about their economic future, so studying at home is not easy, especially for children with low motivation. Such homes often lack the equipment and connectivity that more affluent households take

[29] An article by The Print has pointed out some problems that the students might face dur to the stringency of the educational institutes to compensate on academic loss. Published on 26 April 2020 (https://theprint.in/opinion/pov/indian-educational-institutes-should-break-norms-in-covid-times-not-conduct-online-exams/408660/).

[30] Social Mobility is the phenomenon of shifting from one social position to another either in comparison with family back ground or previous employment (Joye and Falcon 2014).

[31] In an article by *Livemint* the impact of school closure on parents have been highlighted. Published 26 March 2020 (https://bit.ly/3hvSDfs).

for granted, compounding the problem. The urban parents have taken up more of '*digital parenting*[32]' to keep the children engaged. Parents are tapping the internet to add interesting activities to their kids' lockdown. Platforms such as BrainPop, Duo Lingo, Khan Academy, and Scholastic are popular among Indian parents for kids to learn a new language, skill, or just pursue a hobby(Quartz India 2020).

The idea of division and disparity has been brought up many times in the different sections of the chapter. The disparity in how parents deal with education is dependent on the economic viability of the parents. The struggle of the parents is different across their economic structures. While an urban parent can afford digital equipment to help their wards, the struggle of the poor parents is severe. Various reports are available from different corners of the country on how the poor parents are dealing with the education of their children.

- 'In Tripura, a father, who worked as a daily wager, killed himself, disappointed, being unable to buy his daughter a smartphone for online classes.' (NorthEast Now 2020; Times of India 2020a)
- 'A poor man in Himachal was forced to sell his cow to buy a smartphone for online studies of his children. The man sold his cow, also his only sources of income, after he received no financial assistance from banks and private lenders.' (India Today 2020)

These are just a few mentions that clearly bring in the forefront the debate of privileged and underprivileged along with the rural-urban divide. While some parents are turning to digital measures to keep their children engaged, some are making ends meet to provide their children the means to learn. There are unprecedented incidences as well, like The Telegraph reports

> 'Parents of some primary school children are doing their school assignments instead of letting them do it. At times, they are even competing with other parents to be the first to upload such assignments on their common WhatsApp groups'.[33] While the MHRD has issued guideline for digital education (PRAGYTA 2020), the problem is the actual involvement of the students in the whole process.

With the increase in digital involvements, the rate of cybercrime and cyberbullying has also increased. Attacks had soared to 86% in four weeks, roughly between March and April (Economic Times 2020). The parents are forced to deal with the demand of the children. They are to track whether the gadgets are used for a purpose. As India Today (2020) reports,

> A Punjab boy would tell his parents he is studying online on his father's phone but instead would play PUBG Mobile for hours lost 16 lakh rupees from his parents' bank account.[34]

These are anxious times for students, their parents to cope up with the new normal of the education system.

---

[32]The parent who keeps an eye on the young kids and teens online activities in order to protect them from all online dangers is called digital parenting (https://bit.ly/2Fq7fQB).

[33]See, https://www.telegraphindia.com/west-bengal/calcutta/coronavirus-pandemic-when-parents-invade-online-classes-and-do-kids-work/cid/1785886.

[34]See, https://bit.ly/2FhdNkB.

8.3.3 Effect on Teachers, Researchers, and Academicians

According to UNESCO (2020b), over 60 million teachers are at home as COVID-19 school closures are extended. Where in India, the teachers are trained to teach the students face-to-face, and the gesture of the teacher is part and parcel of the classroom teachings. It poses doubts about whether online or virtual classes can be the right substitute for classroom teachings. The former method seldom generates interactions between the students and teachers (Gupta 2020). The Alternative Academic Calendar[35] prepared by the National Council of Education Research and Training (NCERT) provides guidelines to teachers on the use of various technological tools and social media tools available for imparting education in a fun-filled and interesting way. Teachers, Learners as well as parents can use these, and teachers even while at home.

Teachers, especially in private schools, are facing hardship—from arranging whiteboards and smartphones in the middle of lockdown to ensuring that they and their background look presentable enough. Teachers are adjusting in the ways they never had to before (Sahni 2020). The interference of overenthusiastic parents while online classes create tremendous pressure on the teachers as they complain about the quality of teaching, spelling mistakes, improper pronunciation, harsh tone of voice (Deccan Herald 2020). 'Helicopter parenting' (i.e., parents hovering around while the child is taking online classes) and calling out on teachers during online classes surmount to a form of bullying led to teachers' disappointment in e-classes (The Week 2020; Times of India 2020b; Deccan Herald 2020).

The bullying of teachers is not only done by the parents, but the older students resort to means of abusing the teachers on online learning platforms. In an article by The Print (2020), it has been highlighted that:

> One of the biggest issues that female teachers face is online bullying, especially from older children. Students create Zoom IDs in random, unidentifiable names and troll teachers. Some switch off their camera and call teachers' names from these IDs, some use them to send memes.

Moreover, the teachers in private schools face the economic crunch as well. The teachers of the private schools have been, unfortunately, experiencing pay cuts. An exploratory survey by Central Square Foundation (2020) reveals that more than 50% of private schools have uncollected fees, accounting for 13–80% of the annual revenue. Whereas, only less than 20% of teachers in private schools have continued to receive their salaries after March.

The teachers in public schools are dealing with the problem of reaching out to students in the economically weaker sections. In India, state governments are the largest providers of education. However, they are plagued by several challenges,

[35]In this period of Covid-19, which is declared as a global pandemic, the teachers, parents, and students are at homes to prevent its spread in the community. In this situation, NCERT has developed an *Alternative Academic Calendar* for all stages of school education to provide multiple alternative ways of learning at home through interesting activities (https://ncert.nic.in/alternative-academic-calendar.php).

including large student-teacher ratios, infrastructure, and lack of quality training among teachers (Samantaray 2020).

Similarly, researchers have been stalled at home. Research works involving laboratory-based experiments with specialized equipment have remain standstill as there is no 'online' alternative. There is also the trimming of funds to continue or undertake new research and innovation activities worldwide (World Bank 2020). The field-based researches are paused. Scientists working in fields that tend to rely on physical laboratories and time-sensitive experiments of bench sciences[36] reported the largest declines in research time, in the range of 30–40% below pre-pandemic levels. Conversely, fields that are less equipment-intensive—such as mathematics, statistics, computer science is reported to face the lowest declines in research time (Myers et al. 2020).

The research and researchers in the Indian universities have been affected by the sudden hiatus caused by the pandemic. University Grants Commission, while laying down their guidelines in view of the pandemic, has acknowledged that 'research has suffered as the laboratories are closed'.[37] The decision is to be taken by the institutes to allow scholars to carry out their research. The problem is serious for the scholars who are on the verge of completion of their research. The University Grants Commission (2020) in the guideline mentions:

> The M.Phil. or Ph.D. students, whose maximum period for submission of M. Phil./Ph.D. Dissertation/Thesis prescribed under relevant rules/regulations/ordinances of the concerned universities is expired/expiring during the Corona pandemic period, may be allowed to submit their Dissertation/Thesis, including completion of pending formalities, within six months from the date of expiry of regular prescribed period. The extension of six months may also be considered for those students who are yet to submit their Dissertation/Thesis.

The time constraints have been dealt well with by the UGC, but the concerns raised by the scholars is regarding fellowships and grants, which are available only for 5 years. Extension of six months for some research scholars would mean going without any monetary aid. In view of the back draw, a student of Indian Institute of Technology Madras has started a petition a part of which reads:

> Educational institutes remained shut, leaving scholars with no access to research infrastructure. This has resulted in a lot of anxiety among scholars about the expiry of their fellowship. All experimental setups have already either degraded or unusable due to a long delay. Scholars now have their fellowships expiring in the middle of the nationwide lockdown leaving them without any means to survive. The financial pressure is a huge mental stress for us along with their health and future uncertainty.[38]

[36]Bench sciences are subjects like biochemistry, biological sciences, chemistry and chemical engineering.

[37]Mentioned in UGC Guidelines on Examination and Academic Calendar for the Universities in View of COVID-19 pandemic and Subsequent Lockdown (https://www.ugc.ac.in/pdfnews/4276446_UGC-Guidelines-on-Examinations-and-Academic-Calendar.pdf).

[38]The exerpts of petition has been mentioned in an article by Deccan Chronicle, published on 8 July 2020 (https://bit.ly/35A73ZC).

Various monthly scholarships have been blocked or not appropriately disbursed for more than six months in different Indian universities, according to a group of research scholars from across the country.[39]

Besides, there is a persistent gender gap in every aspect of scientific (Blickenstaff 2005; Huang et al. 2020) research a work that is yet again visible in the academics and researchers during the pandemic. Myers et al. (2020) show that female scientists and researchers with young dependents have reported a substantial effect on their ability to devote time to their research. The mothers in academics are the group who during the pandemic have high chances to fall back in their academic works. Instead of research works, they are likely to devote time to homeschooling children and doing household chores (Staniscuaski et al. 2020). Those who are struggling to keep their careers on track, these months of heavier duties may increase the distance between them and their male and childless peers (Huang et al. 2020).

8.3.4 Addressing India's Take on Rights and Sustainability

The Right of Children to Free and Compulsory Education (RTE) Act, 2009 in its Schedule lays down Pupil-Teacher Ratio (PTR) for both primary and upper primary schools. At the primary level, the PTR should be 30:1, and at the upper primary level, it should be 35:1. The Rashtriya Madhyamik Shiksha Abhiyan (RMSA) framework stipulates that the PTR at the secondary level should be 30:1. The kind of teaching practices that have been adopted by India puts to question the inclusiveness, equity, quality, and also future opportunities. There are reported death of young students from parts of the country for being 'incompetent' to meet the needs of Digital classes and assignments. In post-lockdown Karnataka, there is a 'worrying jump' in the number of suicides. The death rose by 23% in May (1127 died) and by 18% in June (1084 died).

As noted by the Regional Director, WHO of South-East Asia Region, it is the stigma related to the COVID-19 infection that may lead to feelings of isolation and depression. Another precipitating factor mentioned could be domestic violence as a factor for impacting mental health with increasing cases of domestic violence globally as lockdown had been imposed. The chairperson has stressed the importance of early identification of mental health conditions, recognition of suicidal behavior, and appropriate management through a multi-sectoral approach.

The head of psychiatry at Victoria Hospital, Bengaluru, says that the reasons for suicide are multifactored, combined, cumulative, un-resolving, and interrelated. People should be watchful of the family member with a history of suicidal attempts and always look for signs (Chetan 2020). It might seem quite unnatural with the low budget smartphones and unlimited data plans, but incidences do say otherwise about India going digital. Rural or urban, there remains a disparity in accessing the digital

[39]The non payement of scholarship has been reported by NDTV Education on 7 August 2020 (https://www.ndtv.com/education/research-scholars-complain-non-payment-of-fellowships).

infrastructure in India. However, a more flexible way of teaching and learning does not end up with infrastructure. Instead, infrastructure is only the first step toward a new paradigm of teaching and learning in post-pandemic time.

This paradigm could represent a shift from traditional, teacher-centered, and lecture-based activities toward more student-centered activities, including group activities, discussions, hands-on learning activities, and limited use of traditional lectures. Full long-term integration of online teaching and learning into university curricula implies further attention to quality (Zhu and Liu 2020). One of the primary challenges to child rights is school dropout and access to proper nutritious food. The chairman of the Delhi Commission for Protection of Child Rights says that the challenge of homelessness and economic constraint shall push children, especially boys, into child labor resulting in reduced school attendance and increased dropouts after the pandemic is over. There would be a rise in cases of early marriage in case of girls and issues related to the Protection of Children from Sexual Offences Act (POSCO) and Juvenile Justice (Baruah 2020).

There has been increased attention toward education for a Sustainable world. There has been a repeated reminder to the various countries by the UN to implement education for a sustainable future. To make higher education a catalyst for creating sustainable, innovative, and equitable societies, governments, and universities have to develop policies to make colleges and universities both affordable and inclusive (UNESCO et al. 2017). The energy and passion of the youth, if utilized properly, can bring substantial positive change to society and progress to the nation. Youth is the creative digital innovators in their communities and participate as active citizens, eager to positively contribute to sustainable development (Government of India 2017). The existing gender disparity in all form of life is an invisible evil which continues even during these times with new major and minor differences. School closures do not just mean that girls are taking on more chores at home; rather, it could also lead to millions of more girls dropping out of school before they complete their education, especially girls living in poverty, girls with disabilities, or living in rural, isolated locations. Even before this pandemic, millions of girls were contending with low-quality education—and millions were not on course to meet minimum proficiency in basic reading and math, nor the secondary level skills, knowledge, and opportunities they need for a productive and fulfilling life. Gender discrimination not only restricts girls' abilities to accumulate human, social and productive assets, limiting their future educational and employment opportunities but also hinders their wellbeing and diminishes their self-belief (UNICEF 2020). Empowering women and girls drive better and quicker sustainable development outcomes for all, support more rapid recovery from the crisis, and place us back on a footing to achieve the Sustainable Development Goals.

The Government of India has launched various programs to reach out to the students as having been mentioned earlier. However, it is very crucial to keep a check that every group is represented well. According to Article 9 of the Convention on the Rights of Persons with Disabilities,[40] the states require to ensure that

---

[40] See, https://www.un.org/disabilities/documents/convention/convention_accessible_pdf.pdf.

persons with disabilities can access information, communication, and technology (ICT) systems. Also, Section 42 of the Rights of Persons with Disabilities Act, 2016[41] directs governments to ensure that 'all contents available in audio, print, and electronic media are in an accessible format.'

School closures have a particularly adverse effect on poorer students, students without stable internet access at home, and children relying on help from their schools in meeting their nutrition and health needs. The situation is especially acute for girls and young women who are excluded from education disproportionately (UNDESA 2020). According to a study by the World Bank (2020), the shock of the closure of educational institutes will lead to learning loss, increased dropouts, and higher inequality, and the economic shock will exacerbate the damage by depressing education demand and supply as it harms household. Together, they will inflict long-run costs on human capital and welfare.

8.4 What Could Be the *'New Normal'* During the Post Reopening of Educational Institutes?

The progression of reopening educational institutes is the biggest challenge to the government of all the countries. The decision shall be sensitive to the political agenda. It has to be decided upon keeping in mind the population in various age groups who shall return to the institutes. The step is not only to bring back the educational spaces to normal but also to ensure help to students who fall behind, proper health facilities, and maintaining physical distance until the pandemic has been wholly eradicated. The new norms to reopen schools should lay utmost importance to mental health, which can see a surge due to the practice of physical distancing among the young student population.

The real situation is, unfortunately, disappointing. In India, around 17.6% of posts for the primary level and 15.7% of posts at the secondary level for government teachers are vacant. Around 108017 schools in India are single-teacher schools, and in about 17% of schools, a teacher has to manage more than 40 children. Besides, presently, around six lakh teachers who are older than 55 years are serving the school education system and are more vulnerable to the threat posed by COVID-19 (Bakshi 2020).

There are not enough options for the teachers as well to maintain social distancing in staff rooms as most of the schools provide a single staff room for all teaching faculty, and in 55% of schools, there is additional room along with staff room for the principal. A survey in 453 schools by 'WaterAid' revealed that only 43.5% of schools have teachers trained on sanitation and hygiene. The preparedness for reopening of schools brings forth the urgency of substantial investment. It has reaffirmed the notion that implementing social distancing norms will require more resources in schools, resources for infrastructure, for filing vacant posts for teachers, for training teachers,

[41] See, http://www.iitg.ac.in/eo/sites/default/files/RPwDAct2016.pdf.

and recruitment of non-teaching staff. On the contrary, the Ministry of Human Resource Development (MHRD) allocations for school education have decreased during the last six years—0.42% of the GDP in 2014–2015 to 0.26% of the GDP in 2020–2021 (Bakshi 2020).

There are a wide number of educational challenges that have been left bare due to the ongoing crisis, which requires immediate management to eradicate the effects the pandemic has brought along. UNESCO and United Nations Office on Drugs and Crime is working to *(i) place rights at the centre of education, (ii) support teachers, and (iii) build bridges between changemakers in education and justice sectors*—all of which is in context of the Global Education Coalition which provides platform to join forces with other multilateral actors, including international organizations and civil society and private sector partners, to ensure the current crisis does not widen learning inequalities and that the rights of children and youth are protected (Giannini and Brandolino 2020).

Critical conditions to assess include access to soap and clean water for hand-washing, and protocols on social distancing. Safety can also mean reducing the number of students on site, through double shifts, prioritizing early grades or particular target groups, or continuing with a blended learning approach. After safety, there must be a focus on the learning recovery process—from assessing learning outcomes during school closures, ensuring their socio-emotional wellbeing, and taking measures to address disparities through remedial approaches. Sufficient manpower should be engaged in the educational institutes that will help in maintaining the necessary prerequisites of the rules and norms that are to be laid down by the government for the safe opening of the educational institutions.

The Ministry of Health and Family Welfare, Government of India, has laid down Standard Operation Procedures (SOP)[42] for the partial reopening of schools for the students of classes 9–12. The guidelines mention hygiene standards to be maintained pre and post-opening of the institutes with proper physical distancing procedures. It has also been mentioned that students attending schools require the consent of the parents to start attending school on 21 September 2020. The issues addressed in the SOP are generic preventive measures, arrangements to be made by the institutes' pre and post reopening, hygiene, and sanitation, risk communication, psychosocial well being and measures to be taken in case of stakeholders in the institutes acquire the COVID syndrome.

8.5 Conclusion

Immediate measures have been taken by the Government of India to cope up with the sudden stop in the education system. However, the responses somehow failed to serve the huge student base of the country with the existing Digital Divide in the

[42]See: https://www.mohfw.gov.in/pdf/FinalSOPonpartialresumptionofactivitiesinschools8092020.pdf.

country. The idea of equity, accessibility, affordability has all been put to question. It has and will directly or indirectly impact all the sections of people in the country. The fundamental human rights and the ideas of Agenda 2030 have also been put to a standstill. And, once the pandemic passes, it shall be put to the test. Of all the shortcomings and violations, the farsightedness is what the future holds, and if this pandemic helps us, 'Build Back Better.' This is a once in a generation opportunity to improve education, alongside economies, to fight the climate crisis. It is a time for education leaders to use this period of disruption to ensure what people learn is truly relevant to their lives and the survival of the planet. The hiatus in all sectors caused can be used to rebuild the systems in a way that the future generation are the ones who shall enjoy the fruits of sustainability and pass it down the lineage. Sustainability and development move in a straight-line path with the catalyzing effects of education and awareness as the key. The rampant exploitation of natural resources by the dominant species on the earth shall help in only short-term developments, which with time, may end its effect and leave nothing for the generations to come. While Agenda 2030 for sustainable development pledges to '*Leave no one behind*' in their endeavor, there is much to do in order to include everyone. The pandemic is just an initial warning to how unprecedented changes in the natural system may look like to develop but with bigger setbacks. The pandemic has laid down a chance and opportunity to correct all that has gone wrong.

References

Ahamed S, Siddiqui Z (2020) Disparity in access to quality education and the digital divide. https://www.ideasforindia.in/topics/macroeconomics/disparity-in-access-to-quality-education-and-the-digital-divide.html. Accessed 11 Sept 2020

AISHE (2019) All India survey on higher education 2018-2019. Ministry of Human Resource Development, Department of Higher Education

Bakshi S (2020) Reopening of schools; is social distancing possible for the majority of schools in India? The Policy Times. https://thepolicytimes.com/re-opening-of-schools-is-social-distancing-possible-for-the-majority-of-schools-in-india/. Accessed 20 June 2020

Baruah S (2020) Covid brings child rights challenges in Delhi: school dropouts, pushed into labour. The Indian Express. https://indianexpress.com/article/india/covid-brings-child-rights-challenges-school-dropouts-pushed-into-labour-6538019/. Accessed 5 Aug 2020

Blickenstaff JC (2005) Women and science careers: leaky pipeline or gender filter? Gend Educ 17(4):369–386

British Council (2014) Indian school education system: an overview. https://www.edcentrum.org/pdf/schooleducation/6_Indian_school_education_system_an_overview.pdf. Accessed 11 Sept 2020

Broadcast India (2018) BARC India universe update (July 2018). https://www.barcindia.co.in/resources/pdf/BARC%20India%20Universe%20Update%20-%202018.pdf. Accessed 10 Sept 2020

Burlutskaia MG (2014) Higher education as means of upward social mobility: the expectation of graduates and the realities of present—day society. Russ Educ Soc 56(4):52–63. https://doi.org/10.2753/RES1060-9393560404

Cambridge Assessment International Education (2018) Global education census report. https://www.cambridgeinternational.org/Images/514611-global-education-census-survey-report.pdf. Accessed 13 Sept 2020

Cauchemez S, Ferguson NM, Wachtel C et al (2009) Closure of schools during an influenza pandemic. The Lancet Infect Dis 473–481
Cawley J (2010) The economics of childhood obesity. https://www.healthaffairs.org/doi/full/10.1377/hlthaff.2009.0721. Accessed 13 Sept 2020
Census of India (2011a) Adolescents and youth in India: highlights from census 2011
Census of India (2011b) Rural-urban distribution of population
Central Square Foundation (2020) Private schools in India: state of the sector report. https://centralsquarefoundation.org/State-of-the-Sector-Report-on-Private-Schools-in-India.pdf. Accessed 14 Sept 2020
Chen WC, Huang AS, Chuang JH (2011) Social and economic impact of school closure resulting from pandemic influenza A/H1N1. J Infect. https://doi.org/10.1016/j.jinf.2011.01.007
Chetan MG (2020) Karnataka sees grave rise in suicides post lockdown. The New Indian Express. https://www.newindianexpress.com/states/karnataka/2020/jul/10/ktaka-sees-grave-rise-in-suicides-post-lockdown-2167809.html. Accessed 15 July 2020
Daniel SJ (2020) Education and the COVID-19 pandemic. Prospects. https://doi.org/10.1007/s11125-020-09464-3
DeBrock L, Scagnoli N, Dutta FT (2020) The human element in online learning. https://www.insidehighered.com/advice/2020/03/18/how-make-online-learning-more-intimate-and-engaging-students-opinion. Accessed 13 Sept 2020
Deccan Herald (2020) Teachers quit e-classrooms due to parent 'bullying'. https://bit.ly/2DV74fg. Accessed 13 Sept 2020
Economic Times (2020) Cybercrime trends, digital safety amidst COVID-19 pandemic. https://ciso.economictimes.indiatimes.com/news/cyber-crime-trends-digital-safety-amidst-covid-19-pandemic/75934732. Accessed 14 Sept 2020
Express New Service (2020) Corp launches tele-counselling for kids. New Indian Express, Thiruvananthapuram. https://www.newindianexpress.com/cities/thiruvananthapuram/2020/jul/14/corp-launches-tele-counselling-for-kids-2169270.html. Accessed 30 July 2020
Giannini S (2020) Build back better: education must change after COVID-19 to meet the climate crisis. UNESCO. https://en.unesco.org/news/build-back-better-education-must-change-after-covid-19-meet-climate-crisis. Accessed 14 July 2020
Giannini S, Brandolino J (2020) COVID-19—education is the bedrock of a just society in the post-COVID world. UNESCO. https://en.unesco.org/news/covid-19-education-bedrock-just-society-post-covid-world. Accessed 4 June 2020
Giannini S, Jenkins R, Saavedra J (2020) Reopening schools: when, where and how? UNESCO. https://en.unesco.org/news/reopening-schools-when-where-and-how. Accessed https://en.unesco.org/news/reopening-schools-when-where-and-how. Accessed 10 July 2020
Government of India (2017) Youth in India. Ministry of Statistics and Programme Implementation, Central Statistics Office, Government of India (Social Statistics Division)
Government of India (2019) Telecom statistics India—2019. https://dot.gov.in/sites/default/files/Telecom%20Statistics%20India-2019.pdf?download=1. Accessed 10 Sept 2020
Gupta S (2020) The impact of SARS-CoV-2 on the education sector in India. Confluence. http://confluence.ias.ac.in/the-impact-of-the-coronavirus-sars-cov-2-on-the-education-sector-in-india/. Accessed 20 June 2020
Halupa CP (2016) Risks: the impact of online learning and technology on student physical, mental, emotional, and social health. https://doi.org/10.21125/iceri.2016.0044
Holsinger D (1974) The elementary school as modernizer: a Brazilian study. In: Inkeles A, Holsinger D (eds) Education and individual modernity in developing countries. Leiden, Brill, pp 24–46
Huang J, Gates AJ, Sinatra R, Barabási AL (2020) Historical comparison of gender inequality in scientific careers across countries and disciplines. Proc Natl Acad Sci 117(9):4609–4616. https://doi.org/10.1073/pnas.1914221117
India Education Report (2005) National Center on Education and the Economy. https://www.ncee.org/wp-content/uploads/2013/10/India-Education-Report.pdf. Accessed 1 Sept 2020

India Today (2020) PUBG player in Punjab reportedly lost 16 lakh from parents' bank account, made to work at scooter repair shop. https://bit.ly/32sdXyo. Accessed 14 Sept 2020
International Labour Organization (2020) Global employment trends for youth 2020: technology and the future of jobs. https://www.ilo.org/wcmsp5/groups/public/—dgreports/—dcomm/—publ/documents/publication/wcms_737648.pdf. Accessed 15 July 2020
Internet and Mobile Association of India (2020) Digital in India 2019—round 2 report. https://cms.iamai.in/Content/ResearchPapers/2286f4d7-424f-4bde-be88-6415fe5021d5.pdf. Accessed 12 Sept 2020
Internet and Mobile Association of India (2019) India internet 2019. https://cms.iamai.in/Content/ResearchPapers/d3654bcc-002f-4fc7-ab39-e1fbeb00005d.pdf. Accessed 12 Sept 2020
Jena PK (2020) Imapct of Covid-19 on higher education in India. Int J Adv Educa Res 5(3):77–81
Jha N, Shenoy V (2016) Digitization of Indian education process: a hope or hype. IOSR J Bus Manag (IOSR-JBM) 18(10):131–139. https://doi.org/10.9790/487X-181003131139
John R (2020) Indian educational institutes should break norms in COVID times, not conduct online exams. https://theprint.in/opinion/pov/indian-educational-institutes-should-break-norms-in-covid-times-not-conduct-online-exams/408660/. Accessed 13 Sept 2020
Joye D, Falcon J (2014) Soc Mobility. https://doi.org/10.1007/978-94-007-0753-5_2775
Kautiainen S, Koivusilta L, Virtanen SM, Rimpela A (2005) Use of information and communication technology and prevalence of overweight and obesity among adolescents. Int J Obes 29:925–933
KMPG India (2017) The 'digital first' journey: how OTT platforms can remain 'on demand ready': 16
Kumar DNS (2020) Impact of COVID-19 on higher education. https://www.highereducationdigest.com/impact-of-covid-19-on-higher-education/. Accessed 13 Sept 2020
Lall M (2005) The challenges for India's education system: briefing paper. https://www.chathamhouse.org/sites/default/files/public/Research/Asia/bpindiaeducation.pdf. Accessed 12 Sept 2020
Lilliard A, Peterson J (2011) The immediate impact of different types of televisions on young children's executive function. Pediatrics 128(4):644–649
Livemint (2020) Every house a school, every parent a teacher as Covid-19 impacts education of 300mn students. https://bit.ly/3hvSDfs. Accessed 1 Sept 2020
Manazir SH, Rubina, Govind M (2020) E-Learning pedagogy in institute of higher education in India post lockdown due to COVID-19 pandemic. Research Based Policy Guidelines on Issues and Constraints of Students
Meeker M (2017) Internet trends 2017—code conference. https://bit.ly/2FfY05z. Accessed 13 Sept 2020
Meyer J (1977) The effects of education as an institution. Am J Sociol 83(1):55–77. http://www.jstor.org/stable/2777763. Accessed 13 Sept 2020
MICA (2019) Indian OTT platforms report 2019: new regional flavours more entertaining content. Centre for Media and Entertainment Studies. https://communicationcrafts.in/wp-content/uploads/2019/12/Indian_ott_report2019.pdf. Accessed 10 Sept 2019
Ministry of Human Resource Development (2020) PRAGYATA: guidelines for digital education. Department of School Education and Literacy, Government of India
Ministry of Statistics and Programme Implementation (2019) Unit level data & report on NSS 75th round for schedule–25.2 (July 2017–June 2018), social consumption: education. Government of India
Mohammadnia Z, Moghadam FD (2019) Textbooks as resources for education for sustainable development: a content analysis. J Teacher Educ Sustain 21(1):103–114. https://doi.org/10.2478/jtes-2019-0008
Myers KR, Tham WY, Yin Y, Cohodes N, Thursby JG, Thursby MC, Schiffer P, Walsh JT, Lakhani KR, Wang D (2020) Unequal effects of COVID 19 on scientists. Nat Hum Behav 4:880–883
National Education Policy (2020) Ministry of Human Resource Development, Government of India. https://www.mhrd.gov.in/sites/upload_files/mhrd/files/NEP_Final_English_0.pdf. Accessed 23 July 2020

National Youth Policy (2014) Ministry of Youth Affairs and Sports, Government of India. https://yas.nic.in/sites/default/files/NYP%20Brochure.pdf. Accessed 27 July 2020

Navarro J, Kohl K, Cetron M et al (2016) A tale of many cities: a contemporary historical study of the implementation of school closures during the 2009 pA (H1N1) influenza pandemic. J Health Polit Policy Law 41(3):393–421

NorthEast Now (2020) Man commits suicide after failing to buy smartphone for his daughter to attend online classes. https://nenow.in/north-east-news/tripura/tripura-man-commits-suicide-after-failing-to-buy-smartphone-for-his-daughter-to-attend-online-classes.html. Accessed 27 July 2020

Office of the High Commissioner for Human Rights (2015) Transforming our world: human rights in the 2030 agenda for sustainable development. https://www.ohchr.org/Documents/Issues/MDGs/Post2015/TransformingOurWorld.pdf. Accessed 29 July 2020

Parsheera S (2019) India's on a digital sprint that is leaving millions behind. BBC. https://www.bbc.com/news/world-asia-india-49085846. Accessed 5 June 2020

Plant W (1965) Longitudinal changes in intolerance and authoritarianism for subjects differing in amount of college education over four years. Genet Psychol Monogr 72:247–287

Press Information Bureau (2020) NCERT and Rotary India digitally sign MoU for e-learning content telecast for class 1-12 over all NCERT TV channels in the presence of Union HRD Minister Shri Ramesh Pokhariyal. Government of India. https://www.mhrd.gov.in/sites/upload_files/mhrd/files/pr_ncert_0.pdf. Accessed 25 June 2020

Quartz India (2020) How urban Indian parents are dealing with coronavirus lockdown. https://qz.com/india/1826430/indias-coronavirus-lockdown-work-from-home-with-kids-is-hard/. Accessed 14 Sept 2020

Razak DA (2020) Sustainable growth and Covid-19. https://www.nst.com.my/opinion/columnists/2020/04/584031/sustainable-growth-and-covid-19. Accessed 1 June 2020

Roesch E (2020) How the shift to remote learning affects students used to face-to-face learning. https://elearningindustry.com/shift-to-remote-learning-affects-students-used-face-to-face. Accessed 13 Sept 2020

Sahni K (2020) Covid-19 and the shift to digital learning: teachers face unprecedented crisis. The Citizen. https://www.thecitizen.in/index.php/en/NewsDetail/index/9/19139/Covid-19-and-the-Shift-to-Digital-Learning-Teachers-Face-Unprecedented-Crisis. Accessed 15 Aug 2020

Samantaray PK (2020) COVID challenges to Indian education system. https://digitallearning.eletsonline.com/2020/06/covid-challenges-to-india-education-system/. Accessed 7 Sept 2020

Saqib NU, Raheem MA, Iqbal M, Salman M, Shahzad T (2018) Effects of extracurricular activities on students. National University of Science and Technology

Savić I (2018) Education for sustainable development and human rights. In: Filho WL, Mifsud M, Pace P (eds) Handbook of lifelong learning for sustainable development. Springer, pp 243–256

Sen A (1999) Development as freedom. Anchor Books, New York

Shaker RR (2015) The spatial distribution of development in Europe and its underlying sustainability correlations. Appl Geogr 63:304–315

Staniscuaski F, Reichert F, Werneck FP, Oliveira L, Mello-Carpes PB, Soletti RC, Almeida CI, Zandona E, Ricachenevsky FK, Neumann A, Schwartz IVD, Tamajusuku ASK, Seixas A, Kmetzsch L (2020) Impact of COVID-19 on academic mothers. Science 368(6492):724. https://doi.org/10.1126/science.abc2740

Tahseen Q (2020) The impact of COVID-19 on education at AMU. Confluence

Thakur GK (2014) ICT and digital divide in Indian school system. IJIMS 2(2):34–38

The Print (2020) Peeping parents, sleeping students, bullying—online classes are a nightmare for teachers. https://theprint.in/india/education/peeping-parents-sleeping-students-bullying-online-classes-are-a-nightmare-for-teachers/458522/. Accessed 10 Sept 2020

The Week (2020) Online classes turning out to be nightmare for teachers amid COVID-19 lockdown. https://www.theweek.in/news/india/2020/05/23/online-classes-turning-out-to-be-nightmare-for-teachers-amid-covid-19-lockdown.html. Accessed 9 Sept 2020

Times of India (2020a) Unable to attend online classes, Tripura student hangs self. Times of India. https://www.google.com/search?q=Unable+to+attend+online+classes%2C+Tripura+student+hangs+self&rlz=1C1CHBD_enIN907IN907&oq=Unable+to+atten&aqs=chrome.1.69i57j69i59j35i39j0l4j69i60.5115j0j4&sourceid=chrome&ie=UTF-8. Accessed 10 June 2020

Times of India (2020b) Helicopter parenting puts pressure on teachers during online classes. https://bit.ly/35wy28m. Accessed 10 Sept 2020

U-DISE (2016) School education in India. U-DISE flash statistics 2016-17. National Institute of Educational Planning and Administration

UIS and GEM (2017) Reducing global poverty through universal primary and secondary education. UNESCO. http://uis.unesco.org/sites/default/files/documents/reducing-global-poverty-through-universal-primary-secondary-education.pdf. Accessed 30 June 2020

UNDESA (2020) Protecting and mobilizing youth in COVID-19 responses. COVID-19 RESPONSE. Policy Brief No. 67. http://www.un.org/development/desa/youth/wp-content/uploads/sites/21/2020/05/PB_67.pdf. Accessed 27 June 2020

UN Economic and Social Council (1999) General comment No. 13: the right to education (article 13 of the Covenant). UN Committee on Economic, Social and Cultural Rights (CESCR). https://documents-ddsny.un.org/doc/UNDOC/GEN/G99/462/16/PDF/G9946216.pdf?OpenElement.Accessed 17 Aug 2020

UN News (2020) Stay the course together to emerge stronger from COVID-19 crisis: UN chief's message to major sustainability forum. United Nations. https://news.un.org/en/story/2020/07/1067622 Accessed 7 July 2020

UNDP (2020) COVID-19: human development on course to decline this year for the first time since 1990. https://www.undp.org/content/undp/en/home/news-centre/news/2020/COVID19_Human_development_on_course_to_decline_for_the_first_time_since_1990.html. Accessed 14 July 2020

UNECOSOC (2020) Progress towards the sustainable development goals: report of the secretary general. https://undocs.org/en/E/2020/57. Accessed 20 May 2020

UNESCO, GEMR, and IIEP (2017) Six ways to ensure higher education leaves no one behind

UNESCO (2011) From green economies to green cities. UNESCO, France. https://www.greengrowthknowledge.org/sites/default/files/downloads/resource/Green_Economy_UNESCO.pdf. Accessed 19 July 2020

UNESCO (2020a) COVID-19 impact on education. https://en.unesco.org/covid19/educationresponse. Accessed 10 September 2020

UNESCO (2020b) The impact of COVID – 19 on women: policy brief. https://reliefweb.int/sites/reliefweb.int/files/resources/policy-brief-the-impact-of-covid-19-on-women-en.pdf. Accessed 17 July 2020

UNESCO (2020c) Reopening schools: when, where and how? https://en.unesco.org/news/reopening-schools-when-where-and-how. Accessed 8 Aug 2020

UNICEF (2020) A new era for girls: taking stock of 25 years of progress. United Nations Children's Fund. https://www.unicef.org/media/65586/file/A-new-era-for-girls-2020.pdf. Accessed 10 May 2020

United Nation (1948) Universal declaration of human rights. UN General Assembly. https://www.ohchr.org/EN/UDHR/Documents/UDHR_Translations/eng.pdf. Accessed 30 May 2020

United Nation (1966) International covenant on civil and political rights. UN General Assembly. https://www.ohchr.org/Documents/ProfessionalInterest/ccpr.pdf. Accessed 9 Aug 2020

United Nations (1966) International covenant on economic, social and cultural rights. UN General Assembly. https://www.ohchr.org/en/professionalinterest/pages/cescr.aspx. Accessed 9 Aug 2020

United Nations (1986) Declaration on the right to development. UN General Assembly. https://www.ohchr.org/Documents/ProfessionalInterest/rtd.pdf. Accessed 8 Aug 2020

United Nations (2015) Transforming our world: the 2030 agenda for sustainable development. Department of Economic and Social Affairs. https://sdgs.un.org/2030agenda. Accessed 9 Aug 2020

United Nations (2020a) COVID-19 and human rights: we are all in this together. United Nations. https://www.un.org/sites/un2.un.org/files/un_policy_brief_on_human_rights_and_covid_23_april_2020.pdf. Accessed 20 June 2020

United Nations (2020b) The impact of COVID-19 on women. United Nations. https://reliefweb.int/sites/reliefweb.int/files/resources/policy-brief-the-impact-of-covid-19-on-women-en.pdf. Accessed 7 Aug 2020

University Grants Commission (2020) UGC guidelines on examinations and academic calendar for the universities in view of COVID-19 pandemic and subsequent lockdown. https://www.ugc.ac.in/pdfnews/4276446_UGC-Guidelines-on-Examinations-and-Academic-Calendar.pdf. Accessed on 10 Aug 2020

World Bank (2020) The COVID-19 pandemic: shocks to education and policy responses. WorldBank. https://openknowledge.worldbank.org/bitstream/handle/10986/33696/148198.pdf?sequence=4&isAllowed=y. Accessed 5 Aug 2020

World Bank Group Education (2020) The COVID-19 crisis response: supporting tertiary education for continuity, adaption and innovation. http://pubdocs.worldbank.org/en/621991586463915490/WB-Tertiary-Ed-and-Covid-19-Crisis-for-public-use-April-9.pdf. Accessed 13 Sept 2020

World Commission on Environment and Development (1987) Our common future. Oxford University Press

Yadav S (2020) Madhya Pradesh's online classes not reaching majority of students. The Hindu. https://bit.ly/3ce9SRx. Accessed 17 Aug 2020

Zhu X, Liu J (2020) Education in and after COVID-19: immediate responses and long-term visions. Post Digital Science and Education. Springer Nature. https://doi.org/10.1007/s42438-020-00126-3

Srashta Chowdhury is a Research Scholar in the Department of Geography and Applied Geography, University of North Bengal. She has completed her masters from the same department and was awarded the University Fellowship. She is pursuing research on the field of Gender Geography and higher education as the core area. Her research interest is cultural geography, population geography, popular and folk lifestyles, and social geography.

Sushma Rohatgi is a Professor in the Department of Geography and Applied Geography, University of North Bengal. She has been teaching in the institute for over thirty years and has supervised several research works. She has served as the director of Women's Studies, the University of North Bengal from 2012 to 2019, and has conducted various consultancy projects with various stakeholders. She is a member of The Geographical Society of India (GSI), Kolkata; Institute of Landscape Ecology and Ekistics (ILEE), Kolkata; Indian Society of Remote Sensing (ISRS), Dehra Dun; Indian National Cartographic Association (INCA), Hyderabad; Association of North Bengal Geographers (ANBG), Siliguri to name a few. Her research interest is significant in Cartography, Remote Sensing & GIS, Population Geography, Gender Studies, and Environmental Management.

Part III
Human, Development and Environment

Chapter 9
Economic Lockdowns and Challenges of Rural Livelihood: Indian Scenario

Sujit Mandal

Abstract India is severely affected by the COVID-19 pandemic worldwide, with the number of confirmed cases and death cases increase at an alarming rate. The COVID-19 pandemic and associated lockdowns affected commercial establishments, education, entertainment, industry, tourism, transport, agriculture, migrants, unemployment, food security, GDP, and recession of the country's economy. Lockdowns in different phases caused a sharp rising of unemployment, stress on supply chains, and the poor's livelihood challenges in India. The present study will address the present situation of the COVID-19 pandemic, lockdowns, and its response to migrant workers, smallholder farmers, the landless workers, and daily wage labourers from the informal sectors and their livelihood crisis. The livelihood crisis differs from one region to another due to regional differences in resource endowment, unemployment, and agrarian distress in India. About 91% of the total of 465 million workers are engaged in informal sectors in India. In India, severely COVID-19 mostly affected states are Maharashtra, Delhi, Gujarat, Tamilnadu, Uttar Pradesh, West Bengal, Rajasthan, Telangana, and Karnataka. Out of 739 districts in India, 139 reported a large number of confirmed cases, 300 reported only a few cases, and 300 districts are not affected by COVID-19. Already lockdowns have created much pressure on the Indian economy and invited livelihood challenges to rural poor. The prolonged lockdowns may further cause the country's income erosion and may push the poor into extreme livelihood crises and poverty.

Keywords COVID-19 · Pandemic · Lockdown · Informal sectors · Income erosion · Rural livelihood crisis

S. Mandal (✉)
Faculty of Science, Diamond Harbour Women's University, Diamond Harbour, West Bengal, India
e-mail: mandalsujit2009@gmail.com

M. Mishra and R. B. Singh, *COVID-19 Pandemic Trajectory in the Developing World*, Advances in Geographical and Environmental Sciences,
https://doi.org/10.1007/978-981-33-6440-0_9

9.1 Introduction

Scientists first identified a human coronavirus in 1965, which caused a common cold. Later on, researchers found a group of same human and animal viruses and named them after their crown-like appearance. Researchers found that seven coronaviruses can infect human bodies. One of them that caused the SARS epidemic in Southern China in 2002, which spread out rapidly to 28 countries, infecting more than 8000 people by July 2003 and raising the death toll of 774. This coronavirus caused fever, headache, and respiratory problems such as cough and shortened breath.[1] Lam et al. (2020) identified SARS-CoV-2-related coronavirus in Malayan pangolins. Lau et al. (2010) and Luk et al. (2019) studied epidemiology, evolution, and Phylogeny of SARS and SARS-related *Rhinolophus bat* coronavirus in China. Huang et al. (2020) propounded the chemical features of patients infected with the 2019 novel coronavirus in Wuhan, China. Van Doremalen et al. (2013), Otter et al. (2016), and Lai et al. (2005) opined regarding the survival of SARS-CoV and various healthcare measures to prevent and check the transmission of the virus from human-to-human and other objects.

Scientists have divided coronaviruses into four sub-groupings, called alpha, beta, gamma, and delta. Seven of these viruses can infect people. The four common ones are 229E (alpha); (ii) NL63 (alpha); (iii) OC43 (beta); and (iv) HKU1 (beta). The three less-common ones are MERS-CoV, a beta virus that causes Middle East respiratory syndrome (MERS); SARS-CoV, a beta virus that causes severe acute respiratory syndrome (SARS); and SARS-CoV-2, which causes COVID-19.[2]

The new or 'novel' coronavirus disease (COVID-19 or 2019-nCoV) has recently been reported from Wuhan (China), and it has also been reported in Thailand, Japan, South Korea, and the USA (WHO). This COVID-19 has caused many fatal cases.[3] Hu et al. (2018) and Zhou et al. (2020) opined that the 2019-nCoV is most closely related to 2 severe acute respiratory syndromes (SARS)-like CoV sequences that were isolated in bats during 2015–2017, suggesting that the bats' CoV and the human 2019-nCoV share a recent common ancestor. Lau et al. (2020) and Huang et al. (2020) suggested that seventeen years later, the severe acute respiratory syndrome (SARS)-like epidemic, an outbreak of pneumonia, presently which is called coronavirus disease (COVID-19) has been reported in Wuhan, China. A Study Group of the International Committee on Taxonomy of Viruses explored that the causative agent is very rapidly isolated from patients and identified to be a coronavirus and named the virus as severe *acute respiratory syndrome coronavirus 2*, i.e., SARS-CoV-2 (Gorbalenya et al. 2020). SARS-CoV-2 has spread out all over the world quickly and affected a large number of the human population with a death toll of 4012 by 10 March 2020 (WHO situation report as of 10 March 2020). The SARS-CoV-2 has spread both in China and outside China and infected people worldwide alarmingly, which means that people are unwittingly catching and passing on the coronavirus.

---

[1] For details see: https://www.webmd.com/lung/coronavirus-history.

[2] View more details at: https://www.webmd.com/lung/coronavirus-strains#1.

[3] Visit: https://academic.oup.com/cid/advance-article-abstract/doi/10.1093/cid/ciaa112/5721420.

This unprecedented and unexpected growing tendency of worldwide transmission is now a pandemic. Riou and Althaus (2020) and Phan et al. (2020) described the pattern of human-to-human transmission of COVID disease from December 2019 to January 2020. The present study aims to demonstrate the temporal and spatial distribution of COVID-19 cases of the world and to suggest some preparedness measures to combat novel coronavirus. The COVID-19 has now affected all the geographical locations of the world.[4]

9.2 Propagation of COVID-19 or 2019-nCoV or SARS-CoV-2

COVID-19 is now a global threat of the twenty-first century. It may be considered as one of the significant factors which may retard all sorts of socio-economic advancement of all affected nations. COVID-19 is an acronym where 'CO' denotes *corona*, 'VI' denotes *virus*, and 'D' denotes *disease*, and 19 marks the year of occurrences. Coronavirus is a single-stranded RNA virus with a diameter ranging from 80 to 120 nm. The first situation report of WHO as of 20 January 2020 revealed that on 31 December 2019, the WHO China Country Office has been informed of cases of pneumonia unknown etiology detected in Wuhan City of Hubei Province, China. However, the causal agent of the cases was not identified.

On 7 January, the Chinese authorities identified a new type of coronavirus. China, 12 January 2020, shared the genetic sequence of the new virus for countries to develop specific diagnostic kits. The Ministry of Public Health, Thailand reported the first imported case of laboratory-confirmed 2019-novel coronavirus (2019-nCoV OR COVID-19) from Wuhan, China, on 13 January 2020. Subsequently, on 15 January, the Ministry of Health, Labour and Welfare, Japan (MHLW), and on 20 January, the Republic of South Korea reported the cases of the novel coronavirus infections. As of 20 January, only four countries of the world, i.e., China, Thailand, the Republic of Korea, and Japan, were affected by novel coronavirus with the confirmed cases of 282 and a total death toll of 6.

Since then, the number started to increase in different countries of the Western Pacific region, Eastern Mediterranean region, European region, American region, Southeast Asia region, and African Region of the world. On 31 January 2020, 19 countries got affected by COVID-19, and it reached over 24 countries by 5 February, 33 countries by 25 February, 53 countries by 29 February, 85 countries by 5 March, 109 countries by 10 March, and 227 on 07 May. Within a span of 3 and a half months (20 January to 7 May 2020) the number of confirmed cases increased from 282

[4]The worldwide distribution of COVID-19 depicts that the USA, Italy, France, Spain, UK, Canada, Iran, India, Japan, South Africa, Switzerland, Brazil, Peru, Mexico, Ecuador, Saudi Arabia, Pakistan, China, Egypt, Russian Federation, Turkey, Germany, Belgium, Netherland, Portugal, Sweden, Ireland, Austria, Poland, Romania, Denmark, Indonesia, Republic of Korea, and Philippines are severely affected by COVID-19.

to 3,672,238, and death cases increased from only 6 to worrying 254,045. As of 7 June 2020, there were 6,799,713 confirmed COVID-19 cases and 397,388 death cases worldwide. On 7 July the number of confirmed COVID-19 cases reached 11,500,302, and death cases reached 535,759. As of 16 August, confirmed COVID-19 cases climbed to 21,294,845, and death cases climbed to 761,779. Such an unprecedented increase of COVID-19 cases shocked the whole world. Almost all the countries of the world are severely affected by the COVID-19 pandemic. At present, all the healthcare personals, government, and non-government organizations and the common public must show the solidarity and fight shoulder to shoulder to combat the pandemic (Yoo 2020). Globally, the American region, European region, and the Southeast Asia region are severely affected by COVID-19 in terms of both confirmed and death cases.

9.3 Status of COVID-19 in India

At present, India has the largest number of confirmed cases in Asia. It ranks third in regard to COVID-19 confirmed cases in the world after the USA and Brazil. Six major cities, i.e., Mumbai, Delhi, Ahmedabad, Chennai, Pune, and Kolkata, account for about half of all reported cases in India. In India, the total confirmed COVID-19 cases have been increasing rapidly: 1397 confirmed cases on 31 March has reached 50,12,269 on 15 September 2020 (Table 9.1).

There is an increasing trend of confirmed cases since 31 March 2020 to till date (21 August 2020). In the month of May, June and July confirmed cases increased at an alarming rate in India. The death cases increased five times only in the month of May, three times in June, and more than two times in July. Severely affected states in India are Maharashtra, Tamilnadu, Delhi, Gujarat, UP, Karnataka, Telengana, West Bengal, Andhra Pradesh, and Rajasthan (Fig. 9.1). A sharp increase in confirmed cases was started from mid of June in major states of Maharashtra, Tamilnadu, and Delhi. Andhra Pradesh, Karnataka, Uttar Pradesh, and West Bengal revealed a sharp increase in the confirmed cases since mid of July 2020.

Table 9.1 Confirmed COVID-19 cases and death cases in India

Date	Total cases	Total deaths
31-Mar	1397	35
30-Apr	33,610	1075
31-May	182,143	5164
30-Jun	566,840	16,893
31-Jul	1,638,870	35,747
20-Aug	2,836,925	53,866
15-Sept	5,012,269	82,010

Source Ministry of Health and Family Welfare, Govt. of India

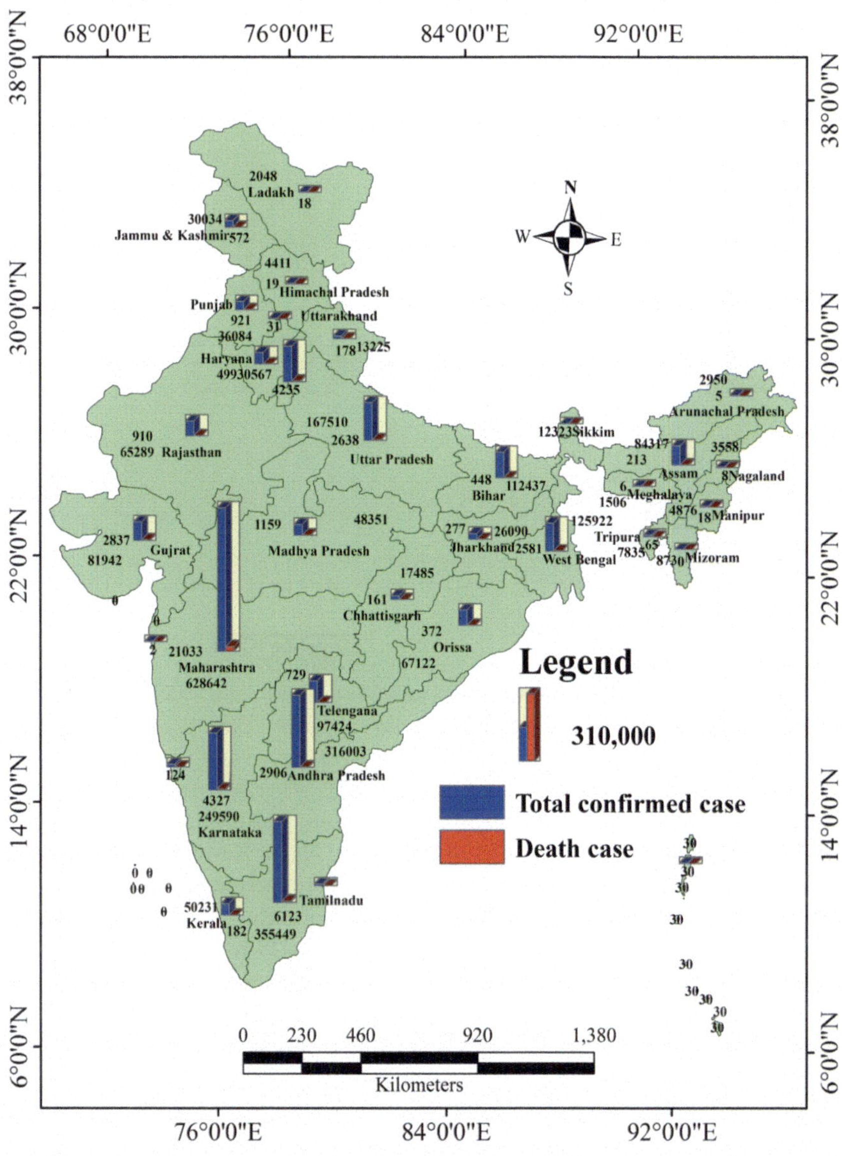

Fig. 9.1 State-wise confirmed COVID-19 cases and death cases as on 20 August 2020 (*Source* Ministry of Health and Family Welfare, Govt. of India)

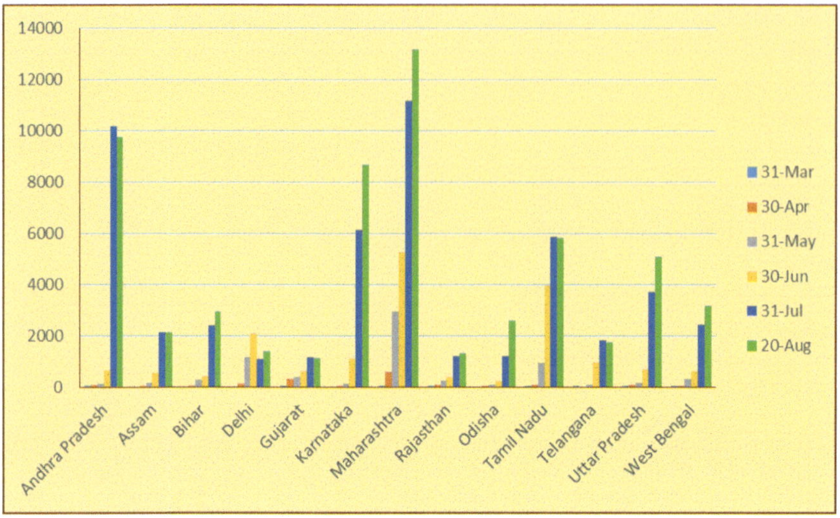

Fig. 9.2 Number of new COVID-19 cases reported on 31 March, 30 April, 31 May, 30 June, 31 July, and 20 August 2020 in different Indian states (Source Ministry of Health and Family Welfare, Govt. of India)

In India, the states of Maharashtra (628,642), Tamilnadu (355,449), Andhra Pradesh (316,003), Karnataka (249,590), Delhi (156,139), Uttar Pradesh (167,510), West Bengal (125,922), and Bihar (112,437) have witnessed a large number of confirmed cases as on 20 August 2020 (Fig. 9.2). Whereas, Arunachal Pradesh, Nagaland, Manipur, Mizoram, Meghalaya, Sikkim, Tripura, Goa, Chhattisgarh, Ladakh, Himachal Pradesh, Chandigarh, Dadra and Nagar Haveli, Daman and Diu, Puducherry, and Andaman and the Nicobar Islands are the Indian states and Union territories with comparatively low infection and mortality caused by the disease. The statistics of infection and death on 20 August 2020 reveals that Tamilnadu, Karnataka, Andhra Pradesh, Delhi, Uttar Pradesh, and West Bengal are severely affected. However, all these states were not so severely affected around two months earlier, i.e., as of 30 June 2020. Another one-month earlier, i.e., during early May, death cases had been mostly concentrated in three Indian states—Maharashtra, Gujarat, and Delhi (Fig. 9.3).

9.4 COVID-19 and Preparedness Measures

Two measures are chiefly taken into account globally to combat COVID-19: clinical measures and non-pharmaceutical measures. WHO has developed interim guidance for laboratory diagnosis, advice on the use of masks during home care and in health-care settings in the context of COVID-19 outbreak, clinical management, infection

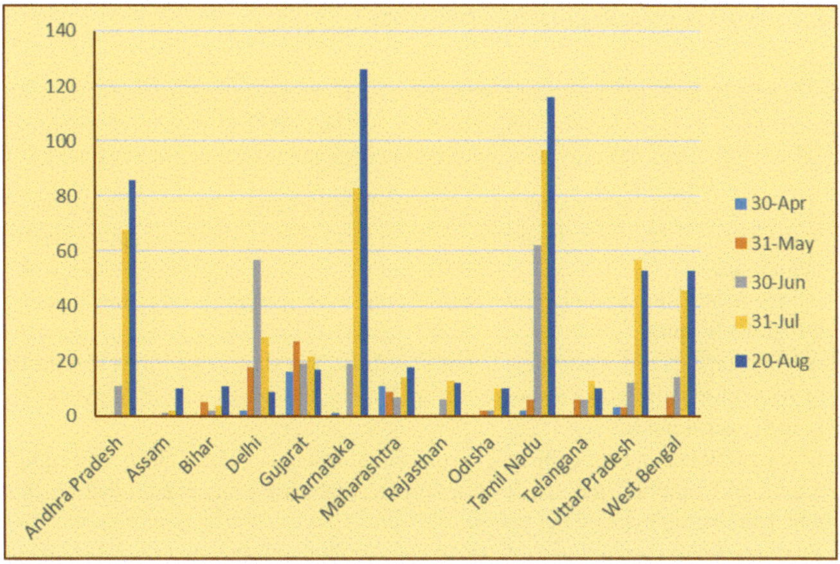

Fig. 9.3 Number of new COVID-19 death cases reported on 30 April, 31 May, 30 June, 31 July, and 20 August (*Source* Ministry of Health and Family Welfare, Govt. of India)

prevention and control in healthcare settings, home care for patients with suspected novel coronavirus, risk communication, and community engagement and Global Surveillance for human infection with COVID-19.

9.4.1 Clinical Measures

It includes laboratory testing and identification of COVID-19 affected confirmed cases for proper care, isolation, and interruption to transmit viruses. Clinical measures will help to specify the locations of affected areas and communities for taking other preventive measures. To promote clinical measures around the world WHO has made an arrangement of a laboratory network in six WHO regions to upgrade the expertise in virology, diagnostics, sequencing, and often viral culture. These laboratories will support the Member States of WHO regarding resting and identifying the COVID-19 cases. Since 24 February 2020 WHO has been providing polymerase chain reaction (PCR) test kits for COVID-19 directly to countries, offering the Member States the capacity to detect cases or clusters before the disease gains a strong foothold.

9.4.2 Non-pharmaceutical Measures

The hospitals and the healthcare system need to be designed properly with all precautionary measures, providing all the staff the adequate protective facilities. The following measures may be suggested to check the infection of COVID-19.

- All the staff who are working in healthcare facilities and who have close contact with COVID-19 patients should always wear scrubs, and it should be appropriately maintained.
- All the health and administrative staff should wear special shoes at work and be left at the hospital.
- All the staff should wash their hands carefully, and shower facilities should be provided for staff to take a shower before leaving the work to avoid transmission to the general public.
- Every candidate in the healthcare system should clean regularly electronic equipment such as mobile, tablets, desktop screen, keyboard, printer, and other objects which are being used by the people.
- Everyone who is attending COVID-19 patients should follow physical distancing and consider sleeping in a separate room and using a separate washroom.

9.4.3 Lockdowns as a Measure for COVID-19 Preparedness

More than 100 countries all over the world introduced either a full or partial lockdown by the end of March 2020 to prevent and check the transmission of coronavirus. Such a decision to lockdown affected billions of people in the world. Some cities, i.e., Stockholm, Jakarta, Seoul, Singapore, Hong Kong, etc., have issued some recommendations of the restrictions of movement and social distancing rather than strict rules. As a result of which the movement of traffic fell down dramatically. South Korea focussed on large-scale testing and contact tracing rather than social distancing and stand still lockdown. In Jakarta and Indonesia, where there was no official lockdown, but the congestion of people dropped. Countries of France, Spain, Iran, Germany, UK, Netherlands, Sweden, Canada, Mexico, Belgium, Ireland, USA, Turkey, China, India, and Switzerland called their strictest lockdown in March 2020. In eight countries, new confirmed COVID-19 cases declined, unlike India. But most of the countries found peak cases of COVID-19 within a month from lockdown. According to WHO situation report, lockdowns are not enough to combat COVID-19. The COVID-19 affected countries must use the opportunity to "find, isolate, test, treat, and trace" all COVID-19 cases. Surely, lockdown will reduce congestion, physical contact, and mass gathering as well as check the transmission of coronaviruses, but strict lockdown should not be followed for a long time. The COVID-19 affected countries should issue guidelines and recommendations in regard to the movement of the people, and it must be followed by being a responsible citizen of the country.

The Government of India confirmed the country's first case of 2019-nCoV on 30 January 2020 in the state of Kerala, when a university student returned from Wuhan, China. The Prime Minister of India declared the "Janata Curfew" on 19 March 2020 to be observed on 22 March 2020 when the 2019-nCoV confirmed cases reached 500 (Business Line, March 22). The government declared a 21 days nationwide lockdown on 24 March 2020 as a preventive measure against the 2019-nCoV pandemic. This lockdown was placed before the entire 1.3 billion population of India when the number of confirmed positive coronavirus cases in India was approximately 500 (Gettleman and Schultz 2020). At the immediate end of the first lockdown, some states, i.e., Odisha and Punjab, extended the lockdown till 1 May 2020, which was followed by West Bengal, Maharashtra, Karnataka, and Telangana. On 14 April, the Prime Minister again declared 2nd phase of lockdown from 15 April to 3 May (Prime Minister Narendra Modi Announcement 14 April 2020). On 1 May, the Government of India again extended the nationwide lockdown further by two weeks until 17 May (NDTV.com).

In the first phase of lockdown, nearly all services and factories were being suspended (Singh et al. 25 March 2020). Approaching the end of the first phase of lockdown, the rate of increase of 2019-nCoV infections had noticeably slowed down in India from a rate of doubling every three days prior to the lockdown to the rate of doubling every eight days on 18 April (Gupta 18 April 2020). During 2nd phase of lockdown from 20 April 2020, the government announced certain relaxations in several sectors such as dairy, agriculture, and plantation, and shops selling farming supplies. Public works departments were also allowed to open maintaining social distancing. Relaxations were also given to Cargo transportation vehicles, including trucks, trains, and planes, Banks, and Government centres.

The lockdown areas were classified into three zones, i.e., *red zone* (indicating the presence of infection hotspots), *orange zone* (indicating some infection), and *green zone* (no infection). There are 130 districts under the red zone, 284 districts under the orange zone, and 310 districts under the green zone in India (Thacker 1 May 2020). On 12 May, the Indian government declared 4th phase of lockdown with some relaxations.

During 'Unlock 1.0' (1–30 June 2020), lockdown restrictions were imposed only in the containment zones, and several activities were allowed in a phased manner. The permission was given to reopen shopping malls, religious places, hotels, and restaurants from 8 June. Under 'Unlock 2.0' (from 1 to 31 July 2020), almost all the activities were permitted in all other areas except containment zones. State governments were allowed to impose proper restrictions. Permission was given in International and interstate travelling. 'Unlock 3.0' started from 1 August 2020, which removed night curfews, permitted gymnasium and yoga centres, permitted Independence Day celebration with social distancing, and declared the closure of educational institutions till 31 August 2020. During 'Unlock 3.0', Tamilnadu and Maharashtra declared lockdowns for the month of August 2020, where West Bengal imposed lockdowns twice a week. Under 'Unlock 4.0' (till 30 September 2020), religious activities, entertainment, political activities, sports, academic functions, and gatherings were permitted with 100 people.

9.5 Impact of COVID-19 and Associated Lockdowns in India

The pandemic invited a sharp rise in unemployment, stress on supply chains, the collapse of the industries, a decrease in Govt. income, etc. All sectors of the economy have been severely affected due to COVID-19 Pandemic in INDIA. More than 45% of households across the nation have reported on income drop (April 2020) as compared to the previous year.[5] During the lockdowns, a large number of people lost employment, and subsequently, the nation has been suffering from economic crisis and livelihood challenges. The pandemic has brought about a large-scale socio-economic changes worldwide at present. It has disrupted lives, created loneliness caused by social distancing; it has led to increased anxiety caused by economic difficulties, fear of illness, and worry about near and dear ones. As a whole, COVID-19 Outbreak has disrupted the lifestyle of people across the world.

9.5.1 Informal Sectors of the Economy

In 2000, 75.2% of employment was in the informal sector and 24.8% in the formal sector. In 2010, employment in the informal sector climbed to 87.4%, and in the formal sector, it was declined to 12.6%.[6] No notable changes have been found in 2013 and 2015–2016 regarding employment share in informal and formal sectors. At present, only 10% of the countries workforce is in the formal sectors. Lockdowns invited suspension of the transport network and all economic activities and services. This brought turmoil in the lives of millions who are primarily involved in informal sectors. They lost their livelihood overnight and got stranded in different pockets of the country. International Labour Organization (ILO) speculated that India is likely to face a severe job crisis because of subsequent lockdowns.[7] All migrant and migrant workers in informal sectors are also severely hit. It may affect their food and nutritional intake, access to healthcare, and education of children worst. The informal sector constitutes about 81–88% of the Indian economy. About 90% of the 500 million workers in India are part of the informal sectors, which contributes half of the national gross domestic product. The sector comprises a vast array of small-to-medium-sized enterprises. The informal sector is the backbone of India's wealth, as the formal sector depends on the goods and services provided by it. It

---

[5]See for details: https://www.statista.com/statistics/1111510/india-coronavirus-impact-on-household-income/.

[6]See the Economic Times report which is available under the URL: https://economictimes.indiatimes.com/news/economy/indicators/national-database-of-workers-in-informal-sector-in-the-works/articleshow/73394732.cms.

[7]For details: https://retail.economictimes.indiatimes.com/news/industry/41-lakh-youth-lose-jobs-in-india-due-to-covid-19-pandemic-ilo-adb-report.

employs the vast majority of India's workers, such as extremely poor, highly mobile, seasonally circulatory, daily-wage, and contract-based labourers.

Central Trade Union had conducted a study that identified 21 trades in India where informal workers are engaged massively. This included the domestic workers, street vendors, home-based workers, homes stay hosts, agricultural labourers, beedi rollers, artisans, construction workers, financial/banking agents, tiffin services, tailoring and production, weavers, animal husbandry, *feri tokri* workers, hospitality staff, factory workers, auto-rickshaw drivers, medical helpers, daily wage labourer, and shop owners.

In India, major informal sectors of the economy are agriculture, forestry, and fishing; construction; trade, repair, accommodation, and food services; real estate, ownership of dwelling, and professional services (Table 9.2). In 2011–2012, 96.8% of unorganized workers were engaged in agriculture, forestry, and fishing, and this percentage increased to 97.1% in 2017–2018. A large number of rural people depend on these sectors of the economy. As of 2017–2018, mining and quarrying (22.5%); manufacturing (22.7%); and transport, storage, communication, and services related to broadcasting (47.7%) also contributed a reasonable share of informal workers.

Lockdowns impacted the income of the workers engaged in all these trades and influenced poor families. Overall income losses already accumulated are significant, and workers are worried about the situation worsening. They are going to face severe everyday problems for getting food, healthcare facilities, children's education, etc. The study depicted that unprecedented lockdowns in India generated some significant insights before informal sector workers, i.e., domestic workers, street vendors, home-based workers, weavers, beedi workers, tendu leave cultivators, factory workers, workers engaged in the hospitality business, workers engaged in constructions, and agricultural workers in terms of job losses and income erosions.

9.5.2 Situation of Migrant Workers

Migrant workers constitute the backbone of the Indian economy because migration is a livelihood strategy of millions of people in India. The high-income states of Goa, Delhi, Haryana, Punjab, Maharashtra, Gujarat, and Karnataka invite workers from other states of the country. Bihar, Odisha, Uttar Pradesh, Jharkhand, and Rajasthan are the states badly affected by the COVID-19 pandemic, from where out-migration takes place frequently. Migrant workers are majorly comprised of daily wage labourers working in the manufacturing and construction industries. They do not possess adequate healthcare, nutrition, housing, and sanitation, since many of them work in the informal sector. Due to the lockdown, more than 300 deaths in India were

Table 9.2 Share of formal and informal sectors across broad sectors to GVA (in percentage)

Sectors	2011–2012		2016–2017		2017–2018	
	Organized/formal	Unorganized/informal	Organized/formal	Unorganized/informal	Organized/formal	Unorganized/informal
Agriculture, forestry, and fishing	3.2	96.8	2.8	97.2	2.9	97.1
Mining and quarrying	77.4	22.6	77.4	22.6	77.5	22.5
Manufacturing	74.5	25.5	76.4	23.6	77.3	22.7
Electricity, gas, water supply, and other utility services	95.7	4.3	95.0	5.0	94.7	5.3
Construction	23.6	76.4	26.6	73.4	25.5	74.5
Trade, repair, Accommodation, and food services	13.4	86.6	13.4	86.6	13.4	86.6
Transport, storage, communication, and services related to broadcasting	53.0	47.0	53.7	46.3	52.3	47.7
Financial services	90.7	9.3	88.1	11.9	88.1	11.9
Real estate, ownership of dwelling and professional services	36.9	63.1	46.8	53.2	47.2	52.8
Public administration and defence	100.0	0.00	100.0	0.00	100.0	0.00
Other services	58.8	41.2	52.7	47.3	52.1	47.9
TOTAL GVA at basic prices	46.1	53.9	47.3	52.7	47.6	52.4

Source Computed from National Accounts Statistics, 2019

being reported till 5 May, caused due to starvation, suicides (Elsa 2020), exhaustion,[8] road and rail accidents (Gettleman et al. 2020), police brutality (Singh 2020), and denial of timely medical care.[9]

Despite government promises and schemes to generate employment in rural areas, some migrant workers began going back to the cities due to lack of employment in their hometowns, as lockdown restrictions were reduced as part of the unlock 1.0 in June. A large number of those were returning to the hubs of interstate migrants, i.e., the Mumbai–Thane–Pune belt, Jaipur and Delhi, Coastal Andhra, and Southern Tamil Nadu. Lockdowns created huge uncertainty about how long this crisis will last and what damage it would cause to the economy, livelihood of people, and basic healthcare services. Considering its size and spread, the management of migrants under lockdowns in different phases represents a massive logistic challenge. The provision of food and basic amenities of the migrant workers, provision of basic income support to the migrants, registrar the migrants family under various Govt. schemes, provision of basic healthcare and preventive kits, counselling and psychological support to the migrants under the distress conditions, and dealing with likely economic stress in the destination areas are emerging as the matters of top priorities.

9.5.3 Impact of Lockdowns on Agriculture and Rural Lives in India

The nationwide lockdown has severely affected lives and livelihoods across rural India. Almost half of India's population depends on agriculture. The rural economy of India is indeed based on agriculture. Agriculture and allied sectors contribute more than half of the workforce in the country. In India, smallholder farmers having less than two hectares of land manage to earn around INR 6000–7000 per month. There are 120 million smallholder farmers in India, and 40% of them are engaged in grain production, and over half of its produce fruits, vegetables, oilseeds, and other crops. Farmers possessing four to ten hectares of land earn around INR 20,000 per month, whereas farmers own more than ten hectares of land earn around INR 40,000 per month. It has been estimated that around 78 million agricultural households possess less than two hectares of land with a monthly average income of INR 6500. Around 85% of farmers are under the category of small and marginal farmers who possesses less than two hectares of land. The monthly expenditure of all these agricultural households exceeds the earning during the non-cropping sessions, which compels them to take loans. Only 11 million agricultural households own more than two hectares of land, out of which 3.65 million households possess more than four hectares of land. During the lockdown period in India, less income

[8]The Economic Times reported on 5 May 2020 entitled 'Suicide leading cause for over 300 lockdown deaths in India, says study'. Retrieved on 13 May 2020.

[9]See the New Indian Express report on 'Two more gas tragedy victims die of COVID-19 in Bhopal; toll reaches seven'. Retrieved on 13 May 2020.

agricultural households started to face severe crises, starvation, and extreme poverty situation. The nationwide lockdowns created a shortage of labourers. They disrupted the production of farm inputs, i.e., fertilizers, seeds, and equipment. It stimulates an increase in price, made inaccessible these inputs to smallholder farmers and marginal farmers.

The Government of India announced several measures, including exemption of agriculture and fisheries from lockdown restrictions in late March, but there have been lacunas in the implementation at the ground level. The Ministry of Home Affairs, Government of India, on 15 April 2020, issued guidelines and exempted agriculture, horticulture, animal husbandry, poultry and fishery, and allied activities from lockdown restrictions. Labourers started to return to work and go to the market for procurement. The agri-input shops and agro-processing centres were also started to function after 15 April 2020. As a whole, the pandemic has disrupted production, declined income from agricultural sectors, increased the number of job losses, and caused income erosion of the labourers, and finally invited livelihood crisis in rural areas.

9.5.4 Lockdowns and Its Impact on Unemployment Status and GDP

The rate of unemployment climbed a peak of 27.1% in the first week of May 2020 just after the first phase of nationwide lockdown. The employment rate in the 3rd week of June was 38.4% and it was 37.8% in the last week of the same month. In the month of March 2020, the unemployment rate was 8.75% and it reached 23.5% during April and May 2020. The labour participation rate in different sectors of economic activities declined significantly as a result of lockdowns. The unemployment rate started to drop first in the first week of June under the unlock 1.0 when it dropped to 17.5%. Subsequently, it fell to 11.6% in the second week and 8.6% in the fourth week of June.[10] The noticeable drops in the unemployment rate in June under Unlock 1.0 signals the revival of economic activities in the country.

The unemployment rate of mostly COVID-19 affected states of Tamilnadu, Andhra Pradesh, Delhi, Uttar Pradesh was below around 15% since 2016–2019 (Fig. 9.4). However, it crossed 20% level in the month of April and May 2020 (Fig. 9.5). The nationwide lockdowns at several phases resulted in an increase in the unemployment rate both in the rural and urban economies in India. From August 2019 to March 2020, there was no such variation in the unemployment rate in the country. It ranged between 7.14 and 8.75%, whereas 8.27–9.71% in urban India and 5.99–8.44% in rural India. However, in April 2020 and May 2020, the unemployment rate climbed to between 23.52 and 23.48%, respectively (Table 9.3). Rural India experienced an unemployment rate between 22.89 and 22.48%, where urban

[10]For details, visit: https://unemploymentinindia.cmie.com/.

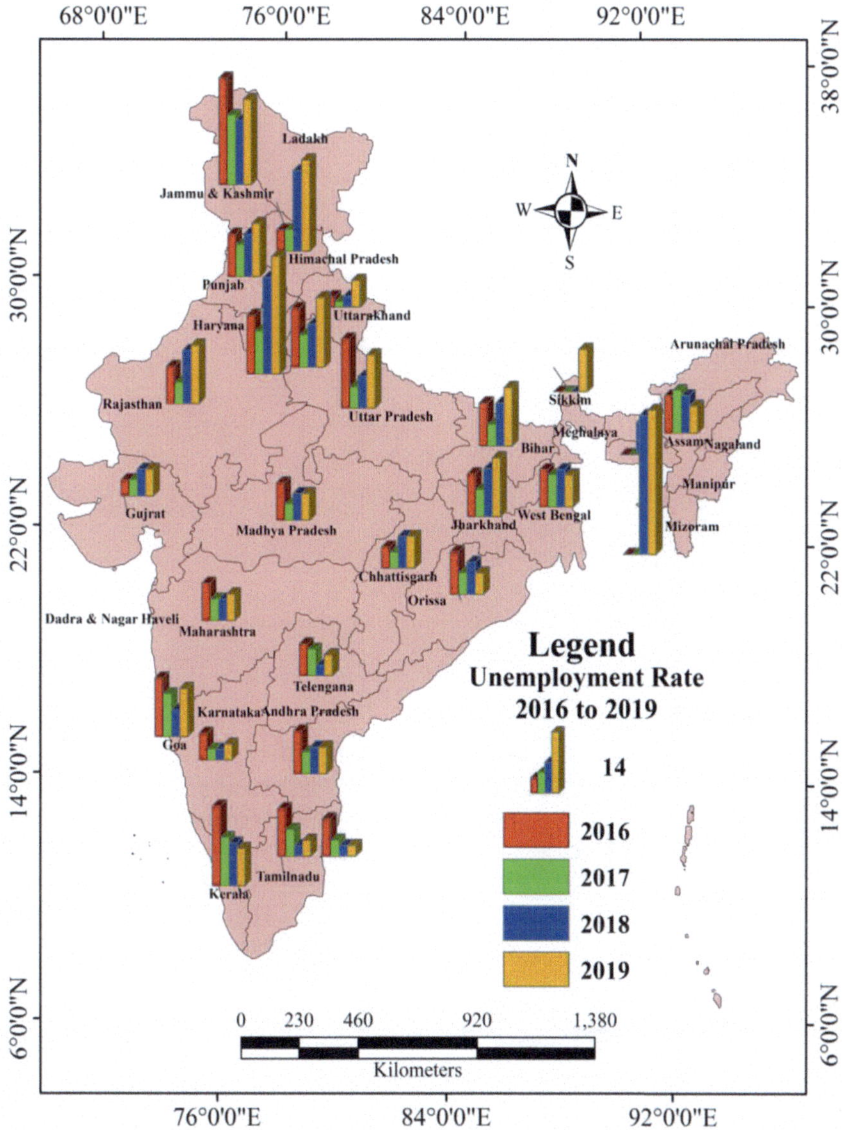

Fig. 9.4 State-wise unemployment rate for the year 2016, 2017, 2018, and 2019 (*Source* Centre for Monitoring Indian Economy CMIE)

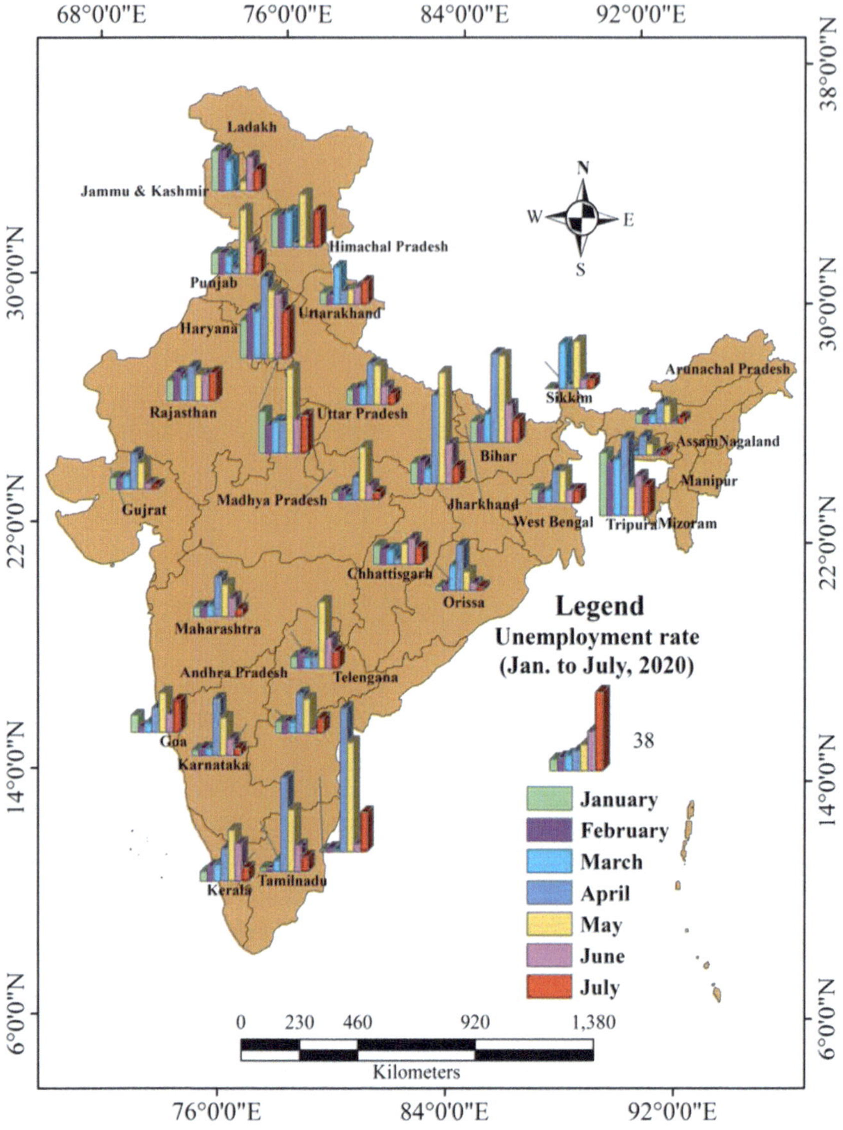

Fig. 9.5 State level unemployment rate from January 2020 to July 2020 (*Source* Centre for Monitoring Indian Economy CMIE)

India observed a slightly higher rate of unemployment—between 24.95 and 25.79%, respectively.[11]

[11]Data is retrieved from the webpage of the Centre for Monitoring Indian Economy CMIE). Visit: https://unemploymentinindia.cmie.com/.

Table 9.3 Rural–urban variation of unemployment rate (%) in India

Months	India	Urban	Rural
Aug 2019	8.19	9.71	7.48
Sep 2019	7.14	9.58	5.99
Oct 2019	8.10	8.27	8.02
Nov 2019	7.23	8.88	6.45
Dec 2019	7.60	9.02	6.93
Jan 2020	7.22	9.70	6.06
Feb 2020	7.76	8.65	7.34
Mar 2020	8.75	9.41	8.44
Apr 2020	23.52	24.95	22.89
May 2020	23.48	25.79	22.48
Jun 2020	10.99	12.02	10.52
Jul 2020	7.43	9.15	6.66

Source Centre for Monitoring Indian Economy CMIE

Due to the introduction of 'unlock 1.0' from 1 June and 'unlock 2.0' from 1 July and impositions of some relaxations in several economic activities, the unemployment rate started to resume towards the previous level. During July 2020, severely COVID-19 affected states of Maharashtra, Tamilnadu, Andhra Pradesh, Karnataka, Delhi, Uttar Pradesh, West Bengal, and Bihar witnessed unemployment rate between 4 and 12%. COVID-19 pandemic and associated lockdowns have resulted in a prolonged economic slowdown in the country. The countries GDP growth fell from 8.2% during January–March 2018 to 3.1% during the same quarter in 2020 (Rajan and Gopalan 2020). In the financial year 2020–2021, the GDP growth rate fell to 23.9% for the quarter of April–June 2020 (Misra and Iqbal 2020). Economic sectors like manufacturing, construction, trade, and hotel industry revealed a negative GDP growth. From April to June 2020, the manufacturing growth rate stood at −39.3%, the mining growth rate at −23.3%, construction growth rate at −50%, and the trade and hotel industry at −47%.[12]

9.5.5 *Lockdowns and Capital Assets*

The pandemic situation has disrupted the basic livelihood strategies of the government, i.e., sustainable use of natural resources, plan for human well-being, income behaviour, and food security of the poor people. COVID-19 induced lockdowns impacted capital assets such as human capital, natural capital, social capital, physical capital, and financial capital in India. It changed the processes such as laws, policies,

[12] Visit: https://www.firstpost.com/business/amid-covid-19-crisis-indias-gdp-contracts-by-23-9-in-april-june-quarter-8772011.html.

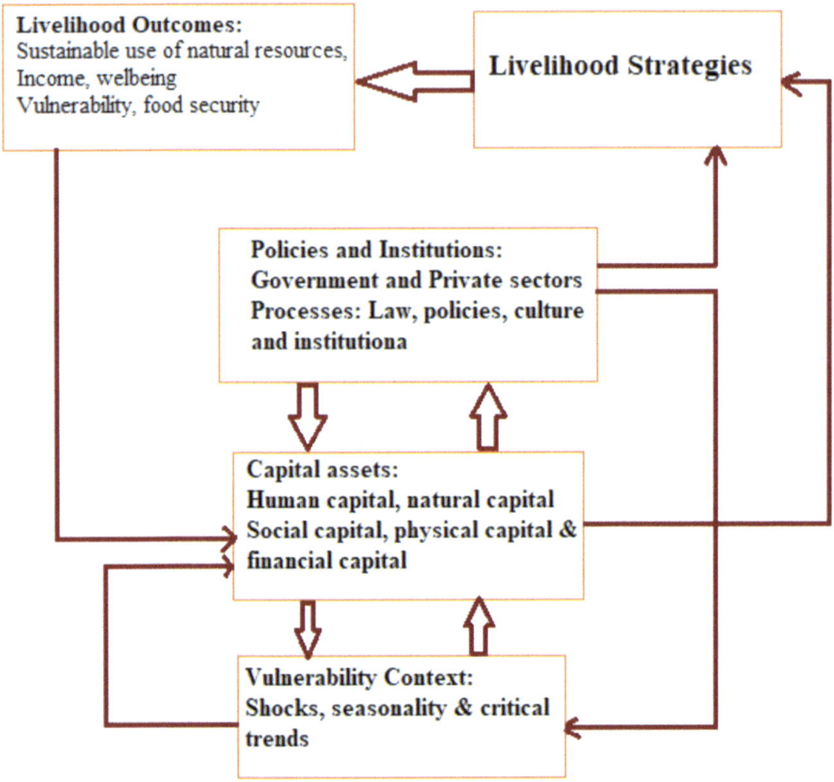

Fig. 9.6 Impact of COVID-19 induced lockdowns on livelihood, capital assets, and human vulnerability (*Source* Prepared by the author)

culture, and institutions in regard to improving the livelihood of the common people and as a result of which erosion in human capital, natural capital, social capital, physical and financial capital has been observed in India (Fig. 9.6). The loss of capital assets has invited a shock and critical trend in society.

9.6 Conclusion

The unexpected COVID-19 pandemic and its rapid expansion worldwide have disturbed the economic, social, religious, political, and financial structures worldwide. The world top most economies are on the verge of collapse. Kristalina Georgieva, the Managing Director of International Monetary Fund (IMF) explained that 'a recession at least as bad as during the Global Financial Crisis or Worse'. At present, the world has been facing a critical economic situation since World War-II. The COVID-19 pandemic is spreading rapidly and causing more economical damages worldwide. The US Official stated that American unemployment would be 30% and its economy would shrink by half. The common people will be losing their jobs, and it has invited a real threat due to the shutting down of industries, supermarkets, and businesses. The stock market of the USA is severely affected by COVID-19, and it is down about 30%. Besides, there is uncertainty and unpredictability regarding the spread of COVID-19. The Organization for Economic Cooperation and Development (OECD) stated that global growth could be cut in half to 1.5% in 2020 if the virus continues to spread. Most of the countries worldwide are passing through the recession and collapse of their economic structure, and 80 countries of the world have already requested the IMF for financial help. Bernard M. Wolf, the renowned Professor of Economics, Schulich School of Business, opines that:

> …it is catastrophic and we have never seen anything like this, we have a huge portion of the economy and people under lockdown that's going to have a huge impact on what can be produced and not produced.

The Director of IMF suggested for things to fight against COVID-19 as:

- *firstly*, continue with essential containment measures and support for the health system,
- *secondly*, shield affected people and firms with large timely targeted fiscal and financial sector measures,
- *thirdly*, reduce stress to the financial system and avoid con tangent, and
- *fourthly* must plan for recovery and must minimize the potential scaring effects of the crisis through policy action.

All the nations, apart from developed and underdeveloped, are poorly affected by COVID-19. All the countries, considering the severity of the pandemic, should extend all kind of cooperation and coordination among themselves to fight against COVID-19 to promote economic and social stability worldwide. Any wrong decisions and policies to combat COVID-19 may lead to a severe impact on other countries' economies as well.

The long run of this crisis does not affect only the workers in the informal sector, but also it looms large for the formal sectors as well. Initially, some states like Bihar, Odisha, Jharkhand, Chhattisgarh, Madhya Pradesh, Assam, Manipur, Arunachal, and West Bengal have a greater likelihood to face livelihood challenges by this pandemic as these states are preconditioned with a 'moderate' to 'high' poverty level. The states like Uttarpradesh, Bihar, Andhrapradesh, Rajasthan, Madhyapradesh, Assam, and West Bengal are the states where workers in the 'below14 years' age group range from 1 lakh to 1 million will face the formidable challenges. If the crisis situation continues, a large number of people in India (mainly with low per capita income) will suffer from extreme poverty, starvation, which is coming to be a severe challenge to the economic and social system of the country. Unfortunately, we could not organize informal sector workers and migrant workers, who are the backbone and wealth of a nation for the well-being and development—both in rural and urban areas. Now, the Lives and livelihood of the poor people are under significant threat.

References

Coronavirus Study Group of the International Committee on Taxonomy of Viruses (2020) The species severe acute respiratory syndrome-related coronavirus: classifying 2019-nCoV and naming it SARS-CoV-2. Nat Microbiol 5:536–544. DOIExternal LinkPubMedExternal Link

Elsa E (15 April 2020) The human cost of India's coronavirus lockdown: deaths by hunger, starvation, suicide and more. Gulf News. https://gulfnews.com/world/asia/india/the-human-cost-of-indias-coronavirus-lockdown-deaths-by-hunger-starvation-suicide-and-more-1.1586956637547. Accessed 13 May 2020

European Centre for Disease Prevention and Control (ECDC) Risk assessment: outbreak of acute respiratory syndrome associated with a novel coronavirus, Wuhan, China; third update 2020. Available from www.ecdc.europa.eu/en/publications-data/risk-assessment-outbreak-acute-respiratory-syndrome-associated-novel-1

Gettleman J, Raj S, Kumar H (8 May 2020). As India reopens, deadly accidents break out. The New York Times. ISSN 0362-4331. https://www.nytimes.com/2020/05/08/world/asia/india-leak-train-reopen.html. Accessed 13 May 2020

Gettleman J, Schultz K (24 March 2020). Modi orders 3-week total lockdown for all 1.3 billion Indians. The New York Times. ISSN 0362-4331

Gorbalenya AE, Baker SC, Baric RS, de Groot RJ, Drosten C, Gulyaeva AA et al (11 February 2020) Severe acute respiratory syndrome-related coronavirus: the species and its vi-ruses—a statement of the Coronavirus Study Group. Bio Rxiv. https://doi.org/10.1101/2020.02.07.937862

Gupta S (18 April 2020) Covid hasn't gone viral in India yet, but some in the world & at home can't accept the truth. ThePrint. Accessed 2 May 2020

Huang C, Wang Y, Li X, Ren L, Zhao J, Hu Y et al. (2020) Clinical features of patients infected with 2019 novel coronavirus in Wuhan, China. Lancet 395:497–506. https://doi.org/10.1016/S0140-6736(20)30183-5

Hu D, Zhu C, Ai L et al (2018) Genomic characterization and infectivity of a novel SARS-like coronavirus in Chinese bats. Emerge Microbes Infect 7:154

India's coronavirus lockdown reveals fresh air, cleaner rivers (22 April 2020) Living Media India Ltd. Accessed 22 Apr 2020

Lam TT, Shum MH, Zhu HC, Tong YG, Ni XB, Liao YS et al (2020) Identifying SARS-CoV-2 related coronaviruses in Malayan pangolins. Nature. Epub ahead of print. DOIExternal LinkPubMedExternal Link

Lai MY, Cheng PK, Lim WW (2005) Survival of severe acute respiratory syndrome coronavirus. Clin Infect Dis 41(7):e67–71

Lau SKP, Li KS, Huang Y, Shek CT, Tse H, Wang M et al (2010) Ecoepidemiology and complete genome comparison of different strains of severe acute respiratory syndrome-related Rhinolophus bat coronavirus in China reveal bats as a reservoir for acute, self-limiting infection that allows recombination events. J Virol 84:2808–2819. DOIExternal LinkPubMedExternal Link

Lau SKP, Luk HKH, Wong ACP, Li KSM, Zhu L, He Z et al (2020) Possible bat origin of severe acute respiratory syndrome coronavirus 2. Emerg Infect Dis. 2020 July [date cited]. https://doi.org/10.3201/eid2607.200092, https://doi.org/10.3201/eid2607.200092

Lockdown makes Ganga water significantly cleaner (4 April 2020). LiveMint. Accessed 22 Apr 2020

Luk HKH, Li X, Fung J, Lau SKP, Woo PCY (2019) Molecular epidemiology, evolution and phylogeny of SARS coronavirus. Infect Genet Evol 71:21–30. DOIExternal LinkPubMedExternal Link

Misra U, Iqbal N (1 September 2020). India's GDP growth contracts 23.9%: what is the economics behind the math? The Indian Express. Accessed 1 Sept 2020

Otter JA, Donskey C, Yezli S, Douthwaite S, Goldenberg SD, Weber DJ (2016) Transmission of SARS and MERS coronaviruses and influenza virus in healthcare settings: the possible role of dry surface contamination. J Hosp Infect 92(3):235–250

Phan LT, Nguyen TV, Luong QC, Nguyen TV, Nguyen HT, Le HQ et al (2020) Im-portation and human-to-human transmission of a novel coronavirus in Vietnam. N Engl J Med 382(9):872–874. https://doi.org/10.1056/NEJMc2001272
Rajan R, Gopalan S (30 March 2020) Covid-19: another blow to India's economy. The Hindu @businessline. https://www.thehindubusinessline.com/opinion/covid-19-another-blow-to-indias-economy/article31208648.ece. Accessed 12 June 2020
Riou J, Althaus CL (2020) Pattern of early human-to-human transmission of Wuhan 2019 novel coronavirus (2019-nCoV), December 2019 to January 2020. Euro Surveill 25(4)
Singh A (20 April 2020). Farmer beaten by cops amid lockdown dies in Madhya Pradesh, six officers suspended, SP transferred. The New Indian Express. Accessed 13 May 2020
Singh KD, Goel V, Kumar H, Gettleman J (25 March 2020) India, day 1: world's largest Coronavirus lockdown begins. The New York Times. ISSN 0362-4331
Thacker T (1 May 2020) Centre issues state-wise division of Covid-19 red, orange & green zones. The Economic Times
van Doremalen N, Bushmaker T, Munster VJ (2013) Stability of Middle East respiratory syndrome coronavirus (MERS-CoV) under different environmental conditions. Euro Surveill 18(38)
World Health Organization (WHO) WHO guidelines on hand hygiene in health care: a summary 2009. Available from: www.who.int/gpsc/5may/tools/who_guidelines-handhygiene_summary.pdf
World Health Organization (WHO) Non-pharmaceutical public health measures for mitigating the risk and impact of epidemic and pandemic influenza. Licence: CC BY-NC-SA 3.0 IGO. 2019. Available from https://apps.who.int/iris/bitstream/handle/10665/329438/9789241516839-eng.pdf?ua=1
World Health Organization (2020) Novel coronavirus. Situation report—4 (2019-nCoV) 24 January 2020. Available at https://www.who.int/docs/defaultsource/coronaviruse/situation-reports/20200124-sitrep-4-2019-ncov.pdf?sfvrsn=9272d086_2. Accessed 24 Jan 2020
Yoo JH (2020) The fight against the 2019-nCoV outbreak: an arduous march has just begun. J Korean Med Sci 35–56. https://doi.org/10.3346/jkms.2020.35.56.7I. Chakraborty PM Science of the Total Environment 728 (2020) 138882
Zhou P, Yang XL, Wang XG, Hu B, Zhang L, Zhang W et al (2020) A pneumonia outbreak associated with a new coronavirus of probable bat origin. Nature 579:270–273. DOIExternal LinkPubMedExternal Link

External Links

https://www.who.int/docs/default-source/coronaviruse/situation-reports/20200121-sitrep-1-2019-ncov.pdf?sfvrsn=20a99c10_4. Accessed 20 Jan 2020
https://www.who.int/docs/default-source/coronaviruse/situation-reports/20200123-sitrep-3-2019-ncov.pdf?sfvrsn=d6d23643_8. Accessed 23 Jan 2020
https://www.who.int/docs/default-source/coronaviruse/situation-reports/20200315-sitrep-55-covid-19.pdf?sfvrsn=33daa5cb_8. Accessed 15 Mar 2020
https://www.who.int/docs/default-source/coronaviruse/situation-reports/20200310-sitrep-50-covid-19.pdf?sfvrsn=55e904fb_2. Accessed 10 Mar 2020
https://www.who.int/docs/default-source/coronaviruse/situation-reports/20200320-sitrep-60-covid-19.pdf?sfvrsn=d2bb4f1f_2. Accessed 20 Mar 2020
https://www.who.int/docs/default-source/coronaviruse/situation-reports/20200325-sitrep-65-covid-19.pdf?sfvrsn=ce13061b_2. Accessed 25 Mar 2020
https://www.who.int/docs/default-source/coronaviruse/situation-reports/20200331-sitrep-71-covid-19.pdf?sfvrsn=4360e92b_8. Accessed 31 Mar 2020

https://www.who.int/docs/default-source/coronaviruse/situation-reports/20200410-sitrep-81-covid-19.pdf?sfvrsn=ca96eb84_2. Accessed 10 Apr 2020
https://www.who.int/docs/default-source/coronaviruse/situation-reports/20200420-sitrep-91-covid-19.pdf?sfvrsn=fcf0670b_4. Accessed 20 Apr 2020
https://www.who.int/docs/default-source/coronaviruse/situation-reports/20200430-sitrep-101-covid-19.pdf?sfvrsn=2ba4e093_2. Accessed 30 Apr 2020
https://www.who.int/docs/default-source/coronaviruse/situation-reports/20200505covid-19-sitrep-106.pdf?sfvrsn=47090f63_2. Accessed 5 May 2020
https://www.who.int/docs/default-source/coronaviruse/situation-reports/20200508covid-19-sitrep-109.pdf?sfvrsn=68f2c632_6. Accessed 8 May 2020
https://www.who.int/docs/default-source/coronaviruse/situation-reports/20200510covid-19-sitrep-111.pdf?sfvrsn=1896976f_2. Accessed 10 May 2020

Sujit Mandal has been serving as a Professor of Geography and Dean of Faculty of Science at Diamond Harbour Women's University in India. He is the former Professor of the University of Gour Banga in India. He has been serving in higher education institutions since 2005. Dr. Mondal has been awarded a Ph.D. from Vidyasagar University. He is specialized in applied geomorphology, environmental hazards, and geoinformatics. He is the author of 70 research articles published in reputed national and international journals. Besides, he has authored three books, published by Springer Nature. He is the Principal Investigators of several research projects funded by UGC and ICSSR; and he is acting as the supervisor for the doctoral program. He is also associated with several academic and administrative bodies in different Universities of India.

Chapter 10
The Contagion Effects of COVID-19 and Public Transportation System: Conceptualizing the Shifting Paradigm in India

Biswajit Paul and Subir Sarkar

Abstract The global outbreak of coronavirus disease 2019 or COVID-19 and the subsequent policies of social distancing and the nationwide lockdown have changed the whole scenario of the world's public transportation system. It has affected the transportation sector in multiple ways. The virus' outbreak has instantly affected the traveling behavior of the people, which has ultimately resulted in complete stagnation of the transport sector worldwide. As most economic activities are directly or indirectly associated with the transportation system, transport-related activities are also severely affected. Thus, it has become a necessary consideration for economists and geographers to understand the overall scenario. This paper explores the impact of COVID-19 on the Indian public transportation system during and post lockdown period using some statistical facts and figures. Analyzes have been done based on the primary and secondary sources of information collected from the primary survey, different news reports, and published research articles. The study reveals a massive reduction in demand for mass transportation modes to get around the fear of spreading the virus and lockdown policy.

Keywords Coronavirus disease · Pandemic · Lockdown · Public transportation · FTA

10.1 Introduction

The transportation system, especially public transportation, is considered the vein of the economy because all the other economic sectors are directly or indirectly associated with it. It provides mobility and access to essential and recreational opportunities. India boasts of having the largest and most diverse transport sectors in the

B. Paul (✉) · S. Sarkar
Department of Geography and Applied Geography, University of North Bengal, Darjeeling, West Bengal, India
e-mail: biswajit.geo@nbu.ac.in

M. Mishra and R. B. Singh, *COVID-19 Pandemic Trajectory in the Developing World*, Advances in Geographical and Environmental Sciences,
https://doi.org/10.1007/978-981-33-6440-0_10

world, which caters to the needs of nearly 1.3 billion people.[1,2] In 2016–2017, the total contribution of the transport sector together with storage, communication, & services related to broadcasting was about 6.76% to India's gross value added (GVA) at the current price. The majority of this is shared by road transportation, with 3.14% of the total share followed by the railways (0.77%), airways (0.15%), and waterways (0.05%).[3] Ministry of Statistics and Program Implementation (GoI) has estimated the growth in GVA (2018–2019) for the Trade, Hotels, Transport, and Communication and Services related to Broadcasting services at 6.9% as against a growth of 8.0% in the previous year (2017–2018).[4] A large proportion of the public assistance recipients do not own a car and rely on public mobility services. Besides, the tourism industry is mostly dependent on public transportation services. The total contribution from travel and tourism to India's economy was 247.37 billion US dollars during 2018. In 2017, India had more than 10 million international tourists generating over 27.31 billion US dollars of foreign exchange earnings.[5] However, the scenario has changed due to the outbreak of coronavirus disease 2019 or COVID-19 or SARS-CoV-2 worldwide (Singh and Neog 2020). Currently, public transportation services are likely to undergo significant changes during these challenging times (TERI 2020). The impetuous expansion of coronavirus disease 2019 has caused a complete halt in the people's daily routines and the extreme cases, taking many lives. The first confirmed case of this newly discovered disease was traced in the Wuhan city of Hubei Province in China on December 31, 2019, and within a concise period, it has extended in other parts of the world (Cui et al. 2019; Lai et al. 2020; Shereen et al. 2020; World Health Organization 2020c). On January 30, 2020, the World Health Organization (WHO) has announced it as a global public health emergency. At the date of March 11, 2020, coronavirus disease caused 80955 infections in China, with 3162 deaths and 43010 infections with 1518 number of deaths outside China (World Health Organization 2020b). On the same date, the novel coronavirus has been proclaimed as a pandemic by WHO. In India, the first case of coronavirus disease 2019 was traced at Kerala's Thrissur district on January 30, 2020.

Interestingly, this is not the first coronavirus-associated epidemic, emerged by a species leap from wild animals to humans. Prior to this pandemic, there was Severe Acute Respiratory Syndrome coronavirus (SARS-CoV) in 2003, and the Middle East Respiratory Syndrome coronavirus (MERS-CoV) in 2012 (Azhar et al. 2019; Cui et al. 2019; Lai et al. 2020; Shereen et al. 2020). SARS-CoV-2 often causes a respiratory disease similar to SARS and MERS, ranging from mild upper respiratory illness to severe interstitial pneumonia, also requiring intensive care (Chen et al.

[1] According to the World Bank, see: https://www.worldbank.org/en/news/feature/2011/09/23/india-transportation.

[2] 2019 Population estimation by the World Bank, see: https://data.worldbank.org/indicator/SP.POP.TOTL?locations=IN.

[3] For detailed statistical tables, see: http://statisticstimes.com/economy/sectorwise-gdp-contribution-of-india.php.

[4] Fore more details, see: http://mospi.nic.in/sites/default/files/press_release/Presss%20note%20for%20first%20advance%20estimates%202018-19.pdf.

[5] For detailed statistical tables, see: www.statista.com.

2020). COVID-19 is poised to become one of the most severe factors hindering public health and the economic crisis world is facing in the twenty-first century.

Along with the sweeping loss of lives and labor productivity, the virus is likely to impact various sectors and human lifestyle over a long period. The combination of the policies regarding social distancing and lockdown is expected to be down, or even a complete stop of production and consumption activities for an uncertain time and cause business in the hospitality and retail sectors to close (Chauhan 2020). Public transportation systems, including shared mobility services like Taxi, Toto, Ola, and Uber, are severely affected by the norms of social distancing. The present study tries to explore the impact of coronavirus disease on the Indian public transportation system during and post lockdown period.

First section of the chapter introduces the concepts of the public transportation system and its vulnerability to COVID-19. The second section deals with the behavioral changes of the citizen, their changing preferences, and the effects of coronavirus disease 2019 on different modes of transportation sectors during and post lockdown periods in India. Subsequently, government initiatives and guidelines toward the public transportation system have been analyzed.

The study is primarily based on the secondary sources of information collected from the World Health Organization, Ministry of Health and Family Welfare (India), International Association of Public Transport (UITP), covid19india.org, and various news reports and published research articles. The primary sources of information are also used for the study, collected through a pre-structured Google form. A total of 212 participants have been surveyed from different parts of India. Different social media platforms and electronic mail have been used to reach the participants of various age cohorts and professions in both the rural and urban areas.

10.2 Public Transportation Under the "*New*" Threat

Public transportation services are one of the worst-affected economic sectors by the coronavirus disease 2019 or COVID-19. The sector has experienced a demand-cum-supply shock. There is an unprecedented reduction in travel and commercial activities. Consequently, the impacts have surged over transport activities more severely. In India, the Gross Value Addition (GVA) from the transport sector accounts for 4.8%, where a significant proportion is covered by road transportation (3.1%) followed by the railways (0.8%), airways (0.2%), and waterways (0.1%). Besides these, the motor transport records more than 1.3 million employment in 2017 (8.7% of the total industrial employment) (Kumar and Sankar 2020).

The transportation system availed by the general public based on requisite fare and ply with scheduled times is called the Public Transportation System (PTS), which includes both the freight and mass transportation services (Robert Wood Johnson Foundation 2017). The present chapter mainly focuses on mass or passenger transportation as it primarily depends on the demand of the people, unlike the freight transportation system, which is dependent on the movement of the essential goods and commodities. The rate of virus transmission is less in freight transportation compared to mass transportation because mass transportations are directly associated with the carriage of people between the places. According to WHO, the coronavirus spreads primarily through direct contact or saliva droplets from an infected person while sneezing, talking, singing, or coughing (World Health Organization 2020d).

The higher risk of disease transmission in PTS is sourced by its nature of accommodating a large number of people in a confined space with limited ventilation. There is no infrastructural access to control or identify potentially sick persons and a variety of common surfaces to touch, such as money, tickets, handrails, windows, seats, and doorknobs (I & I Editorial Board 2020). WHO, together with the scientific community, has declared that there is a possibility of coronavirus transmission through aerosols in the absence of aerosol-generating procedures, particularly in an indoor setting with poor airspace (World Health Organization 2020a, d). Public transport is an essential service to provide mobility, also in times of pandemics, not least to provide access to health care facilities (Leigh 2020). The key objective for public transport operators, therefore, has to be continuing the operation. Consequently, public transport operators should focus their pandemic plan efforts on staff, trying to protect them, and preparing to deal with absenteeism (UITP 2020).

The exact span and severity of the impact of the present pandemic on the economy are still unknown. However, Evan and Over (2020) has provided some potential economic channels of COVID-19 for the low and middle-income group of countries. The "aversion behaviour"[6] of the citizen during an epidemic or pandemic is likely to affect most on the economic condition, and this behavior comes from the preferences and choices of the individuals for travel, going out, and other social activities (Fig. 10.1). To avoid the chance of virus infection, people have no alternative but to try minimize the outing (as the vaccine or the medicinal support is still unknown). Government mandates (lockdown and social distancing) and institutional protective decisions to avoid the infection also result in the shortfall in revenue collection of every sector, including trade and transportation.

10.3 The Pandemic Trapped Indian PTS

The negative impact of coronavirus outbreak exhibits its severe face in developing economies like India by disrupting almost all existing human-centric sectors. The transportation sector is not an exception. The transportation system in India has long been struggling with insufficiencies such as high congestion level, limited multimodal integration, inadequate public transportation system, degraded footpaths, an exploding growth in the number of private vehicles (especially motorcycles), and the non-existence of dedicated cycle tracks. All of these together have impacted

[6] Aversion behavior is behavior of people that comes with a feeling of dislike or repugnance toward something with a desire to avoid or turn from it. See the meaning of aversion: https://www.merriam-webster.com/.

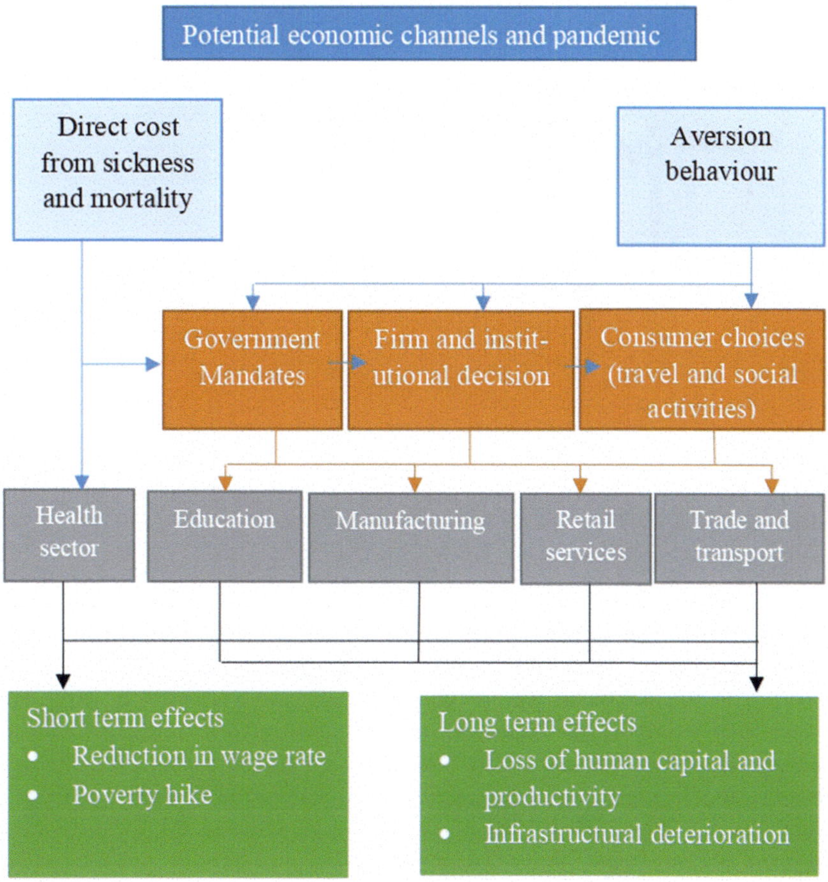

Fig. 10.1 Broad channels of pandemic impact on the economy (Drawn by the author based on Evan and Over 2020)

the natural environment in the long term, causing a significant increase in carbon emission and noise pollution levels being the most rampant. The imposition of the lockdown has allowed the natural environment to revive. Besides, the loss of human resources to accidents and mishaps have reduced drastically (Fig. 10.2). The revenue, however, from the transport sector and the economic dependency on the sector has been significantly compromised. The government in the pre-pandemic world had been working and trying to improve the condition in order to implement different developmental measures to make the road transportation system more sustainable. However, the sudden disruption is likely to act as a hindrance to the existing inclusive policies concerning PTS. Also, the coronavirus outbreak has created new challenges for this sector, especially in the cities, where high travel demand is prevailed (Shakti Foundation 2016; TERI 2020).

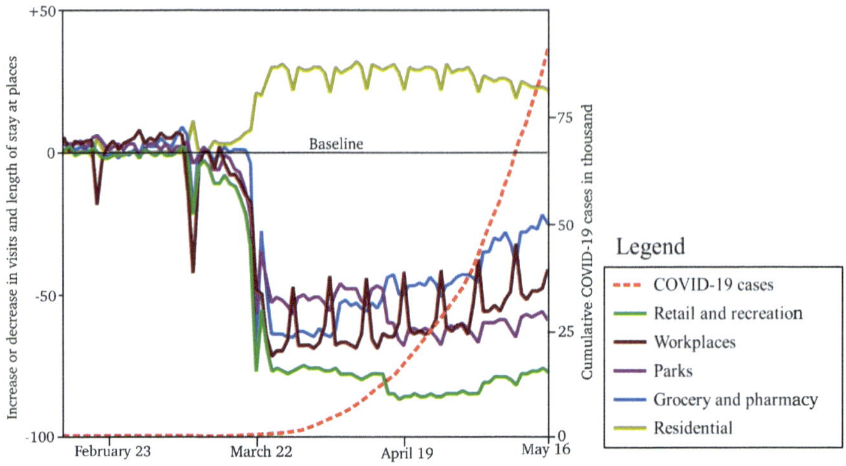

Fig. 10.2 Mobility Trends in India during coronavirus 2019 (Drawn by the authors following Radhakrishnan et al. 2020)

The mobility behavior of the population is dynamic, and its magnitude and pattern are subject to change in response to socioeconomic security. Figure 10.2 shows the changing trend in visits and length of stay at places during the pre-lockdown, lockdown, and partial lockdown period from February to May 2020 in India. The baseline is taken as the median value for the corresponding day of the week during the five weeks (January 3–February 6, 2020). Prior to the removal of complete lockdown, certain sectors had already started work so, and there were relaxations in specific essential sectors. The graph makes it clear that the visit to the workplaces, grocery, and pharmacy increases. The visit to transit stations had yet not shown a sudden increase as people were availing private mode of transport, visiting the retail and recreation spaces; however, were visited the least as the relaxations to reopen these spaces have yet to be given by the government.

10.3.1 Effect During the Lockdown Period

India earns considerable revenue from the major transportation modes (Table 10.1). Coronavirus 2019 outbreak, imposed lockdown, and “stay at home approach” has directly affected the supply and demand in the economy (Singh and Neog 2020). Disruption in the transport system is put to a standstill as the demand in the service sector is crucial for the Micro, Small, and Medium Enterprises (MSMEs), which are more fragile to market demand (Bouey 2020). The total exports including reexports in India during the month of July 2020 were 23.64 billion US$. As compared to July 2019, the overall exports in July 2020 registering a negative growth of −10.21%. In terms of Indian Rupee, the total exports in July 2019 were Rs. 181190.34 crore,

Table 10.1 Gross revenue trends over the years in India

Years	Railways (US$ billion)	Roadways (Toll fees) (Million rupees)	Airways[a] (Million rupees)
FY17	25.02	69379.2	25094.2
FY18	24.64	86306.9	29834.9
FY19	25.56	91876.7	27648.5
FY20 (P)	24.78	NA	NA

NA: Data not available. (P): Provisional
[a]Revenue earned from international airlines and flights passing through the Indian air zone
Data Source Singh and Neog (2020), www.ibef.org and www.indiastat.com

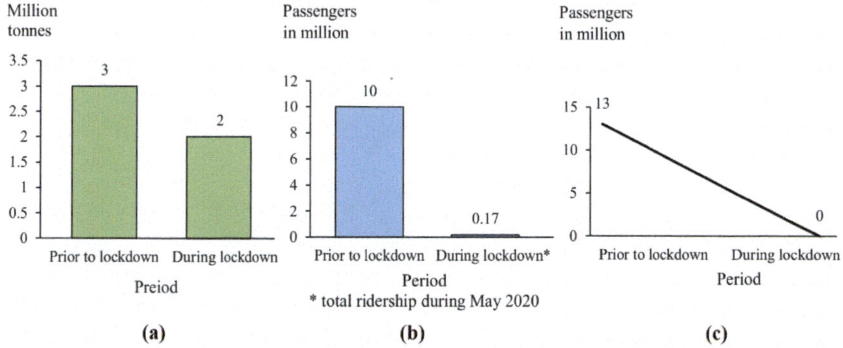

Fig. 10.3 **a** Daily freight flow; **b** daily passenger non-suburban (intercity) special migrant rail ridership, and **c** daily passenger suburban rail ridership before and during the lockdown in India (Drawn by the authors following Kumar and Sankar 2020)

whereas in July 2020 the figure came down to Rs. 177305.79 crore, registering a negative growth of −2.14%. A negative growth rate of −28.40% is also found in case of total imports in India during July 2020 (28.47 billion US$), as compared to July 2019 (39.76 billion US$).[7]

In India, passengers and cargo movement during the lockdown period was less than half of the pre-coronavirus level. The number of people availing the public road transportation via buses per day is also decreased to 4 million (during May 2020) during the lockdown period compared to the pre-lockdown period's figure (i.e., 70 million) (Kumar and Sankar 2020). Figure 10.3a shows the declining trend of freight flow from the pre-lockdown period to the lockdown period in India. The total volume daily freight flow has been decreased by 1 million tons from the pre-lockdown period to lockdown. In the case of daily passenger volume in both the suburban (Fig. 10.3c) and

[7]Press release by the Ministry of Commerce and Industry, Department of Commerce, Economic Division (GoI) on 14th August 2020. See: https://commerce.gov.in/writereaddata/UploadedFile/NTESCL_637330240112020361_Press_Release_July_2020.pdf.

the non-suburban (intercity) special migrant rail ridership (Fig. 10.3b), a decreasing trend has been found. The satellite view of the national highway of Delhi near the India gate, for instance, shows a drastic reduction of vehicular movement from the pre-lockdown period to the lockdown period (Fig. 10.4).

Mishra (2020) has estimated the sector-wise expected loss in gross value added (GVA), published in *Business Today (2020)*. According to the estimation (Table 10.2), there is a possibility of 8.76 lakh crore rupees loss in GVA. Nevertheless, the trade,

Fig. 10.4 Satellite view of Delhi near India Gate. **a** Before lockdown December 7, 2019 and **b** during lockdown March 28, 2020 (*Source* Google Earth)

Table 10.2 Estimated loss in gross value added (sector-wise) due to COVID-19 outbreak

Sectors	Disruption severity	Loss estimated in Rs. Lakh Cr.
Mining and quarrying	Complete	0.31
Manufacturing	Near-complete excluding medicine	2.16
Construction	Complete	1.06
Trade, hotel, transport, communication, and relating broadcasting	Near-complete excluding broadcasting	2.42
Financial, real estate, and professional services	Near-complete excluding banking and healthcare	2.81
Total		8.76

Data source Mishra (2020)

hotel, and transport sector together could register 2.42 lakh crore rupees loss due to COVID-19, which could augment the total loss to 27.63% (Mishra 2020).

10.3.1.1 The Rail Transport System at Stake

India boasts for the world's largest rail network extended over 123236 km, with 9141 freight trains and 13452 passenger trains transporting 23 million travelers daily from 7349 stations.[8] The rail transport network of India is also recognized as one of the largest single management railway systems in the world. A major proportion of the Indian economy and industrial sector is directly dependent on the railway network because of its vast network and connectivity apart from being an energy-efficient mode of transport. Before the pandemic situation, train travel remained the prime preference for the majority of the people in India while traveling a longer distance. The outbreak of coronavirus disease in 2019 has severely affected the rail transportation system, causing a rapid decline in passenger volume from 8.44 billion in FY 2019 to 8.1 billion in FY 20 at the end of March (Fig. 10.5). After the nationwide lockdown, announced in the country, almost all the passenger train services have been suspended for an indefinite time, which ultimately results in the decrease in revenue collection from the passenger segment for the upcoming financial year. The railways' projections estimated that the lockdown might result in a drastic fall in the estimated total earnings up to 1.48 lakh crore rupees for FY2020–21, increasing the earnings gap by 63,000 crore rupees. If it is added with the earlier resource gap of over 28,000 crore rupees for FY 2019–2020, the national transporter is looking at a deficit of a whopping 90,000 crore rupees (Sharma 2020).

[8]See the website of India Brand Equity Foundation, available under the URL: www.ibef.org.

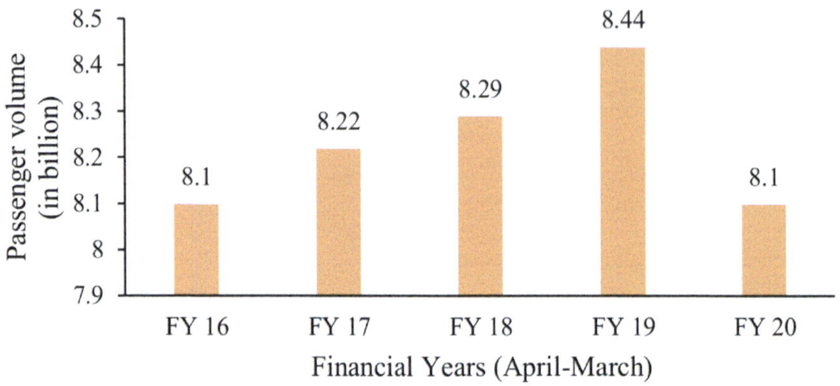

Fig. 10.5 Trend of passenger volume in Indian railways (as of April 2015–March 2020) (Drawn by the authors with the d*ata from the* Ministry of Railways, Govt. of India)

10.3.1.2 Air Transport in a Fix

The aviation industry has emerged as one of the fastest-growing industries in India since the initial years of the last decade. The International Air Transport Association (IATA) predicts that India is expected to be the third-largest air passenger (international and domestic) market in the world by 2024.[9] However, the situation has already changed with the coronavirus 2019 outbreak. The IATA forecast is likely to be incomplete, and it will take much more time, reaching the third position in the air passenger market. During the lockdown period, all the passenger carriages had stopped flying. Only the "Lifeline Udan" flights were operated by the Ministry of Civil Aviation (GoI) to transport essential medical cargo to remote parts of the country to support India's fight against COVID-19. As a result, the Indian aviation industry is expected to suffer 24 thousand crore rupees' revenue loss due to the pandemic (Manju 2020).

10.3.1.3 The Bottleneck of the Road Transport System

Similarly, public road transport services have been limited to essential services only as a part of the outbreak strategies. The ban on operations during lockdown has caused significant financial losses to operators. In a study by UITP and World Bank Group (2020) it is predicted that, when services resume, demand may not return to pre-lockdown levels.

The report says that around 20 lakh people have lost their employment. The Bus & Car Operators Confederation of India worries that approximately equal numbers are on the verge of losing their jobs as private bus and tourist taxi operators have seized the services due to the lockdown (The Economic Times 2020a).

[9]See the website of India Brand Equity Foundation, available under the URL: www.ibef.org.

National Highway Authority India (NHAI) had suspended toll collection on all national highways as the 21 days nationwide lockdown was announced, which results in a massive impact on the revenue collection from the road transport sector (Livement 2020). The 57-day nationwide lockdown due to COVID-19 resulted in a sharp 13% downfall in toll collections and remittances, according to CRISIL Research (The Economic Times 2020b). As per the International Credit Rating Agency (ICRA) estimation, toll collections would fall to 6.5–8.0% in FY21, and there would be a significant impact on those industries and firms where employees are dependent on the revenue from public transportation (Businessline 2020).

10.3.1.4 The Gasping Tourism Sector

The transportation sector is not a standalone industry; it possesses many direct impacts on the other allied activities. The travel and tourism industry is one of the major industries, having significant concerns with the public transportation system. The coronavirus outbreak has also severely affected the tourism sector in India. The arrivals of foreign tourists in the country have reduced to below two million in 2020 till April (Fig. 10.6a). The statistics show an increase over a span of 12 years; if not for the virus, the graph would have a positive growth during the latest year.

According to the Ministry of Tourism (Govt. of India), there was around 25% growth in the total domestic tourist, whereas 8.9% growth in foreign tourist arrival during 2018 and 2019 (Table 10.3). However, this figure has definitely receded (Fig. 10.6a). A remarkable negative growth rate (−22.65) is found of Foreign Tourist Arrival (FTA) on e-Tourist Visa from January to March 2020. Only for March, the growth rate remains the lowest in the last few decades, i.e., −66.4% (FTAs in March 2020 counts 3,28,462 as compared to 9,78,236 in March 2019) (Table 10.4). There is an almost identical scenario regarding the foreign exchange earnings (FEE) from tourists in India during 2019 and 2020. There is a sharp decrease in FEEs during March 2020 (Rs. 5,833 crores) as compared to March 2019 (Rs. 16,214 crores). The growth rate was −15.6% in FEE for January to March 2020 and −64.0% for March 2019 to March 2020 (Table 10.5).

The FTA has suddenly witnessed a decline that results in a drastic lowering of the FEE from 29.96 million US $ in 2019 to 6.15 million US $ till April 2020 (Fig. 10.6b). This trend has made it clear that forthcoming months in the current financial year have a greater likelihood to experience more decline due to the ongoing situation.

10.3.2 The Post Lockdown Scenario

The second phase of unlocking has been initiated by the Government of India, with effect from July 1, 2020. However, the challenges are persistent to contain the spread of coronavirus disease. In May, the World Bank Group and UITP, an International Association of public transport, have conducted a survey among the private and

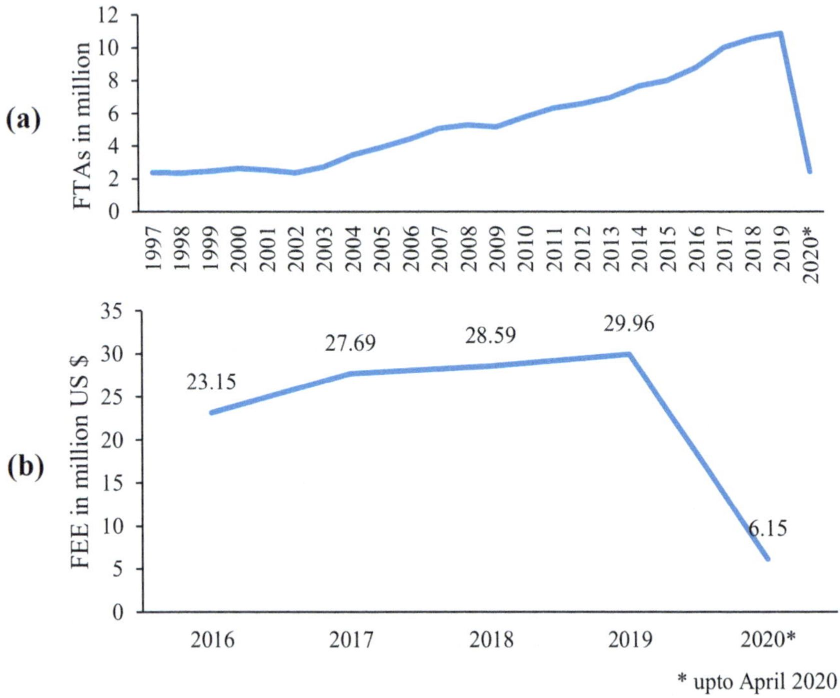

Fig. 10.6 **a** Trend of Foreign Tourist Arrival (FTAs) (*Data source* Ministry of Tourism, Govt. of India 2020) and **b** Foreign Exchange Earnings (FEE) from tourists (*Data source* Bureau of Immigration, Govt. of India 2020)

Table 10.3 Foreign tourist arrival and number of domestic tourist (2018–2019)

2018 (R)		2019 (P)		Growth rate	
Domestic	Foreign	Domestic	Foreign	DTV 2018–19	FTV 2018–19
1853787719	28851130	2321982663	31408666	25.3%	8.9%

P: Provisional, R: Revised: Data for 2017 is repeated for 2018, and 2019 data is estimated by applying all India growth rate for 2019/18 on 2018 data (*Data source* State/Union Territory Tourism Departments, Govt. of India)

Table 10.4 FTAs on e-Tourist Visa during March 2020

2019		2020		Growth rate
January to March	3179792	January to March	2462244	−22.6%
March	978236	March	328462	−66.4

Data source Ministry of Tourism, Govt. of India, 2020

Table 10.5 Foreign Exchange Earnings (FEEs) from tourism in rupees terms

2019 (in crore rupees)		2020 (in crore rupees)		Growth rate (%)
January to March	52378	January to March	44203	−15.6
March	16214	March	5833	−64.0

Data source Ministry of Tourism, Govt. of India, 2020

public bus service operators from India. Their report shows that over 60% of PTS operators believe that the service and demand will not be more than 50% of the pre-lockdown period. While 12% believe that the service levels will resume to 75–100%, keeping only four percent as very optimistic to believe that that the demand will return to what it was before. Seventy-eight percent expect that it will take more than six months from the end of lockdown for the demand levels for buses to reach pre-COVID levels (UITP 2020).

Various research works have stated that work from home will be an integral part of work culture after the lockdown. There is a 377% jump in web search regarding remote work in India during February to May amid the COVID-19 crisis.[10] In the post lockdown period, people will switch to prefer personal vehicles as the mode of commutation to workplaces in case of restarting work from the office space. People have been cycling miles, driving their four-wheelers, and commuting with their two-wheelers instead of getting into crowded public transport vehicles. The number of public vehicles plying on the roads during this post lockdown phase are so inadequate in number as well as the frequency that people who have no other means and are in a desperate need to go to the workplace are boarding on the public vehicles. It exposes the passengers to susceptible to acquiring the virus for not being able to maintain proper physical distance in a crowded riding space. In India, most of the state-run buses have started plying out in small numbers, whereas the railways and metros within the city have started operation in very few states of India.

We conducted a survey to examine the respondents' changing lifestyles and preferences to portray the impact of lockdown on the public transportation system. The participants considered in the study have been categorized with their age, and gender characteristics, educational qualification, occupational structure, and their place of residence. 82.08% of participants belong to the age group 15–30 years, and 50.94% of male participants have responded to the survey. People mostly engaged in Higher Education and academics have responded as can be seen that 46.23% have a post-graduate degree or above, and 23.58% are graduates. 43.40% of respondents are students, and 16.98% are teachers by profession. Most of the respondents are from the urban area (Table 10.6).

This survey has mirrored that almost every aspect of human life has taken a shift from the real space of being physically present to a virtual space. We find that only 15.09% of the respondents use public transportation in order to commute to their workplaces (Table 10.7). An almost identical result is observed in the case of

[10]See the website of edex live: https://www.edexlive.com/news/2020/may/24/work-from-home-job-searches-see-a-377-per-cent-spike-from-feb-to-may-says-report-12211.html.

Table 10.6 Characteristics of the respondents ($n = 212$)

Characteristics	Numbers (f)	Percentage
Age composition		
<15	14	6.60
15–30	174	82.08
30–45	20	9.43
>45	4	1.89
Sex composition		
Male	108	50.94
Female	102	48.11
Preferred not to say	2	0.94
Educational qualification		
Primary	8	3.77
Secondary	44	20.75
Higher Secondary	12	5.66
Graduate	50	23.58
Post-Graduate & Above	98	46.23
Occupational structure		
Unemployed	26	12.26
Students	92	43.40
Researchers	20	9.43
Teachers	36	16.98
Homemaker	12	5.66
Business	8	3.77
Others (Army, Geologist, Agricultural supervisor, Accountant, and IT Officer)	18	8.49
Place of residence		
Rural	102	48.11
Urban	110	51.89

Data source Primary survey, 2020

Table 10.7 Mode of transportation used to commute workplaces and market

Workplaces			Market and shopping centers		
Alternatives	f	%	Alternatives	f	%
Work from home	80	37.74	Work from home	72	33.96
Personal vehicle	100	47.17	Personal vehicle	124	58.49
Public transport	32	15.09	Public transport	16	7.55

Data source Primary survey, 2020

Table 10.8 Online meeting and conferences during and post lockdown period

During lockdown			Post Lockdown		
Alternatives	f	%	Alternatives	f	%
Remained the same as pre-lockdown	68	32.08	Remained the same	180	84.91
Increased	118	55.66	Decreased	32	15.09
Does not use	26	12.26			

Data source Primary survey, 2020

marketing and shopping. Maximum respondents desire to commute to the market place by walking or by personal vehicle to avoid the chance of transmission of the virus in public transportation or shared mobility services. The same tendency is found for the transit preferences in commuting to the place of work. The Energy and Resources Institute (TERI) opines that the share of Delhi metro services has decreased by 13% due to the outbreak of COVID-19. There is a need to promote alternate low emission modes of travel in Delhi NCR. In Bengaluru city, 50% of the users of public transportation services will switch to personal vehicles, 25% will opt work from home approach, and the remaining 25% will use carpool during the pandemic situation. In the city of Kolkata, the share of personal cars has increased by 10% during the COVID-19 outbreak (TERI 2020).

The survey report shows that 55% of respondents prefer to attend the meeting and conferences in virtual mode during the lockdown. In the post lockdown period, the preference remains the same as any specific medical treatment and vaccination is yet to be operational. With the opening of several workplaces, many have stopped working from home; 15% of respondents are reported to have shifted to the offline mode during post lockdown (Table 10.8, Fig. 10.7).

Similar is the case regarding online classes and education. During the lockdown period, only 5.66% of the respondents did not avail of any online classes. Whereas, the use of online learning platforms had increased for 75.47% respondents as compared to the pre-lockdown period. The proportion decreased by 2.83% of the respondents. 16.04% had availed online classes as same as the pre-lockdown period. In the post lockdown scenario, only 4.72% had stopped using this online mode of education either because their private coaching classes have started in physical mode, or they have a lack of resources. The majority of the respondents (i.e., 84.91%) continue learning through online mode (Table 10.9).

Coronavirus disease 2019 has also triggered the "virtual tourism"[11] in the form of a *virtual walk* and *video-based virtual tour*. In West Bengal, the Heritage Walk

[11]Virtual tourism has become a trendy and popular term in tourism industry, also known as virtual reality tourism or VRT. It is essentially a hybrid concept combining both the notions of virtual reality (VR) and tourism. Virtual tourism facilitates a tourism experience, without actually having to travel anywhere. In its simplest form, virtual tourism may comprise of a video of a tourism destination. The "tourist" watches the video, utilizing their hearing and sight senses. For more detail, see: https://tourismteacher.com/virtual-tourism/#0-what-is-virtual-tourism. This is also termed as smart torism and remote tourism.

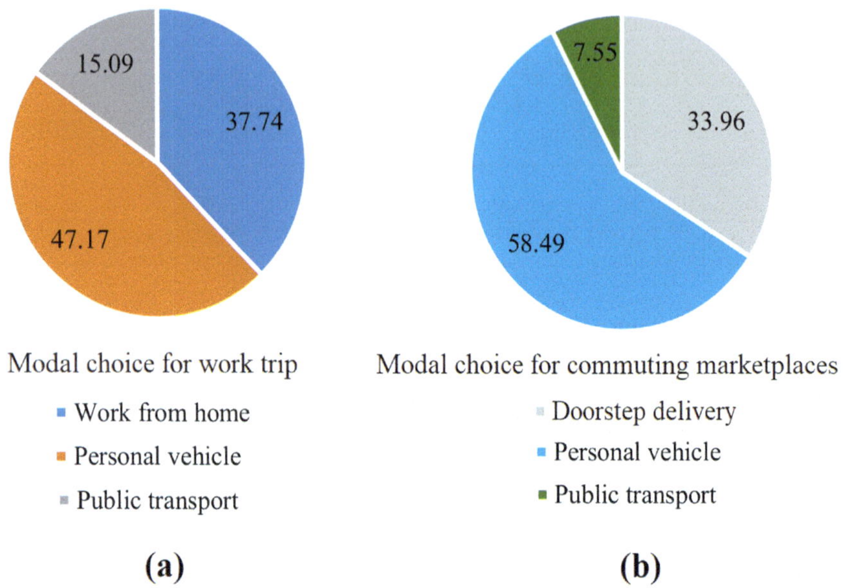

Fig. 10.7 The choice of people regarding the transit for **a** a work trip in post lockdown period and **b** commuting marketplaces in post lockdown period (*Source* Authors)

Table 10.9 Use of online education platforms during and post lockdown period

During lockdown			Post lockdown		
Alternatives	f	%	Alternatives	f	%
Does not use	12	5.66	Stopped using	10	4.72
Using prior to lockdown	34	16.04	Continuing the use	180	84.91
Increased	160	75.47	Decreased	22	10.38
Decreased	6	2.83			

Data source Primary survey, 2020

Calcutta team have arranged such a virtual tour to experience different historical and architectural sites using modern technologies.[12] Despite the existing digital divide in India, the whole education system has shifted to a digital platform, and people prefer personal vehicles instead of public transport, which will result in a huge ebb in demand for a very long period.

Besides the impacts mentioned above of COVID-19 on the public transportation sector, passenger and freight activities are affected. Per day road accidents during the lockdown period have decreased many times compared to pre-lockdown (Kumar and Sankar 2020). COVID-19 outbreak and lockdown have resulted in a significant reduction in the average emission level, contributed by the transportation sector. The

[12] See details in their official website: https://www.heritagewalkcalcutta.com/.

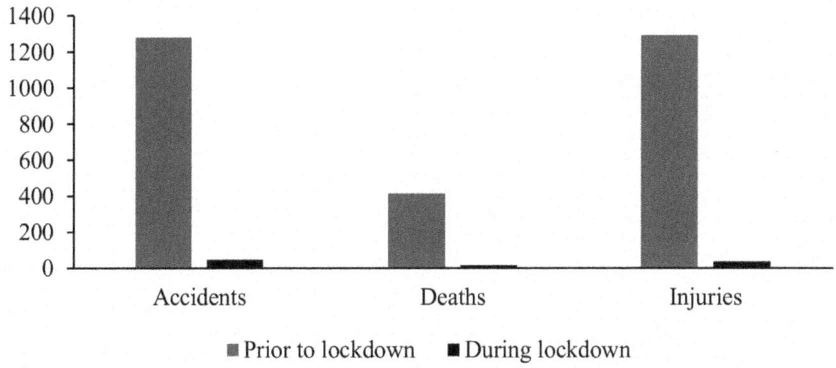

Fig. 10.8 Daily road accidents in India (*Data source* Kumar and Sankar 2020)

air pollution level has dropped to levels last seen in 2006. Particulate matter pollution in the major cities of India exhibits a similar declining trend during the lockdown period. However, the recent relaxation during the unlock period is causing the level to become re-rising significantly (Radhakrishnan et al. 2020) (Fig. 10.8).

10.4 Coping with the Crisis

We should also look into the world's strategies and planning for public transportation during the ongoing pandemic situation. It will help in giving a critical insight to understand the Indian situation in transportation planning during COVID-19.

In the USA, many of the bus transport agencies have temporarily waived their fares to promote the norm of social distancing and to keep safe the riders and operators. This practice may also be a financial cushion to the riders during the difficult situation of pandemic (WSP 2020a). The Australian government has identified three scenarios based on varying levels of physical distancing to understand the impact of the transition stage of coronavirus outbreak on public transportation capacity—*Scenario 1, i.e., strict distancing* (1.5 m between people, no standing); *Scenario 2, i.e., moderate distancing (no person sitting directly next to, behind, or diagonally across from another)*; and *Scenario 3, i.e., relaxed distancing (allows for a gap between people on each seat and 1 m distance between the people facing each other).* The public transportation system of Australia is following the three phases depending upon the situation and mode of transport (WSP 2020b).

In Seoul city of South Korea, the authority has introduced live updates of the congestion level show up on every bus top of the city. Passengers can also check out the level of congestion by the mobile application, which is allowing the analysis of bus passenger trends. The city authority is plying extra buses to balance the crowd level and maintain physical distancing. The city authority has also established a detailed guideline depending on different levels of congestion in the metro rails for managing the norm of social distancing (Lee 2020).

In the city of Mexico, the government is planning to quadruple its cycle lanes to cut down the pressure on its metros, maintaining the norm of social distancing. The same is the case of Bogota, the capital city of Colombia. The city authority has converted 100 km of city streets into cycle lanes to cut down the load on their bus services. Budapest (Capital of Hungary), is also creating cycle lanes as a safe alternative for commuting work and market places (Bhatt 2020).

To mobilize the public transport system in India, the Ministry of Health and Family Welfare (Govt. of India) has released detailed guidelines for domestic travel on 24 May 2020. The nutshell of the said guidelines is as follows

- *Do*s and *Don't*s shall be provided along with tickets to the travelers by the agencies concerned.
- All passengers shall be advised to download the Arogya Setu App[13] on their mobile devices.
- Suitable announcements about COVID-19, including precautionary measures to be followed, shall be made displayed at airports/railway station/bus terminals and in flights, trains, and buses.
- The States/UTs shall ensure that all passengers shall undergo thermal screening at the point of departure, and only asymptomatic passengers are allowed to board the flight/train/bus.
- During boarding and travel, all passengers shall use face covers/mask. They will also follow hand hygiene, respiratory hygiene, and maintain environmental hygiene.
- At airports, railway stations, and bus terminals, required measures to be taken, ensuring social distancing protocols to follow.
- Airports/railway stations/bus terminals should be regularly sanitized/disinfected, and the availability of soaps and sanitizers shall be ensured.
- Thermal screening at the exit point shall be arranged.

Institute for Transportation and Development Policy (ITDP-India Program) has also provided comprehensive and detailed guidelines for the public bus services in India. The guidelines include a list of recommendations to ensure the smooth and safe functioning of the bus services in the post lockdown period. Some of the highlights of the guidelines are[14]:

- Only those commuters must be allowed to board the bus who are wearing masks.
- Conductors must ensure that the buses do not get overcrowded.

[13]A mobile application developed by Government of India, for helping augment the efforts of limiting the spread of COVID-19, with an objective of enabling Bluetooth based contact tracing, mapping of likely hotspots and dissemination of relevant information about COVID-19 to the general public. It was released on April 2, 2020. See: https://static.mygov.in/rest/s3fs-public/mygov_159050700051307401.pdf.

[14]For detailed guideline, see: https://www.itdp.in/tag/public-transport/.

- Contactless ticketing systems and digital payments must be developed and promoted by the authorities and bus operators.
- Agencies must communicate relevant and updated passenger information to avoid overcrowding at the bus terminals. Terminals, bus stops, and buses must be sanitized regularly.

The Council of Scientific and Industrial Research (CSIR) and the Central Road Research Institute (CRRI) have prepared few guidelines based on the multidimensional approach to managing the public transportation system of India by maintaining social distancing norms.[15]

- *Approach A: Redesigning the facilities suiting to social distancing*
- *Approach B: Reducing the demand and Capacity Enhancement*

It is important to acknowledge that pandemic planning is not a standalone project, but should be integrated with existing crisis management structures and procedures in order to be effective. Along with the guideline provided by the government, the following steps could be helpful to manage the situation:

- Maximize contactless payment for shared mobility services to avoid any physical interaction during the journey and to establish best practices to keep riders safe.
- Online booking of tickets to travel which will ensure the minimization of money and ticket transfer by hand;
- Video instruction of *Do*s and *Don't*s can be shown during the journey;
- Cleaning and sanitizing the vehicle after every trip or at layovers during the route;
- Incentive-cum-awareness programs may be helpful to gain and rebuild ridership in the public transportation system;
- An integrated plan for public transport and financial help from the state and central government to the transport operators;
- Engagement of the government and stakeholders to determine the requirements of the transportation system in the country;
- Adaptation of changing demand. (Conversion of passenger vehicles into carrying goods.)
- Riders should avoid unnecessary conversation and interaction with the fellow riders during the journey.

10.5 The Gap Between the "Cup and the Lip"

The mammoth volume of passengers that the Indian transport carries is the critical concern of worries whether the system could run adopting the protocols of social distancing. There are constant efforts by the government to prepare safe spaces of travel for the people with guidelines issued. Sanitizing the vehicles, seating arrangement, the number of passengers to be carried, ticketing system, distancing, passenger,

[15]For detailed guidelines, see: https://crridom.gov.in/sites/default/files/SocialDistancing_DraftFINAL%203.5.2020.pdf.

and employee hygiene: are the few measures that have been repeatedly mentioned. Despite all measures, the major concern is whether it can be executed in reality and how. *HuffPost India*'s interview with a Senior Associate of Urban Planning and Design at Urban Works Institute (a network partner in India for the New York-based Institute for Transportation and Development Policy, or ITDP) brings forth the real scenario. It is a big question whether the country could reorganize the transport system that would be well equipped to deal with the overwhelming demand as it enforces physical distancing, and how the changes should be implemented (Jain 2020). The gap between ambition and execution of strategies and planning is vast. There are various reasons behind that; one of those is an enormous gap between the required number of vehicles and operational vehicles. A study of Climate Trends, a body of strategic communication of Delhi, says that during this COVID-19 pandemic, the country faces a massive shortage of bus services. As per the report, India would need nearly 6.6 lakh buses[16] for 25 million daily commuters, and currently, the country has only 25 thousand operational bus services. Thus, we have twenty-four times fewer buses in its public transport fleet than it needs for people to follow social distancing (Yadav 2020; Gandhiok 2020; Intelligent Transport 2020; Sachdev 2020). Due to this inadequacy, maintaining social distancing would be very difficult, and could increase the risk of infection.

The government-owned public transports can still have checks and balances to maintain the rules in order to assure the safe carriage of passengers utilizing regular audits by the officials. On the other hand, privately owned vehicles, especially city buses, autos, and e-rickshaws can be the real problem creators. With no formal announcements regarding the timings, there are chances of crowding in the pickup stands of the local modes of transport. A contactless journey is somehow not a very practical choice to travel short distances. Passengers traveling short distances within the city may be reluctant to go cashless for a journey costing between Rs. 5–50. Introducing e-ticketing is the solution that may be the solution; however, it may raise travel costs for short hauls. While the government-owned transport sector can still afford to go digital, the private short distance transport sectors may not intend to shift digital (Fig. 10.9).

The regular sanitization of the vehicles is another big issue. There is a need to ensure that the products used are not harmful and are enlisted by the government in the list of chemicals allowed to be used. The operators may be more prone to using pesticides, due to their cost-effectiveness, instead of enlisted alcohol-based sanitizer. Using toxic chemicals as disinfectants can adversely affect the health of the people on board.

As distancing has been made necessary, the public may have to opt for alternate commuting. The Kolkata Metro has laid guidelines where children and older people cannot avail of the service. So in case, a passenger has to be accompanied by a child or an older person may have to think of availing some other means that might require

[16]As per Mumbai's BEST (Brihanmumbai Electricity Supply and Transport) guidelines, i.e., 30 passengers per bus (25 seated and five standing).

Fig. 10.9 Commuters are waiting to enter Howrah Station and flouting social distancing norms after lockdown restrictions were eased in Kolkata (*Source* Samir Jana, Hindusthan Times 2020)

a higher cost of travel. The public, despite the resume in transport services, is pushed to forced alternative choices.

In the previous discussion, it was seen that the daily commuters are shifting to personal vehicles to avoid the risk of virus transmission (Table 10.7). However, it is also the fact that buying a car or availing personal mobility is not the option for many Indians, and there must be a ramping up of public transportation infrastructure (Yadav 2020).

In most of the Indian megacities, footpaths are occupied by vendors and shopkeepers. After the unlock period has started, most of the footpaths are getting back to the pre-lockdown situation. Thus re-imagination of our city streets design where there will be a separate lane for cycling and walking would be very difficult.

While the government is trying to make sure that the least number of people are affected by sharing the transportation space, the public availing services are equally responsible for practicing the rules and regulations while traveling. While the government has to ensure spreading awareness among the masses continually, it is on the masses to acquire awareness and maintain the system during the pandemic.

10.6 Conclusion

Public transport is one of the essential services to be maintained as long as reasonable and viable. It is also a very high-risk environment for transmissible diseases, including novel coronavirus disease. Therefore, it is necessary to follow and adopt the guidelines given by national health organizations and authorities. It also can be mentioned that the imagination of modern society without any mobility and transport was impossible just a few months back, but the outbreak of coronavirus disease in 2019 brings a complete halt in the public transportation system. The whole world is rethinking mobility alternatives and trying to abstain from community transmission. A massive shift has been noticed from transportation base activities to virtual communication.

On the one hand, the increase in the use of bicycles will be beneficial for both the natural environment and individuals' health, but the sudden increase of personal cars and motorcycles will cause an adverse effect on the environmental and human health by increasing noise and emission gases. Public transport services also provide a sustainable alternative to private vehicles. The suspension of the public/rental services and reduction in demand can put a big question on the practice of sustainability in times to come. For the time being, the consumption of fossil fuels has decreased, but in the near future, the consumption rate will go up.

Where we could build better, we are gearing up for much chaos with no one to blame. The policy interventions should be more concrete. The post-pandemic public transportation system may not be the same as it always has been in connecting cities, towns, and villages but to sustain the future environment and reduce the risk of degradation. Strict policy measures are to be thought of by the government to bring back the limping transport sector to normal.

References

Azhar EI, Hui DS, Memish ZA et al (2019) The middle east respiratory syndrome (MERS). Infect Dis Clin 33:891–905. https://doi.org/10.1016/j.idc.2019.08.001

Bhatt A (2020) Post Covid-19 lockdown, will India's public transport systems be able to maintain social distancing? Scroll.in. https://scroll.in/article/960063/post-covid-19-lockdown-will-indias-public-transport-systems-be-able-to-maintain-social-distancing. Accessed 20 Aug 2020

Bouey J (2020) Assessment of COVID-19's impact on small and medium-sized enterprises: Implications from China. https://doi.org/10.7249/CT524

Businessline (2020) ICRA sees highway toll collection dipping. The Hindu Businessline. https://www.thehindubusinessline.com/economy/icra-sees-highway-toll-collection-dipping/article31302349.ece/amp/. Accessed 2 June 2020

Chauhan MC (2020) COVID-19 and its effect on rail and transport industry. Urban Transport News. https://www.urbantransportnews.com/covid-19-and-its-effects-on-rail-transport-industry/. Accessed 24 June 2020

Chen N, Zhou M, Dong X et al (2020) Epidemiological and clinical characteristics of 99 cases of 2019 novel coronavirus pneumonia in Wuhan, China: a descriptive study. The Lancet 395:507–513. https://doi.org/10.1016/S0140-6736(20)30211-7

Cui J, Li F, Shi Z-L (2019) Origin and evolution of pathogenic coronaviruses. Nat Rev Microbiol 17:181–192. https://doi.org/10.1038/s41579-018-0118-9

Evan D, Over M (2020) The economic impact of COVID-19 in low and middle—income countries. Centre for Global Development. https://www.cgdev.org/blog/the-economic-impact-of-COVID-19-in-low-and-middle-income-countries. Accessed 25 June 2020

Gandhiok J (2020) Social distancing: Delhi needs to triple bus fleet. The Times of India. http://timesofindia.indiatimes.com/articleshow/76719678.cms?utm_source=contentofinterest&utm_medium=text&utm_campaign=cppst. Accessed 28 Aug 2020

Hindusthan Times (2020) Social distancing norms flouted across Bengal on first day of Unlock 1. Hindusthan Times. https://www.hindustantimes.com/india-news/social-distancing-norms-flouted-across-bengal-on-first-day-of-unlock-1/story-JDC54WNBhw5ouhCfVNJTmI.html. Accessed 27 August 2020

I & I Editorial Board (2020) Mass transit, the pandemic petri dish the left loves. Retrieved from Issues & Insights. https://issuesinsights.com/2020/03/20/mass-transit-the-pandemic-petri-dish-the-left-loves/. Accessed 25 June 2020

Intelligent Transport (2020) India requires 24 times more buses to allow commuters to social distance effectively. Intelligent Transport. https://www.intelligenttransport.com/transport-news/100881/india-requires-24-times-more-buses-to-allow-commuters-to-social-distance-effectively/. Accessed 26 Aug 2020

Jain A (2020) How can India Reopen public transport post lockdown? An expert explains. Huffpost. https://www.huffingtonpost.in/entry/public-transport-lockdown-bus-train-masks_in_5ebbc585c5b687aa20a05128. Accessed 25 Aug 2020

Kumar A, Sankar TP (2020) Impact of Covid-19 on India's transport sector. https://www.itef-oecd.org/covid-19-impact-india-transport-sector. Accessed 28 June 2020

Lai C-C, Shih T-P, Ko W-C et al (2020) Severe acute respiratory syndrome coronavirus 2 (SARS-CoV-2) and corona virus disease-2019 (COVID-19): The epidemic and the challenges. Int J Antimicrob Agents 55(3):105924. https://doi.org/10.1016/j.ijantimicag.2020.105924

Lee S (2020) How South Korea is managing public transportation under COVID-19. IDB https://blogs.iadb.org/transport/en/how-south-korea-is-managing-public-transportation-under-covid-19/. Accessed 25 July 2020

Leigh G (2020) Public transport in a pandemic. Good idea or best avoided? Forbes. https://www.forbes.com/sites/gabrielleigh/2020/03/17/public-transport-in-a-pandemic-good-idea-or-best-avoided/#2f6c2a4d3047. Accessed 20 June 2020

Livement (2020) Road developers see toll revenues pick up as highway traffic returns. https://www.livemint.com/industry/infrastructure/road-developers-see-toll-revenues-pick-up-as-highway-traffic-returns-11592306919740.html. Accessed 28 July 2020

Manju (2020) Indian aviation industry could face Rs 24 k crore revenue loss due to COVID-19. Times of India. https://timesofindia.indiatimes.com/business/india-business/indian-aviation-industry-could-face-rs-24k-crore-revenue-loss-due-to-covid-19-lockdown/articleshow/75595340.cms. Accessed 25 July 2020

Mishra HH (2020) Coronavirus in India: COVID-19 lockdown may cost the economy Rs 8.76 lakh crore; here's how. Business Today. https://www.businesstoday.in/opinion/columns/coronavirus-in-india-covid-19-india-lockdown-economy-cost-gdp-gva-nationwide-shutdown/story/399477.html. Accessed 28 May 2020

Radhakrishnan V, Sen S, Singaravelu N (2020) Pollution levels decline in many Indian cities due to COVID-19 lockdown. The Hindu. https://www.thehindu.com/data/data-pollution-levels-decline-in-many-indian-cities-du-to-covid-19-lockdown/article31667020.ece. Accessed 18 June 2020

Robert Wood Johnson Foundation (2017) Public transportation systems. County Health Rankings & Roadmaps. https://www.countyhealthrankings.org/take-action-to-improve-health/what-works-for-health/strategies/public-transportation-systems. Accessed 25 June 2020

Sachdev A (2020) Existing number of buses inadequate to ensure social distancing, will need 6,00,000 more: Report. CNBCTV18. https://www.cnbctv18.com/healthcare/exsiting-number-of-buses-inadequate-to-ensure-social-distancing-will-need-600000-more-report-says-6186191.htm. Accessed 26 August 2020

Shakti Foundation (2016) City public transportation development in India. www.intelligenttransport.com. Accessed 10 Aug 2020

Sharma M (2020) Lockdown effect: Indian Railways staring at Rs 90,000 crore shortfalls in earnings. India Toady. https://www.indiatoday.in/india/story/lockdown-effect-indian-railways-staring-at-rs-90-000-crore-shortfalls-in-earnings-1675200-2020-05-07#:~:text=The%20railways'%20initial%20projections%20estimated,lakh%20crore%20for%20FY2020%2D21.&text=If%20it%20is%20ad. Accessed 25 June 2020

Shereen MA, Khan S, Kazmi A et al (2020) COVID-19 infection: origin, transmission, and characteristics of human coronaviruses. J Adv Res 24:91–98
Singh MK, Neog Y (2020) Contagion effect of COVID-19 outbreak: another recipe for disaster on Indian economy. J Public Aff, e2171:1–8. https://doi.org/10.1002/pa.2171
TERI (2020) Impact of COVID-19 on Urban Mobility in India: Evidence from a perception study. The Energy and Resources Institute
The Economic Times (2020a) COVID-19 has led to 20 lakh job losses in bus, taxi sector; more on anvil: Industry Body. The Economic Times: https://economictimes.indiatimes.com/jobs/covid-19-has-led-to-20-lakh-job-losses-in-bus-taxi-sector-more-on-anvil-industry-body/articleshow/76492675.cms?from=mdr. Accessed 1 July 2020
The Economic Times (2020b) Toll collections to see 13% drop due to 57-day coronavirus lockdown: CRISIL Research. https://economictimes.indiatimes.com/news/economy/finance/toll-collections-to-see-13-drop-due-to-57-day-coronavirus-lockdown-crisil-research/articleshow/75576705.cms. Accessed 26 June 2020
UITP (2020) Management of COVID-19: Guidelines for public transport operators. Int Assoc Publ Trans. https://www.uitp.org/sites/default/files/cck-focus-papers-files/Corona%20Virus_EN.pdf. Accessed 10 July 2020
UITP and World Bank Group (2020) Impact of COVID-19 on Indian bus operators. https://cms.uitp.org/wp/wp-content/uploads/2020/07/Statistics-Brief-India-Bus.pdf. Accessed 14 June 2020
World Health Organization (2020a) Considerations for public health and social measures in the workplace in the context of COVID-19: annex to considerations in adjusting public health and social measures in the context of COVID-19, 10th May 2020. https://www.who.int/publications/i/item/considerations-for-public-health-and-social-measures-in-the-workplace-in-the-context-of-covid-19. Accessed 24 June 2020
World Health Organization (2020b) Coronavirus. Health topic. https://www.who.int/health-topics/coronavirus#tab=tab_1. Accessed 24 June 2020
World Health Organization (2020c) Laboratory testing for coronavirus disease 2019 (COVID-19) in suspected human cases: interim guidance, 2nd March 2020. World Health Organization. https://apps.who.int/iris/handle/10665/331329. Accessed 25 June 2020
World Health Organization (2020d) Transmission of SARS-CoV-2: implications for infection prevention precautions. https://www.who.int/publications/i/item/modes-of-transmission-of-virus-causing-covid-19-implications-for-ipc-precaution-recommendations. Accessed 20 July 2020
WSP (2020a) Public transportation and COVID-19 funding and finance resiliency: considerations when planning in an unprecedented realm of unknowns. https://www.wsp.com. Accessed 22 June 2020
WSP (2020b) Public transportation and COVID-19: how to transition from response to recovery. https://www.wsp.com. Accessed 22 June 2020
Yadav N (2020) India needs over 600,000 buses for 25 million commuters daily to follow social distancing norms, according to a study. Business Insider. https://www.businessinsider.in/india/news/india-needs-over-600000-buses-for-25-million-commuters-daily-to-follow-social-distancing-norms-in-times-of-coronavirus/articleshow/76417090.cms. Accessed 26 Aug 2020

Biswajit Paul is a Research Fellow in the Department of Geography and Applied Geography, University of North Bengal. He has completed his master's from the same department and is a university medal holder. He has been awarded the UGC fellowship in the year 2019. He had been awarded the Junior Research Fellowship under the National Geospatial Chair Professor Scheme, DST (Govt. of India). His area of research focuses on transportation geography and planning. His other areas of interest are population geography, geospatial science, regional development, and planning.

Subir Sarkar is a Professor in the Department of Geography and Applied Geography of the University of North Bengal. He has been awarded a Ph.D. degree from the University of North Bengal in the year 1989. Dr. Sarkar has written more than 30 book chapters and published more than 36 journal articles. He has also authored/edited 9 books. Prof. Sarkar is continuing the collaborative research with scientists from the Polish Academy of Sciences since 1990 under the INSA—PAS Exchange program. He has also completed 6 International and 31 national-level projects, and his major areas of research interest are Geomorphology, environmental hazards, geo-informatics, hydrology and watershed dynamics, decentralized development planning.

Chapter 11
Inter-state Labor Migration in India: The Normal and Reverse Phase

Rupai Hembram and Uttam Garai

Abstract The novel coronavirus has started spreading from Wuhan city of mainland China in December 2019, and it spread out over several countries throughout the world. To stop the spread of this virus, the Indian Government has announced for countrywide lockdown. It has a significant impact on the Indian economy, especially in the unorganized sector. The workers of this sector, especially the migrant workers, are the worst affected section of our society. The present study tries to explore the socioeconomic shock of this crisis on interstate migrant workers. We have divided the whole discussion into two major parts. The first one is the pre-pandemic situation of interstate migration in the country and the second one is the pandemic situation. For the first one, as there is no accurate data on migrant workers countrywide, we have considered the D3 series data of the Census of India 2011. Whereas, for the second one, various study reports and news reports have been analyzed. It is revealed that migrants were going toward high developed states from the low developed states during the pre-pandemic situation. However, the pandemic shock changed the direction of migration; a reverse migration was witnessed toward their native places. This chapter brings to fore the forces of detracton at the Indian megacities, which acted as the push factors; the challenging phase of transit from cities to native villages; and the graved scenario of social stigmatization over migrants workforce in their "homeland."

Keywords Migration · Spatial pattern · Unemployment · COVID-19 crisis · Stigma · Joblessness

---

R. Hembram (✉)
Department of Geography, Dr. Meghnad Saha College, Itahar, Uttar Dinajpur, West Bengal, India
e-mail: rupaihembram13@gmail.com

U. Garai
Department of Geography, Sidho-Kanho-Birsha University, Purulia, West Bengal, India

M. Mishra and R. B. Singh, *COVID-19 Pandemic Trajectory in the Developing World*, Advances in Geographical and Environmental Sciences,
https://doi.org/10.1007/978-981-33-6440-0_11

11.1 Introduction

The first COVID-19 case was reported from Wuhan city of mainland China in December 2019. The case was from a market where wild animals were consumed (Xu et al. 2020), but in India, the first case was reported on January 30, 2020. The patient was a student studying at Wuhan University. She came back home to spend her vacation. Within 41 days, the number of COVID-19 positive patients increased to 50 (Rawat 2020). The World Health Organization (WHO) has announced the COVID-19 disease as a pandemic on March 11, 2020 (WHO 2020). After two weeks later, the Government of India has taken strict measures to control the health crisis. As an initial controlling measure, the government has announced 21 days countrywide lockdown. It imposes all kinds of human mobilities to be restricted, and all kinds of transport facilities to be paused. In his address to the nation, the prime minister of India requested people to stay wherever they were (Ghosh 2020).

This chapter will not discuss the health crisis the Indian has faced. Rather, it will be more concerned with how the initial health crisis has been transformed into the socioeconomic shock. The pandemic shock has proved itself to be absolutely asymmetric if the socioeconomic shock is concerned; it has not similarly impacted everyone. Out of this health crisis, the perpetual debate started arising as the phases of lockdowns were extended. The citizens became dissected about whether to prioritize lives or livelihoods.

Thanks to the lockdown, the screen-time most Indians spend on their mobile phones has gone up enormously. Peoples now tend to spend far more time checking out the incessant flow of WhatsApp traffic than they had ever done before the pandemic. If it is not the mobile phone people are browsing to pass their time during the lockdown, it is the television screen. The streaming services have enormously expanded the viewing menu. However, this is just half the story.

However, the "theory" of "life is not bigger than the business" simply poses "meaningless" to those whose bread and butter depend upon daily earnings. They sought that they would die if the lockdown would be extended (Arya and Ahmed 2020). The lockdown was not working for the poor. The slum dwellers who live in congested lanes, for them the precautions like social distancing, work from home, self-isolation nothing works.

The lockdown meant the shutting down of the doors of economic activities. All the factories were closed. The activities in the construction sector were discontinued (Mishra 2020b). The workers who came from different corners of the country to the Metros in search of jobs were the worst affected section. This crisis degraded their socioeconomic status (Sengupta and Jha 2020). Lockdown reached the workers of the unorganized sector into the darkness of unemployment (Sundar 2020). The unorganized sector has a significant contribution to our Gross domestic production (GDP) (GOI 2019), and not only that, but it also engages a large number of laborers (Kishore and Jha 2020). For the migrant workers, their livelihood has been "robbed." The lockdown kept the *haves* at home to minimize infection risk while the *have-nots* migrant workers were left on the highways, exposed to the infection (Hannant

2020). In this chapter, we will try to explore the socioeconomic shock of this crisis on interstate migrant workers.

We have divided the whole discussion into two major parts. The first one is the pre-pandemic situation of interstate migration in the country and the second one is the pandemic situation. For the first one, as there is no authentic data on migrant workers countrywide, we have considered the D3 series data of the Census of India 2011. Whereas, for the second one, various study reports and news reports have been analyzed.

11.2 InterState Migration in Pre-pandemic India: Glance Back to the *Normal*

In this section, we will focus on the spatial pattern of short-term interstate migration. Based on the rate of migration, India's whole is divided into five distinctive zones (Table 11.1). Uttar Pradesh and Bihar comprised 35.55% out-migrants out of the total migrants in India. These two states come under the very high out-migrating zone. These states are located on the great Gangetic plain. Most of the north-Indian states exhibit a low contribution to out-migration, excluding these two states. The western part of India, including Madhya Pradesh and West Bengal in eastern India, is at the next tier—i.e., the "High" out-migration zone. Chandigarh and Puducherry also come under this category. Alongside this, the Moderate zone comprises Andhra Pradesh, Jharkhand, Tamil Nadu, NCT of Delhi, Gujrat, Odisha, Haryana, Chhattisgarh, Punjab, and Kerala. On the other side, the north-eastern states contribute very

Table 11.1 Share of migrants by source states

States (share of migrants as percentage to total)
Very high share Uttar Pradesh (20.64); Bihar (14.90)
High share Maharashtra (6.35); Madhya Pradesh (6.25); Rajasthan (5.67); West Bengal (5.04); Karnataka (4.78)
Moderate share Andhra Pradesh (3.99); Jharkhand (3.79); Tamil Nadu (3.24); NCT of Delhi (3.20); Gujarat (3.06); Odisha (3.05); Haryana (2.74); Chhattisgarh (2.64); Punjab (2.23); Kerala (2.15)
Low share Assam (1.55); Uttarakhand (1.30); Jammu & Kashmir (0.68); Himachal Pradesh (0.67); Chandigarh (0.58); Puducherry (0.38)
Very low share Goa (0.21); Manipur (0.21); Tripura (0.14); Meghalaya (0.12); Nagaland (0.11); Arunachal Pradesh (0.10); Andaman & Nicobar Islands (0.06); Mizoram (0.05); Sikkim (0.04); Daman & Diu (0.03); Dadra & Nagar Haveli (0.03); Lakshadweep (0.02)

Source Computed by the authors based on Census of India 2011 datasets

insignificantly to this out-migration process. Andaman and Nicobar, Daman, and Diu and Goa have also reported very low migration (Table 11.1; Fig. 11.1).

Work or employment opportunity dominates male migration in the country. However, four states and two union territories are the exceptions of this. For Madhya Pradesh, Punjab, Kerala, and Lakshadweep, the primary reason is the movement with the household (Fig. 11.2a). The secondary reason for male migration is to move with the households for most of the Indian state. However, the migrants of Madhya Pradesh, Manipur, Meghalaya, Andaman, and the Nicobar Islands, Dadra and Nagar Haveli, and NCT of Delhi moved for other reasons. Education was the second dominating reason for Mizoram and Haryana (Fig. 11.2c).

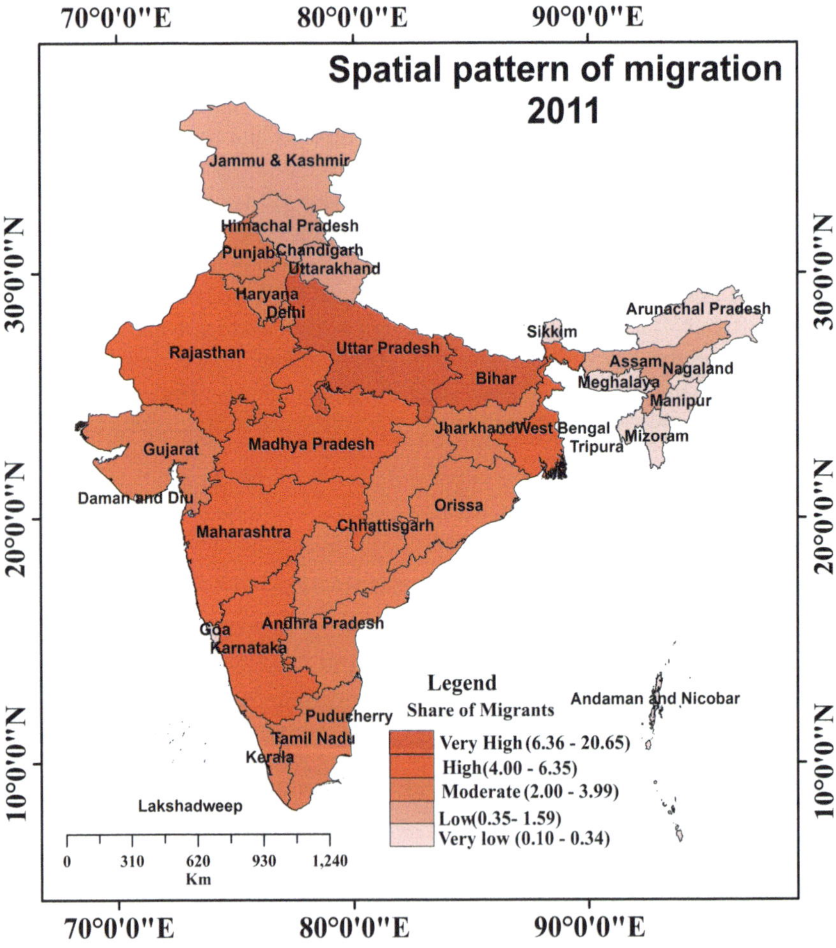

Fig. 11.1 Spatial pattern of migration in India (*Source* Prepared by authors, based on Census of India 2011 datasets)

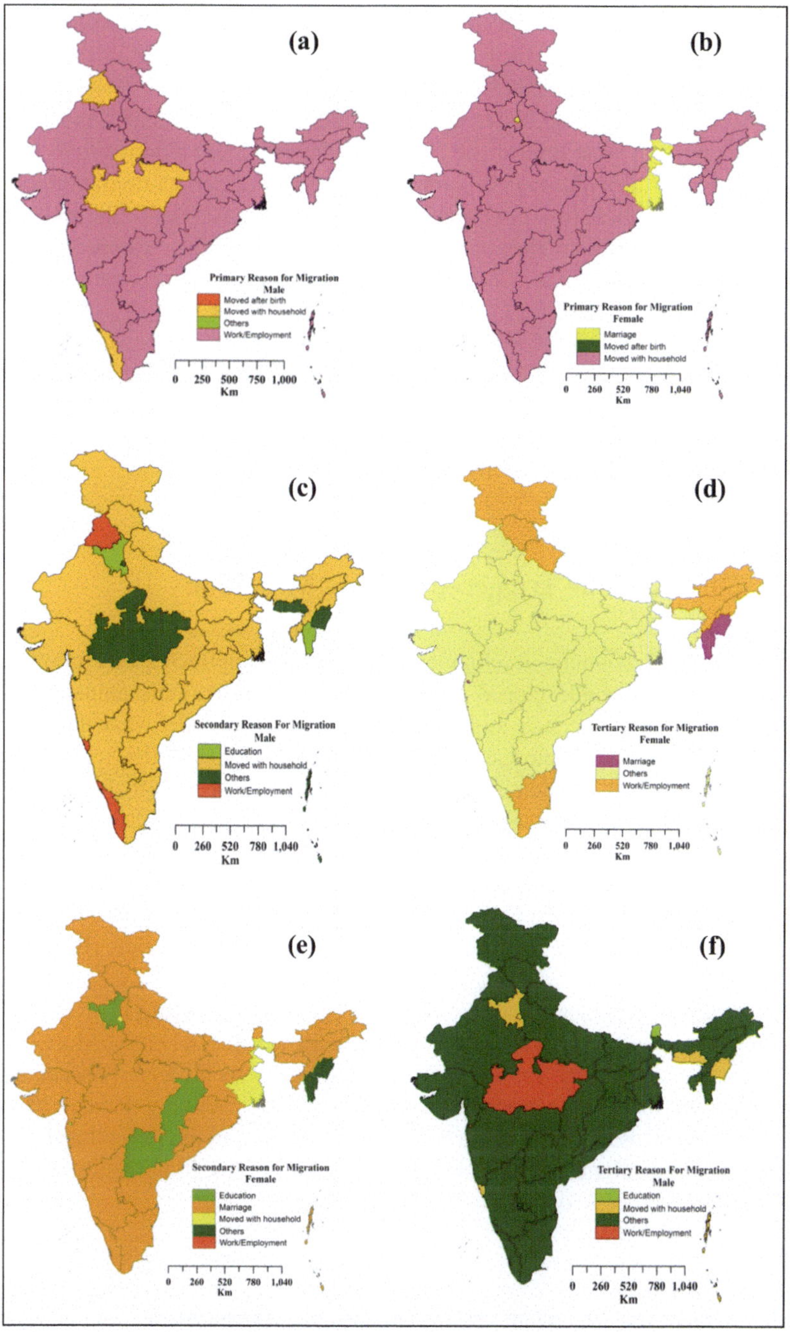

Fig. 11.2 Dominating reasons for migration in India (*Source* Prepared by the authors based on Census of India 2011 datasets)

Migrants of Madhya Pradesh are reported to move for work or employment, and migrants of Sikkim moved for education, where for the other states, the tertiary reason is seemingly undefined (Fig. 11.2f).

Females are mostly migrated to move with the household. Whereas, most of the female migrants of West Bengal and NCT of Delhi have been reported to move for marriage (Fig. 11.2b). Marriage becomes the secondary reason for migration for the other states except these two. For the states of West Bengal, NCT of Delhi, and Dadra and Nagar Haveli, the secondary reason is posed by moving with the household. The second-largest stream of the out-migration of Haryana and Chhattisgarh is moved for education (Fig. 11.2d).

The tertiary reason for female migration is undefined for most of the states. The female migrants of Tamil Nadu, Assam, Uttar Pradesh, Puducherry, Nagaland, Arunachal Pradesh shows the tertiary reason as work or employment. Whereas, the tertiary streams of female out-migrants of Chandigarh, Manipur, Mizoram, Dadra, and Nagar Haveli is sourced by marriage (Fig. 11.2e).

11.3 On Pandemic Triggered *Reverse* Migration

The inequality in economic development is the key driving force for the movement of migrants from underdeveloped states to developed states (Panchamukhi 2013). The increased rate of interstate migration has been seen after 1991 when the government enacted the new economic liberalization policy (Bhagat 2010). The female members of society play a dominating role in migration, as they shift after marriage. Now contrary to the traditional trend, females are migrating for other reasons also (Panchamukhi 2013), and Male migration is chiefly due to unemployment at the source and job opportunities at the destination (Skeldon 1986). However, data reveals that the temporary migration is seven times larger than the permanent one, and the former is dominated by the economically improvised section of society (Bhagat and Keshri 2018). The short-term migrants have low income; however, they balance it with their low per capita expenditure (Handral et al. 2018). It keeps them devoid of savings or investing in insurance or other financial security schemes. Most of the migrant workers have not enrolled their name in public social security schemes as well; moreover, those who enrolled were getting contingencies only (Pandey 2020a). Altogether, the closure of the "source of bread" posed a tremendous threat to the livelihoods of the migrants in the pandemic captured Metros. It ultimately arose the question of their access to the basic needs to live on.

11.3.1 The Sea of "Migrants" Set Out for Homeland

The worst affected section is the workers of the unorganized sector at the time of lockdown, especially the migrant workers. The situation of this section of workers during the COVID-19 crisis is pathetic throughout the world as they do not have any social protection (Sengupta and Jha 2020). In India, 92% of the workers are engaged in the unorganized sector (Tiwary et al 2012), and workers of this sector are poor, and their situation is like they are on the epicenter of the volcano. The number of jobless people is much higher in the construction sector compared to the financial sector (Kishore and Jha 2020). An estimation made by CSDS and Azim Premji University (2019) says that about 29% of the population of Megacities are daily wage earners (Sing and Magazine 2020). Another estimation by Kundu (2020) says that there are 65 million interstate migrants in India, and out of this figure, around 33% are daily wage based workers. The total estimated figure is 12–18 million, including casual workers, workers of the informal sector, street vendors, and other vulnerable communities (Sing and Magazine 2020). The exact volume of return migrants is unknown because there is no official record at the pan-India level. However, it is clear that the pandemic has pushed lakhs of migrant workers to the darkness of joblessness; they were headed to return back (Fig. 11.3a), they faced many problems on the way, and when they reached their home, no warm welcome was there (The Hindu Net Desk March 31, 2020). Such a mammoth volume of migration has not been seen on the road since partition. On May 4, officials of Maharashtra reported that 35000 migrants were sent to their native place after completing the primary treatment (PTI 2020a). Uttar Pradesh and Bihar witnessed a high volume of migrants who wanted to return to their homes (Sing and Magazine 2020).

a b

Fig. 11.3 Two cities witnessing the same scenario when **a** the desperate migrant workers gathered at Anand Vihar bus terminal, Delhi (*Source* ANI [view the Newsroom Post, dated 29 March 2020: https://newsroompost.com/gallery/amid-lockdown-a-sea-of-migrants-in-delhi-try-to-find-their-way-back-homeamid-lockdown-a-sea-of-migrants-in-delhi-try-to-find-their-way-back-home/511214.html. Accessed on 12 August 2020]) and **b** the migrants as thousands gather at Bandra station, Mumbai (*Source* India Today [see the India Today dated 14 April 2020: https://www.indiatoday.in/india/story/lockdown-woes-mumbai-stations-flooded-with-migrant-labourers-hoping-to-get-back-home-1666908-2020-04-14. Accessed on 15 August 2020])

11.3.2 The Force of Detraction

The COVID-19 induced lockdown seized the opportunities of work and employment, which had remained the critical force of attraction to the workers to migrate in the cities. As the lockdown was announced, the daily wage earners lost their job as an immediate effect. The World Bank reports that the COVID-19 had impacted the livelihood of 40 million migrations. It has caused unemployment, and the protocols of social distancing created a more complex situation for migrant laborers (PTI 2020e). According to an estimation by SWAN (2020), about 12 crores of migrant workers suffered from livelihood crises due to lockdown (Edwin 2020). SWAN conducted a study on migrant workers, involving 11159 individuals as samples. They found that 89% of the migrant workers got their wage unpaid during the 21 days of nationwide lockdown in the country. It left them in a pathetic situation with no money and no food. The desperate struggle of the famished migrants started protesting in different pockets of the country (Edwin 2020). PTI reported that 100 migrant workers were protesting in Surat. They were suffering from an acute food crisis, and they wanted to return to their homeland (PTI 15 April 2020g). However, strict lockdown protocols did not allow them to return. Against this decision, lawyer-activist Prasanta Bhusan filed a complaint, where he wrote those who were negative to COVID-19 should be allowed to return, and the government should provide the necessary transport facilities (Pandey 2020b). The migrant workers left for home not in fear of the virus but in fear of hunger (Pandey 2020b).

Many industrial hubs, factories shut their door. In Lucknow, the plywood factory was forced to close. Surrendra Panday, a labor of that factory, started his journey toward home on bare feet, as no food and transport facilities were available (Sharma and Khanna 2020). All the vehicles, including trains and buses, were restricted from carrying passengers except transporting essential goods.

Pandey reported on April 21, hundreds of migrant workers were found under a bridge near Delhi. They had not got any meal for the last three days. They were placed in a shelter, but unfortunately, the shelter got burnt. Thus, they came out from there (Pandey 2020b). The workers who were stranded in other states not only suffering from food and money but they were also the victims of rumor. On April 14, a few hours after the announcement of the second phase of lockdown, thousands of people gathered outside the Bandra railway station (Fig. 11.3b). There was a rumor that after the end of the first phase of lockdown, they could avail trains to return back to their home. They were protesting with the demand to run the train. In response, the police force was introduced to disperse the crowd (Mengle 2020; Joshi 2020; Pandey 2020a).

The daily wage earners, stranded at the workplace during the lockdown, were worried about the harvesting of the crop at their home. Anil Yadav, a 28-year-old man, told the Economic Times on April 15, that he had lost one-acre areas of sugarcane. He used to earn Rs. 7000–8000 per month, but all has gone. Another worker Asha Devi was staying in a shelter with her 13 other group members. She has four young children. They went for work in a brick kiln in Ghaziabad. After the announcement of the countrywide lockdown, the broker (locally called 'contractor') promised them to pay money, but one day suddenly, the broker disappeared, leaving them helpless (PTI 2020d). Manoj Ahriwal, a 25-year-old worker, came to Delhi to join his family members just three days before the lockdown has been announced. They were worried about the crop. His mother worried if they could not be able to harvest the crop, they would have nothing to eat throughout the year (Pandey 2020b).

The above incidents are a few of such immeasurable plights faced by migrant workers in different megacities. However, the interesting is the "reverse" direction of the migration. It is unusual in terms of the usual rural-to-urban migration trend of the country, which has been expedited for the last three decades. The understanding of the force of detraction makes it indispensable to discuss the force of attraction, which has been causing this mammoth volume of migrants to gather in the megacities for long.

So, who migrates, and why migrates? An initial *push factor* is prevalent in Indian scenario. It transfers people from rural-to-urban areas, and often such flows are called the mobility "*by default*." The growing rural distress with agriculture failing to offer sustenance for the majority of cultivators is a critical concern, and this mainly influences the decision of those with small landholdings and those face dispossessed of land due to factors which include heavy debt burdens to set out for alternative job opportunities from the cities. State policies have often proved both inadequate and ineffective to support the ailing rural economy. With steady losses of sustainable livelihoods, there has been a continuing stream of *out-migration* from the rural economy.

This initial force has created the initial stream. Some migrate on a 'permanent' basis, having no plan to return; some are the 'seasonal' migrants who temporarily return from urban areas to their villages during the harvest times, and then go back to urban centers; and the rest, the 'footloose' who have no means to decide on their future plans. Both the 'seasonal' and 'footloose' are often escorted by local contractors on the basis of payments to cover the initial travel costs and also as cuts from the meager wages they receive by migrant workers in the urban centers. Migration has also been facilitated by the prevailing familial links between the rural folks and the urban workmen.

Out of all these things, the irony is that the inflows of migrants to urban areas provided only cheap labor having no legal compulsion for employers. So, the current miseries are not all about an accident; instead, it unfolded a long history of deprivation and injustice to the migrant workers on the part of the employers to enjoy the benefit from the "footloose" migrants. They had never been given any legal status as the working population. The factories, construction sites, and other labor-intensive activities made use of the migrants in their cost-cutting exercises as they could be hired at wages, which could dip lower than the statutory minimum. The surplus migrants who were not absorbed in the formal or informal workplaces became self-employed in different activities, ranging from vendors to shop-keepers and petty services. So, the rural migrants accomplished the urban economy significantly, providing cheap labor to factories, cheap services to households, and in various other forms, keeping their job securities at acutely vulnerable condition consciously or unconsciously.

It was worrying but not astonishing that this vulnerable community lost their livelihood and shelter in their workplaces with the four-hour notice from the state for a complete shutdown. The sufferings by those uprooted people—lacking access to shelter, food, or sources of income, or even means of transport to return to their places of origin continued to follow them throughout their entire trajectory of *reverse* migration, which will be the subject matter of the following columns.

11.3.3 Problems in Transit

The Home Ministry of the Indian Government, at the onset of the crisis, had issued an advisory to all state governments to control the exodus of migrants; and, the permission was given to utilize the National Disaster Response Fund (NDRF) for arranging their food and shelter (PTI May 23, 2020b). However, the reverse migration stream could not be managed. The issues of insecurity loomed large in the cognition of the migrants, which drove their thought that if they could reach their home, they could survive.

Throughout the month of April, Indian highways witnessed the plightful procession of the thousands of migrants heading toward their homeland by walking a few hundreds of kilometers. Many of them were walking on bare feet, and those who could afford traveled through hired vehicles (Fig. 11.4a) (PTI 10 May 2020f). The situation was out of control, and no one could stop the migrants. Hunger, cashlessness, panic, and rumors captured the distressed migrants. Confusions made them undertake desperate attempts in heading toward homeland like walking along highways, along rail tracks, by cycling with managing all their belongings, minors, babies, old family members, and pets (Fig. 11.4b).

a b

Fig. 11.4 **a** Migrant workers are returning their homeland being piled on a truck (*Source* PTI [see the India Today dated 18 September 2020: https://www.indiatoday.in/india/story/fake-news-triggered-large-scale-migration-during-lockdown-mha-1722102-2020-09-15. Accessed on 20 September 2020]). **b** Migrant workers returning back by walking along the rail track (*Source* ANI [see the Newsroom Post dated 29 March 2020, https://newsroompost.com/gallery/amid-lockdown-a-sea-of-migrants-in-delhi-try-to-find-their-way-back-homeamid-lockdown-a-sea-of-migrants-in-delhi-try-to-find-their-way-back-home/511214.html. Accessed on 20 August 2020])

Under this circumstance, the railway authority, from May 1, 2020, started plying hundreds of "Shramik Special Trains" for transporting stranded migrants. The number of trains plied for this purpose came inadequate as everyone became desperate to quit the cities. Many could not register themselves for the train (Babu et al. 2020). Besides, the trains were not maintaining th scheduled time, and there was no availability of food and water on board due to COVID-19 protocol. Seven people were reported to have died on the train (PTI 28 May 2020c).

Suraj Bhan Sing was working at Surat in Gujrat, failed to register himself for travel because he had no Aadhaar card. So, there were only two options left either to walk or to die hungry (Babu et al. 2020). Mohammad Imran, a daily wage earner, started his 600 km long, unpredictable journey to reach home by walking. He accompanied his parents, children, and wife (Babu et al. 2020). Some of them could not fulfill their desire; they breathed their last on the way. Fifteen migrants, returning along the railway track, were hit by a goods train and died on the spot at Aurangabad at early in the morning on May 8. It was reported that the fatigue due to the long journey made them taking rest along the track to face the absolute fatality, keeping their journey to home unfulfilled (Sharma 2020). Six persons were found dead and 14 injured badly in a truck accident in Narsinghpur district of Madhya Pradesh. All of them were coming from Hydrabad and wanted to go to Uttar Pradesh (PTI 10 May 2020f).

> Sanju Yadav and her family went to dream city Mumbai 10 years ago to fulfill their dream. Her husband was working in a factory, and besides that, they were selling snacks in the evening time from 4 pm to 10 pm. They applied for tickets for 'Shramik special' trains, but they had not got the chance. Without any options, they started their journey on 9 May towards home with the e-rickshaw they owned. Her husband drove from 5 am to 11 am and 6 pm to 11 pm. They ate whatever dry food they had, and for sleeping, they used the *footpath*. In this way, they overcome 1300 km out of 1500 km, but suddenly one truck hit their vehicle from behind. The accident snatched away, Sanju Yadav and her daughter's life. (Pandey 2020c)

Road accidents, heatstroke, fatigue caused several fatalities, which are merely a "number" rather, they are the terminations of the source of bread for the families concerned. Woefully, none of these fatalities were attributed by the coronavirus, rather the failure of the system to giving protection to the economically vulnerable groups of the society.

11.3.4 The Welcome at Homeland Was Not "Warm" Always

The coronavirus pandemic dealt yet another blow to vulnerable migrants. Blamed for leaving their homes in defiance of the lockdown, hungry, and cash-strapped, migrants struggled, while those who managed to reach their native places face hostility. Reaching their homeland after a long journey, those people discovered themselves being pointed with raising fingers toward them as the bearers of the virus.

a b

Fig. 11.5 **a** Villagers barricading the entry to Durdih village in Lakhisarai district of Bihar on March 25, 2020 (*Source* India Spend [see the India Today dated 01 April 2020: https://www.indiaspend.com/bihars-migrants-return-to-face-stigma-under-prepared-medical-facilities/. Accessed on 25 August 2020]). **b** Migrant workers quarantined themselves on the branches of trees (*Source* ANI [see the Times of India dated 28 March 2020, available under the URL: https://timesofindia.indiatimes.com/city/kolkata/west-bengal-labourers-who-recently-returned-to-village-quarantine-themselves-on-tree/articleshow/74866464.cms. Accessed on 21 August 2020])

Isolating the migrating laborers at destination points was imposed as a part of the strategic measures of limiting the spread of the disease. However, the local administration was not ready to manage the situation. Under this mismanaged situation, ostracism and stigmatization emerged as the vital issues in most parts of the states receiving a higher volume of migrants.

In many villages, inhabitants and local leaders barricaded the entry and exit points of villages and placed posters saying something like "Outsiders are not allowed"; they were mandating those who return to self-report (Fig. 11.5a). If the self-report was not made, the villagers reported it to the authorities. The fear that outsiders will bring COVID-19 was rampant across Indian villages. People frequently called up hospitals, police stations, and district government officials to report COVID-19-related issues (Agrawal 2020).

The state of West Bengal was the receiver of mammoth volumes of migrating workers returning from Delhi, Gujrat, Kerala, Tamilnadu, and Maharastra. The state government has arranged the quarantine centers almost at each of the C.D. Blocks where the school and college premises, community halls were temporarily converted to quarantine centers with the supervision of the government health officials at the district level. However, the lack of proper infrastructures, adequate PPE kits, insufficient SWAB tests, and the non-experience of the health workers remained the key constraints. Besides, the issues of nonfulfillment of basic services, e.g., electricity, food, water, cleanliness in toilets, are often raised by the borders of different quarantine centers in the state (Kundu 2020).

It was also reported from different corners of the state that the local villagers had opposed opening the quarantine centers in the school premises within the village. Under that circumstance, the returned workers were compelled to spend the nights under the sky until the local administration managed the situation (Kundu 2020). The fake news, rumors, and unscientific information spread rapidly amongst the villagers, getting facilitated by the low literacy level of the vast rural areas. Migrant workers who returned from Chennai to their home at Purulia were not allowed to enter the home. The villagers asked them to stay in quarantine on the branches of trees (Fig. 11.5b) (Rai and Ramashankar 2020).

Gita Sen, an expert in public health, says three types of people are more vulnerable to stigmatization. People who are in quarantine, health workers, and those who are suffering from discrimination throughout ages like migrant workers, religious minorities, and people of north-eastern states (Perappadan 2020). The resistance of the society toward the returning migrants also posed challenges to their livelihood in their homeland. After coming back, they faced difficulties with running their families. This problem was particularly critical for those who had no cultivable land in their possession. Returning migrants those were under home quarantine, government officials pasted poster on their wall, this became a matter of stigmatization. They were not accepted by society to getting involved in the wage-earning system (Mishra 2020a).

Even amidst those pandemic times, the incidences of caste discrimination on migrant workers were reported. The incidences of lynching or attempt to lynching on migrants belonging to tribal communities questions on the social security issues, even at the homeland, which had seemed to be "secured" to them as they attempted leaving the cities. Some migrants of Uttar Pradesh was reported to face caste discrimination. "Upper caste" people harassed Ravi Maury and his fellow workers from Prayagraj after returning home (Kumar and Mohanty 2020). Samu Munda from West Singhbhum district of Jharkhand returned home, but the villagers did not allow them to enter the village. His family members were also restricted from meeting with him. He claimed that he had no symptoms of COVID-19, and he had even agreed to the COVID-19 test, but no one listened (Kumar and Mohanty 2020). Surin Soren from Jharkhand's Dumka district came from Varanasi faced social boycott. The villagers barricaded his house. They declared that all the members of his family had been infected (Kumar and Mohanty 2020). Kishore Behra, who was a mason, came back from Mumbai, forced to take shelter under a culvert with his co-workers near Birpali village of Bolangir district of Odisha (Barik 2020).

11.4 Legislations: Facts and Trivia

There are several legislative measures, but most of them are confined in papers. The state of Indian has long been witnessing the non-implementation and weak enforcement of different laws, which makes them non-functionals on the ground. Migrant worker issues faced the same reality. There are several laws that are not appropriately enforced (Table 11.2). In 1970 an act was launched to regulate and abolish the contract labor whatever necessary. It states about the registration of establishment, licensing the contractors and make provisions for the welfare of the contract laborers and provide legal support to them. The disobey of this law is a punishable offence (GOI 1970). The experts thought this law was not sufficient to protect the interest

Table 11.2 Relevant legislative measures available that might work

Act	Year of enactment
Contract Labor (Regulation & Abolition) Act	1970
Inter-state Migrant Workmen Act	1979
Disaster Management Act	2005
Unorganized Workers Social Security Act	2008
The Street Vendors Act	2014

Source Compiled by the authors from different sources

of migrant workers (Sinha 2020). Besides, the interstate migrant workmen law 1979 was enacted to protect the migrant workers from exploitation. The act says every migrant worker should be registered by the contractor in both the home state and host state. There should be more detailed records about the terms and conditions of work. The most interesting point the article says that there should be a database for all migrant workers (Sinha 2020).

The national disaster management act, 2005, made provisions for managing any disaster. It says that the government should be prepared for any upcoming disaster. Disaster management is a continuous process that demands proper planning about it, organizing citizens, making coordination with various departments, and implementing various measures, which are essentials (Phukan 2020). Preparedness is the tool that is possible to be tested only during a disaster. The present crisis speaks about it.

The unorganized workers' social security act of 2008 has been introduced to secure the livelihood of the working class, including the landless poor. Although, some of the welfare schemes dedicated to unorganized workers were already in vogue (Mishra 2018). Similarly, the vendors' act, 2014, has been launched to protect their livelihood and to regulate the street-vending. It highlighted the measures like conduct a survey of street vendors every five years, issuance of certificates (GOI 2014).

All the legal provisions could have provided an excellent shield to the rights and security of the migrant workers; however, the real situation speaks something different. One can ask about the legal status of the migrants and about the authorized contractors and employers, as mentioned under the 1979 Act. Unfortunately, the long forty years since enactment has turned out to be inadequate to implement it properly (Sinha 2020).

Have the rights of the weaker sections always been safeguarded? The answer is "no," unfortunately. Instead, the forces of liberal capitalism worked critically. The greatest possible number of economic decisions, having been organized on individual lines, are made by individuals or households rather than collective institutions or organizations. It appeared as true for the cases of legislations that ere enacted to safeguard the rights of the vulnerable labor forces. The 2nd National Commission on Labor (2002) suggested minimal social protection for workers in the unorganized sector (PRS 2005). In 2006 the National Commission on Enterprises in Unorganized Sector (NCEUS) published a report where it was mentioned that the employers of the unorganized sector would get legal protection. However, it is also said to give social security to unorganized workers. The commission highlighted issues like health and education, low wages, and employment opportunities (GOI 2007).

Amid this pandemic situation, the ordinances of state governments ultimately favor the employers and worsen the condition of workers. Four states of India—Uttar Pradesh, Madhya Pradesh, Rajasthan, and Assam passed the ordinances which derogate the rights of workers. The government of Uttar Pradesh suspended all labor rights for the next three years (Gaur 2020). Except for some basic laws, the Madhya Pradesh government also freezes the labor laws (Haq et al. 2020). The new manufacturing units in Gujrat are exempted from prevailing labor laws for at least 1200 days. The state of Rajasthan also revoked the existing labor laws. The ordinances replaced the existing 8 h working norms per day to 12 h per day. It has snatched away the earned right of the working class, which has a long history of bloody struggle. The ordinance also says that the laborers will not get any remuneration as overtime duty for the mentioned 12 h (Sen 2020). India is readily a labor surplus country where there is every possibility for this section to be deprived if not safeguarded by the state. Still, the arguments from the corporate sectors and employers often corroborate their faints about the slip of cheap labors from their grips. The government of India needs serious intervention.

11.5 Conclusion

The pandemic has changed the world. However, the shock is not equal to everyone. For some people, it is only a health emergency; for others, it is a threat to their livelihoods. It has badly impacted the lower economic section of the society, especially the migrant workers. Amid this crisis and the frightful ambiance for the migrants, the Supreme Court of India has sincerely intervened with the governments after looking at the pathetic condition of migrant workers. On March 30, the Court asked the central government a status report regarding the measures taken for migrant laborers. The apex court also expressed that the big problem for a migrant is fear and panic, not the virus (Bureau 2020). On May 26, the Supreme Court questioned the role of central and state government. The three judges bench ordered both governments to take necessary steps as soon as possible to tackle the migrant crisis (Rajagopal 2020).

On May 28, three judges bench directed the state and union territories:

- to reach stranded workers to their respective home within 15 days;
- to smoothen the registration process for traveling of the migrant workers by decentralizing the process;
- to ensure the information related to transport options should be made available on local media;
- to provide essentials facilities that are needed at the destinations (Venkatesan 2020).

The present paper tries to trace how the initial health crisis of the pandemic has transformed into a severe socioeconomic shock with particular reference to migrant workers in India. Economic insecurity in their homeland has forced them to migrate to metros and cities; the social insecurity during the pandemic carries the panic for them. When these peoples desperately return to their homeland for survival during the crisis, their homelands treat them as "exotic" and stigmatize them as the "carrier of the virus." The pandemic exposed the hackneyed condition of migrant workers. These peoples are the people of "nowhere" in reality. When will the civil society think for this most economically and socially vulnerable workforce of the nation?

References

Agrawal P (2020) COVID 19 lockdown: Bihar migrants who fled cities faced ostracism at home. Scroll.in

Arya S, Ahmed S (2020) Haves want lockdown, have nots fear they will starve. Times of India

Babu V, Saini S, Swaroop V (2020) Across the country, migrants still forced to walk thousands of miles. Hindustan Times

Barik S (2020) A rude welcome for Odisha's migrants as villages lack basic facilities. The Hindu

Bhagat RB (2010) Internal migration in India: are the underprivileged migrating more. Asia Pac Population J 25(1):27–45

Bhagat RB, Keshri K (2018) Internal migration in India: Intensity, flows and impact. Paper presented in the workshop on "Comparing Internal Migration in the Countries of Asia", July 9–10, 2018, Asian Demographic Research Institute, Sanghai University, Sanghai, Chaina

Bureau OW (2020) 'Panic, fear bigger problem than Coronavirus': SC On Plea On Migrant Workers' Exodus. Outlook

Edwin T (2020) Why migrant workers are protesting: no money to buy essentials, limited access to cooked food. The Hindu Business Line

Gaur V (2020) Uttar Pradesh brings ordinance to suspend most labour laws for 3 years. Economic Times

Ghosh D (2020) Total lockdown from today, stay at home for 21 days: PM on corona virus. NDTV, March 25

GOI (1970) The contract labour (regulation and abolition) act, 1970. http://legislative.gov.in/sites/default/files/A1970-37_0.pdf. Accessed 12 Aug 2020

GOI (2007) Report on social security for unorganized workers: summary and recommendations. Indian J Hum Dev 1(1):187–195

GOI (2014) The street vendors (protection of livelihood and regulation of street vending) act, 2014. http://legislative.gov.in/sites/default/files/A2014-7.pdf. Accessed 10 Aug 2020

GOI (2019). Periodic labour force survey 2017–18. Retrieve from http://www.mospi.gov.in/sites/default/files/publication_reports/Annual%20Report%2C%20PLFS%202017-18_31052019.pdf. Accessed 7 Aug 2020

Handral AR, Anbukkani P, Kumar P (2018) Internal labour migration in India: recent trends and pattern. Ann Agrc Res 39(2):1–8

Hannant M (2020) In India lockdown protects the rich-and exposes poor a humanitarian and economic disaster. Scroll.in

Haq et al (2020) Some states put freeze on labour laws to get business going. Hindustan Times. https://timesofindia.indiatimes.com/business/india-business/remittances-to-india-likely-to-decline-by-23-in-2020-due-to-covid-19-world-bank/articleshow/75311447.cms

Joshi S (2020) Cops lathicharge migrants as thousands gather at Bandra station to leave Mumbai, defy lockdown orders. India Today

Kishore R, Jha A (2020) Economic impact of COVID-19 pandemic to vary in sectors. Hindustan Times
Kumar C, Mohanty D (2020). Migrant workers battle stigma, bias back home. Hindustan Times
Kundu I (2020) No end to migrant woes: after arduous journey, social stigma hits them at home. India Today
Mengle SG (2020) Coronavirus: Migrant workers protest outside Bandra railway station. The Hindu
Mishra S (2018) Social security for unorganised workers in India. J Soc Sci 53(2):73–80
Mishra D (2020a) 'No one wants to go near them'—returning migrant workers in Bihar face social boycott. The Print
Mishra L (2020b) Bricks fall apart for construction sector. The Hindu, March 30
Panchamukhi VR (2013) Changing pattern of Internal Migration in India 1999/00–2007/08: Issues and challenges. Indian Econ J 61(1):129–148
Pandey S (2020a) Social security for migrant workers during COVID-19. Economic and Political Weekly 55 (26–27)
Pandey G (2020b) Corona virus in India: Desperate migrant workers trapped in lockdown. BBC News
Pandey V (2020c) Coronavirus lockdown: the Indian migrants dying to get home. BBC News
Perappadan BS (2020) Preventing stigma related to COVID-19 requires full-throated campaign, says expert Gita Sen. The Hindu
Phukan S (2020). Disaster Management Act is a good law: Sibal. The Hindu
PRS (2005) Report of the national commission on labour. https://www.prsindia.org/uploads/media/1237548159/NLCII-report.pdf. Accessed 15 Aug
PTI (2020a) 35000 migrant workers sent to hometowns so far: Maharastra official. NDTV
PTI (2020b) 4 crore migrant workers in India, 75 lakhs return home so far: MHA. The Tribune
PTI (2020c) 7 death reported on-board migrant special trains on May 27. NDTV
PTI (2020d) Lockdown 2.0: Money and jobs gone, stranded desperate to return home. Economic Times
PTI (2020e) Lockdown in India has impacted 40 million internal migrants: World Bank. The Economic Times
PTI (2020f) Migrant resort to desperate means to reach home; centre asks states to permit more special trains. The Times of India
PTI (2020g) Migrant workers stage another protest in Surat, say they want to go home. NDTV
Rai S, Ramashankar (2020) What awaits migrants when they reach home. Times of India
Rajagopal K (2020) Supreme Court orders Centre and States to immediately provide transport, food and shelter free of cost to stranded migrant workers. The Hindu
Rawat M (2020). Corona virus in India: Tracking country's first 50 COVID-19 cases; what numbers tell. India Today, March 12
Sen S (2020) Rethinking migration and the informal Indian Economy in the time of a pandemic. The Wire
Sengupta S, Jha MK (2020) Social policy, COVID-19 and impoverished migrants: challenges and prospects in locked down India. Int J Commun Soc Develop 2(2):152–172
Sharma T (2020) "Extremely anguished": PM Modi on death of migrants hit by train in Maharastra. NDTV
Sharma S, Khanna S (2020) India's migrant workers face long walk aimed coronavirus lockdown. Reuters
Sing S, Magazine A (2020) Explained: Indian migrants, across India. The Indian EXPRESS
Sinha B (2020) This 1979 inter-state workmen law could help avert migrant crisis if enforced strictly. The Print
Skeldon R (1986) On migration patterns in India during the 1970s. Population and Development Review, pp 759–779
Sundar KRS (2020) Unorganised workers left high and dry by lockdown. The Hindu Business Line
The Hindu Net Desk (2020) Lockdown displaces lakhs of migrants. The Hindu

Tiwary G, Gangopadhyay PK, Biswas S, Nayak K, Chatterjee MK, Chakraborty D, Mukherjee S (2012) Socio-economic status of workers of building construction industry. Indian J Occup Environ Med 16(2):66
Venkatesan V (2020) Have SC's directions to redress migrants' suffering brought necessary course-correction? The Wire
WHO (2020) World Health Organization. WHO announces COVID-19 outbreak a pandemic. https://www.euro.who.int/en/health-topics/health-emergencies/coronavirus-covid-19/news/news/2020/3/who-announces-covid-19-outbreak-a-pandemic. Accessed 2 Aug 2020
Xu Z, Shi L, Wang Y, Zhang J, Huang L, Zhang C, Liu S, Zhao P, Liu H, Zhu L, Tai Y (2020) Pathological findings of COVID-19 associated with acute respiratory distress syndrome. Lancet Respir Med 8(4):420–422

Rupai Hembram is an assistant professor in the Department of Geography at Dr. Meghnad Saha College. He has completed his masters in Geography from Sidho-Kanho-Birsha University in the year 2016. Before engaging himself in teaching, he was a Junior Research Fellow at the Department of Geography at Sidho-Kandho-Birsha University. His areas of research interest are Population Geography, Social Geography, and Geography of transport.

Uttam Garai is an independent researcher. He has completed his masters in Geography from Sidho-Kanho-Birsha University in the year 2016. His areas of research interest are Population Geography, Social Geography, and Geomorphology.

Chapter 12
Human Discourses in Action: Community Health Workers' Contribution to Health Security and Pandemic Preparedness

Moumita Mondal

Abstract Ensuring healthy lives and promoting well-being at all ages is essential to sustainable development. Health emergencies, such as the increasing frequency of different non-communicable and communicable diseases, like the COVID-19, pose a global risk and have shown the critical need for preparedness. Community health workers have traditionally been used to improve community health initiatives, manage the risk of infectious diseases, and fill the healthcare system gap by extending health services to the last people. They can act as community-level educators and mobilizers, contributing to surveillance, monitoring, and healthcare referral systems. The present study discusses why developing countries like India should utilize community health workers' services to ensure the universal health of the Sustainable Development Goals-3, which seeks to improve community-level resilience against non-communicable and communicable diseases, notably the present pandemic.

Keywords Community health worker · COVID-19 · Pandemic · Universal health care system · SDG · Resilient

12.1 Introduction

A common proverb says, '*health is wealth.*' However, millions have no access to basic health care facilities globally. Health should be considered as one of the fundamental rights of the human being. Health does not mean the absence of disease or frailty, but it is a state of complete physical, mental, and social well-being (Alma-Ata declaration 1978).[1] The promotion and protection of people's health are essential to sustain economic and social development and contribute to a better quality of life and world

[1]The Declaration of Alma-Ata was first international Conference on Primary Health Care which was held in Alma-Ata, USSR, 6–12 September 1978. Available under the URL: https://www.unicef.org/about/history/files/Alma_Ata_conference_1978_report.pdf. Accessed on 10 August 2020.

M. Mondal (✉)
Department of Geography, Rammohan College, Kolkata, India
e-mail: moumitamondal1423@gmail.com

M. Mishra and R. B. Singh, *COVID-19 Pandemic Trajectory in the Developing World*, Advances in Geographical and Environmental Sciences,
https://doi.org/10.1007/978-981-33-6440-0_12

peace. Primary health care facilities are essential, especially in developing regions, for social justice attainment as part of development.

The declaration of Alma-Ata formally adopted primary health care (PHC) as the means for providing a comprehensive universal and affordable health care service for all countries (Hall and Taylor 2003). Primary health care services' significant objectives are the prevention of illness and promoting good health among all. Primary health care addresses health care in the community, providing promotive, preventive, curative, rehabilitative services accordingly.

Safe, people-centered, and integrated health services are critical for moving toward universal health coverage. According to WHO, people-centered care should ensure reaching the last person with a basic healthcare facility. It includes attention to people's health in their communities. Community health workers play a crucial role in shaping health policy and health services in remote areas. WHO supports countries in moving toward universal health coverage by improving their health service delivery system's efficiency and effectiveness. Primary health care is an interface between the basic healthcare system and common people. In developing countries, a large group of people stays out of the healthcare system. They cannot access basic healthcare facilities. To reach the marginal group of people, the primary health care system is a useful tool.

A primary healthcare system is inclusive in nature and interconnected with different people, institutions, and actions. Scientifically sound primary health care systems ensure universal health care accessible to all in a community. Primary Health Care is an approach to health beyond the traditional health care system that focuses on 'health equity producing' social policy. Health is an integrated part of society. Primary health care helps to minimize the negative impact of social inequalities on health.

Primary Health Care System cannot achieve its full potentiality without the participation of Community Health Workers (CHWs). Communities in remote areas may have difficulties in accessing basic health services. This problem is more in low-income countries where improvised governance, lack of accountability in the health-care system, and poor healthcare accessibility are general features. Recruitment of health workers from the community may ease the accessibility of the community. So, a well-structured health system with a strong workforce is crucial to improve primary health care. A large number of Community Health Workers with shorter training can be transformed into human resources. They can provide a wide range of health-related services and assistance to the community by creating a bridge between the health service system and communities. The formal health system supports the CHWs with their shorter training period than professional health workers. So, they can be called the instant and super flexible health workforce, which can be stretched out at emergency times.

Disease patterns have been changing due to the changing environment and lifestyle. Moreover, premature death has increased due to the increase of communicable and non-communicable diseases (Global status report on non-communicable diseases, 2010). Non-communicable diseases (NCDs), such as heart disease, stroke, cancer, chronic respiratory diseases, and diabetes, are the leading cause of mortality in the world. These non-communicable diseases are mostly overlooked in low-income countries (Gowshall and Taylor-Robinson 2018). The number of patients with these NCDs is increasing in those countries at an alarming rate. These NCDs are taking the form of an invisible epidemic that passively influences the vicious cycle of poverty in underdeveloped countries. Non-communicable diseases need long-term care and support from the healthcare system. An organized and robust primary health care integrated with secondary and tertiary care levels may reduce diseases' burden. Community health workers can be recruited to monitor and monitor non-communicable diseases so that emergency cases can be referred to as the next level of care. Surveillance and record-keeping of the diseased people help countries have the information they need to fight epidemics. They are fundamental pillars of public health.

Community health workers are known as different names in different countries; in India, they are known as ASHA (Accredited Social Health Activist), family welfare assistant (FWA) in Bangladesh. They shoulder the implementation of the government's health-related policies at the ground level. Their persistent efforts have helped to run universal vaccination successfully and improve mother and child health in developing countries. They are also prime health workers in different health-related programs, e.g., malaria and dengue eradication programs, TB eradication programs, diarrhea eradication programs, etc. In this global pandemic situation, community health workers have been working tirelessly as frontline workers. The workers visit houses and advice symptomatic patients to stay isolated and check their status every day. Ground-level surveillance has become possible for this workforce.

The pandemic has put the most prominent question before the policymakers while they will start to rethink health structure. Health-related disasters are comparatively undermined in international dialogue. Disaster and 'disaster risk,' which are unexpected events implying serious health, economic, and political threats, pose challenges to any country's development process. It requires special consideration and management skills beyond routine procedures to minimize the country's negative impact on Gross Domestic Production (GDP). Large-scale communicable diseases outbreak (like the present COVID-19 pandemic) represent such a disaster (Boyce and Katz 2019). Importantly, these communicable disease outbreaks are frequent, and this is usual. So global leaders need to rethink resilience building against those diseases. A huge trained workforce is urgently needed to cope up with the situation. Community health workers can be deployed for that purpose. On the other hand, growing numbers of non-communicable diseases are subjects of significant concern worldwide. Children also have a greater likelihood of facing fatalities from non-communicable diseases (such as rheumatic heart disease, type-1 diabetes, asthma, etc.) if comprehensive care for disease prevention is not provided. According to WHO's projection, the total annual number of deaths from non-communicable diseases will increase to 55 million by 2030 if no intervention occurs against these diseases (Global status report on non-communicable diseases, 2010).

The Alma-Ata declaration (1978), for the first time, considered the importance of the community health workers for 'universal health coverage' achievement. The Alma-Ata declaration is committed to achieving comprehensive primary health care and emphasizes CHWs as the cornerstone of this effort. The conference recommended that the governments prioritize the full utilization of human resources by defining the technical role, supportive skills, and attitudes required for each category of health workers according to the functions that need to be carried out to ensure effective primary health care. They should be trained and retrained based on the area's problem as they can play a more significant role in providing primary health care. Primary health care is to be most effective if it employs means that are understood and accepted by the community and applied by the community health workers at a cost the community and the country can afford.

Health did not get important in disaster risk reduction-related conferences until Sendai Framework for Disaster Risk Reduction (SFDRR 2015) was held.[2] The need to focus more on people's health and livelihoods gave birth to Sendai Framework for Disaster Risk Reduction (Opemo 2020). The *Hyogo Framework* (2005–2015) recommended integrating disaster risk reduction into the health sector and emphasized on hospital-centric treatment systems.[3] Community people's engagement in the health system was not mentioned there. The Sendai Framework has encouraged the national health system's resilience building by integrating disaster risk management into primary, secondary, and tertiary health care at the local level. The *Sendai Framework* (2015) emphasized promoting and training the community health group in disaster risk reduction approaches in health programs. Health is a fundamental resource that enables the human being to cope with an adverse environment. Basic preparedness in health structure can reduce the intensity of the outbreak effectively. Risk is always associated with disasters, so resilience capacity building is important to minimize the risk. A disaster such as a pandemic causes illness of a large number of people within a shorter time that causes an acute shortage of human resources and tremendous pressure on health institutions to serve the patients. Resilience building against the epidemics or pandemic is possible by scientific perspective with a humanitarian approach.

The 2030 agenda for sustainable development goals was aimed to end poverty and lead the world toward prosperity and opportunity. Sustainable development goals talk about equality and responsibilities over the environment, economy, and society beyond the border.[4] Discrimination, in terms of resource distribution, is a major hindrance to sustainable development goals. Billions of people worldwide shared

---

[2]Sendai Framework for Disaster Risk Reduction is an international document which was adopted by UN member states between 14 and 18 March 2015 at the world conference on Disaster Risk Reduction held in Sendai, Japan. Available under the URL: https://www.preventionweb.net/files/43291_sendaiframeworkfordrren.pdf. Accessed on 15 July 2020.

[3]Global Action Plan for The Prevention and Control of Noncommunicable Diseases (2013–2020). Available under the URL: https://apps.who.int/iris/bitstream/handle/10665/94384/9789241506236_eng.pdf;jsessionid=933C6568A5B68EE1CDF7348590B19ABE?sequence=1. Accessed on 10 September 2020.

[4]UN system task team on the post—2015 UN Development Agenda; (review of the contribution of the MDG Agenda to foster development: Lessons for the post—2015 UN development agenda); March 2012. Available under the URL: https://sustainabledevelopment.un.org/content/documents/843taskteam.pdf. Accessed on 31 August 2020.

their hope for a better future. For a better future, 17 goals were adopted by the global leaders in 2015. The third goal of SDGs ensures healthy lives and promotes well-being for all ages. Target 3.8 of Sustainable Development Goals (SDGs 2015) speaks for 'universal health coverage' and 'health security' and encourages to build up a strong health workforce, especially for developing nations. World nations must need to ensure public health at an affordable cost to achieve the targets. Financial crises associated with weak health systems jeopardize other SDGs (Mackey et al. 2018). Health is essential for achieving other goals, such as poverty and hunger eradication, inclusive and sustainable economic growth. Indeed, many of the goals are interrelated with health. Outbreaks of any kind of epidemic or pandemic may challenge the achievement of the goals. Development is a continuous and inclusive process. Any uncertainty may slow down the process of development.[5]

12.2 Pandemic and Health: Communicable Disease and Non-communicable Diseases

The pandemic is an extraordinary situation, while the fast-spreading of diseases occurs across a wide geographic area over a shorter period. Pandemic affects society, economy, and the polity of the region negatively. A considerable number of people also get affected at the time of the pandemic. It is basically a 'health emergency' when special attention in health care is required to overcome the situation. But a healthcare system without prior preparedness cannot be able to handle the pandemic situation. Most of the time, pandemic concepts are associated with communicable or very fast-spreading diseases. On the contrary, non-communicable diseases which are non-infectious in nature but alarming levels of these disease occurrences are a matter of concern nowadays.

'Communicable diseases' are caused by microorganisms (e.g., viruses, bacteria, parasites, etc.) that quickly spread from one person to another through the air, water, blood, or contaminated fomites, etc. (WHO, Regional Office for Africa 2020). Sometimes insects play an important role in the transmission of these diseases (WHO, Regional Office for Africa 2020). Surveillance and immediately reporting of cases are important in communicable disease management as communicable diseases take the form of pandemic quickly, if not managed properly. Sustainable development goal-3, specifically target 3.3, concerns infectious diseases. A large number of people from low-income and lower-middle-income countries are suffering from diseases like AIDS, tuberculosis, malaria, and neglected tropical diseases and waterborne diseases, including other communicable diseases. Collectively, these diseases accounted for an estimated 4.3 million deaths in 2016.[6]

---

[5]See: Sustainable Development Goals were adopted by UNGA on 25 September 2015; UNO. Available under the URL: https://sdgs.un.org. Accessed on 15 July 2020.

[6]For details, view the World Health Statistics 2019, Available under the URL: https://apps.who.int/iris/bitstream/handle/10665/324835/9789241565707-eng.pdf. Accessed on 24 September 2020.

Now we are living in the era of globalization when integration and free movement has increased. We need to depend on and interconnect with each other for daily livings. An uninterrupted movement of people and goods among different regions, high population density due to urbanization, and the exploitation of natural resources increase the transmission of communicable diseases. Hygienic practices and precautionary measures can check communicable disease transmission only.

On the other hand, non-communicable diseases proliferate, especially in developing countries, due to changing lifestyles and environments. The burden of non-communicable diseases results in the backwardness of socio-economic conditions in those countries (Global status report on non-communicable diseases, 2010). People are compelled to expense more for health care instead of better lives. High expenditure for non-communicable diseases' treatment is considered as a threat to poverty reduction efforts in developing nations. Unless serious action is taken to address these diseases timely, the burden of non-communicable diseases (NCDs) will reach the levels beyond all stakeholders' capacity to manage.

Non-communicable diseases (NCDs), such as cardiovascular diseases, cancer, diabetes, and chronic respiratory diseases, are the leading global causes of death and are responsible for just over 70% of death worldwide.[7] These non-communicable diseases (NCDs) share key modifiable behavioural risk factors like consumption of tobacco and alcohol, unhealthy diet (a diet high in saturated fat and trans fat, low intake of fruits and vegetables), and physical inactivity, leading to overweight and obesity, high blood pressure, and ultimately disease (Non-communicable diseases progress monitor 2020) (see Footnote 8). They continue to be an important public health challenge in all countries, including low- and middle-income countries where more than three-quarters of NCD deaths occur.[8]

The magnitude of the epidemic due to non-communicable diseases has been rising in recent times though the diseases are preventable to a great extent. Surveillance and preventable measures against these diseases are essential components in health care to reverse these diseases' advances. NCD risk factors are spread throughout society, and they often begin early in life and continue through adulthood. Both prevention and treatment interventions are necessary to decline the rate of NCDs cases (Global status report on non-communicable diseases, 2010). Achievement of this goal is also possible for low-income countries through the primary health care system reform. Improvement in health care system performance should be implemented to improve NCD control outcomes. As increasing NCDs tends to impact the development process of developing countries negatively; therefore, those countries should prioritize NCDs prevention and control efforts. Community health workers can be deployed in surveillance and primary treatment intervention of NCD patients.

---

[7]Global status report on noncommunicable diseases 2010. Available under the URL: https://apps.who.int/iris/bitstream/handle/10665/44579/9789240686458_eng.pdf?sequence=1. Accessed on 19 September 2020.

[8]Noncommunicable diseases progress monitor 2020. Available under the URL: https://www.who.int/publications/i/item/ncd-progress-monitor-2020. Accessed on 24 September 2020.

12.3 Pandemic and Community Health Worker

Pandemic is an emergency health situation, while a significant proportion of the population gets affected over a wide geographical area. Suddenly demand for health-care workers increases due to pandemic as the number of patients hikes in a shorter period. We have seen that most advanced countries in the healthcare system also have disrupted due to the outbreak of COVID-19. So for developing countries, it is a difficult challenge. There is always a considerable deficit in the actual demand and availability of trained health workers in developing countries.[9] It becomes more challenging to cope with the situation for low-income countries at the time of outbreaks.

Pandemic affects the health workers not only physically but also psychologically. They get frequent contact with the infected persons that raises their vulnerability to communicable diseases. The shortage of healthcare equipment and trained health workers creates tremendous work pressure on the existing workforce that causes physical as well as psychological stress on them.

In most countries, most physicians and pharmacists in the health workforce are male, while female health workers dominate in the nursing and midwifery work-force. The *Labour Force Surveys* (LFS) data from 57 countries confirm this trend (Fig. 12.1). Women's representation in the health workforce has increased in developing countries, too, as the primary health system has been fueled up through community health workers recruitment in those countries. In most countries, including the Indian subcontinent, community health workers are basically all women's health workforce. In pandemic situations, they face biases and violence in their work field as well as in domestic spaces. The community health workers are deployed to field visits where they often get assaulted by the locals. On the other side, they also face social restrictions from their neighbors (sometimes from inside the family) as they have a higher chance of infection during field visits.

Community Health Workers deliver essential preventive, promotive, and limited curative primary health care services to the community normally. However, at the pandemic, they are tasked to perform several activities related to the health crisis and their mandated works. Amidst this pandemic, they have ensured expecting and new mothers and child health in India (Boniol et al. 2019).

Communicable diseases can result in pandemic very quickly. So, early awareness creation and contact tracing are important to prevent the spread of diseases. If information spreads faster than the disease, then pandemic may be prevented. We live in a time when misinformation and lack of understanding can have devastating consequences in low-income communities. Here community health workers experience connecting people and spreading information are very much useful. The community health workers are able to influence community through sharing insights and field experiences with local decision-makers, providing advice and helping people to access primary health care services (Kane et al. 2016).

[9] View details at: https://www.who.int/mediacentre/news/releases/2013/health-workforce-shortage/en/. Accessed on 30 July 2020.

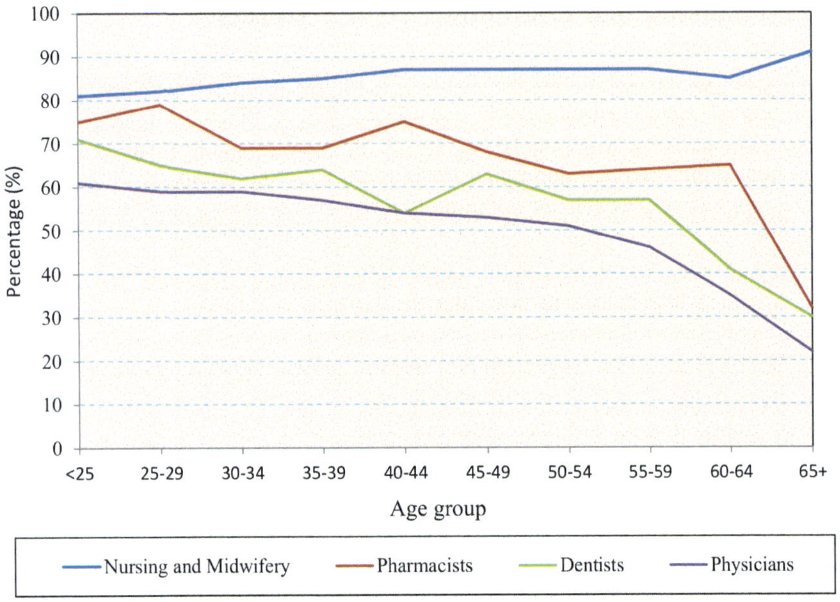

Fig. 12.1 Share of women health workers in different roles by age group (*Source* Prepared by the author from Boniol et al. 2019 with the LFS Survey datasets [see: https://apps.who.int/iris/bitstream/handle/10665/311314/WHO-HIS-HWF-Gender-WP1-2019.1-%20eng.pdf?ua=1])

Developing nations are struggling to mitigate the basic needs of their people. Awareness of sanitation and hygienic practices are rare in those countries. However, these are fundamental requirements to battle against communicable diseases. So, to educate the people, community health workers may play an important role. They serve like community messengers who inform locals about where to access for Covid-19 testing in this pandemic or health services or vaccination, etc. Their outreach skill is very useful for contact tracing in the pandemic time.

Outbreaks cannot be prevented from occurring, but they can be instrumental in more quickly containing these health threats by reporting sentinel cases. It is estimated that up to $750 million economic losses per year could be avoided through CHW scale-up, considering community health workers' potential capabilities in containing public health emergencies (Dahn et al. 2015). The SFDRR has recommended public and private investment in training the community health workers to mitigate disease outbreaks risk.

So, an emerging consensus among global health leaders is that building stronger health delivery systems, with special emphasis on community-based primary health care, will be required in the future to ensure adequate preparedness against future epidemics and universal health coverage (Dahn et al. 2015). Community health workers should be integrated with the existing health system infrastructure for better functioning.

World Health Organization organized the Global Conference on Primary Health Care in Asthana, Kazakhstan, in October 2018. The conference endorsed a new declaration emphasizing the crucial functions of primary health care around the world (Global Conference on Primary Health Care, 25–26 October 2018, Astana, Kazakhstan). It aims to refocus efforts on primary health care to ensure that everyone can enjoy the highest possible attainable standard of health. Community health workers were recognized as vital to achieving that goal.

12.4 Importance of Health in Sustainable Development Goals

Sustainable Development Goals were adopted on 25 September 2015 by the UN General Assembly, where 17 goals and 169 associated targets were taken for a better and more sustainable future (Table 12.1). We cannot aim to achieve just one goal as they are all interconnected. There must be an integration in their achievement to transform the world. These goals are aimed to be achieved by 2030 and need multi-stakeholder involvement for implementation.[10] For that, an integrated insight is essential for holistic progress in multiple goals. It aims to end inequality in resource distribution and establish justice and a sustainable environment in both developed and developing countries.

Health is closely linked to the other 16 goals of sustainable development, and is dependent on the their acheivements (Bueno de Mesquita et al. 2018). Sustainable development goal 3 speaks to ensure healthy lives and promote well-being for all ages, including the commitment to ending the epidemics of AIDS, TB, Malaria, and other communicable diseases by 2030. More specifically, to achieve targets 3.1 and 3.2, the security of essential nutrients, safe drinking water, affordable, effective medicines, and universal vaccination is required for all.[11] To fight against communicable and non-communicable diseases, the spread of awareness is important, and it is possible through education, empowerment, and collaboration at different levels only. The member states must ensure access to health services by introducing affordable essential medicine and vaccine for all and providing protection against catastrophic health expenditure to achieve target 3.8 of sustainable development goals.

Obstacles in management and financial hardships are generally associated with weak health systems. Inadequate financing mechanisms intensify inequities and injustice in health, which ultimately jeopardizes the achievement of other SDGs. For instance, according to the Global Status Report on non-communicable diseases 2010, almost 100 million people have pushed into poverty annually because of catastrophic health expenditure. More than half of the world's population is hardly able to access essential health services. Globally increasing non-communicable diseases

[10]Available at the URL: https://sustainabledevelopment.un.org/partnerships/about.

[11]Available under the URL: https://www.who.int/sdg/targets/en/. Accessed on 30 September 2020.

Table 12.1 A glimpse on the Sustainable Development Goals

• Goal 1: End poverty in all its forms everywhere
• Goal 2: End hunger, achieve food security and improved nutrition, and promote sustainable agriculture
• Goal 3: Ensure healthy lives and promote well-being for all at all ages
• Goal 4: Ensure inclusive and equitable quality education and promote lifelong learning opportunities for all
• Goal 5: Achieve gender equality and empower all women and girls
• Goal 6: Ensure availability and sustainable management of water and sanitation for all
• Goal 7: Ensure access to affordable, reliable, sustainable, and modern energy for all
• Goal 8: Promote sustained, inclusive, and sustainable economic growth, full and productive employment, and decent work for all
• Goal 9: Build resilient infrastructure, promote inclusive and sustainable industrialization, and foster innovation
• Goal 10: Reduce inequality within and among countries
• Goal 11: Make cities and human settlements inclusive, safe, resilient, and sustainable
• Goal 12: Ensure sustainable consumption and production patterns
• Goal 13: Take urgent action to combat climate change and its impacts
• Goal 14: Conserve and sustainably use the oceans, seas, and marine resources for sustainable development
• Goal 15: Protect, restore, and promote sustainable use of terrestrial ecosystems, sustainably manage forests, combat desertification, and halt and reverse land degradation and halt biodiversity loss
• Goal 16: Promote peaceful and inclusive societies for sustainable development, provide access to justice for all, and build effective, accountable, and inclusive institutions at all levels
• Goal 17: Strengthen the means of implementation and revitalize the Global Partnership for Sustainable Development

Source UNDP's sustainable development report (2019)

which need long-term treatment and high medication costs result in a catastrophic expenditure.

Therefore, universal health coverage could contribute to achieving the SDGs by producing equitable and sustainable health outcomes. Gaps in promising governance compound many health disparities between people with different socioeconomic statuses. For instance, poor people are more vulnerable to access to health services, medication, and information. Corruption in the system enhances their vulnerability. In addition, factors such as ethnicity, gender, age, disability, and location can further exacerbate these health disparities. Hence, it is essential for tracking indicators that measure the health of vulnerable groups. Monitoring the status of equitable access to health care could also illuminate the status of human rights and social equality within the states. Those people not receiving adequate health services are probably also disadvantaged in other social aspects. A better understanding of factors contributing to access to health services will help shape policies to attain SDG-3 and support the achievement of other SDGs such as attaining gender equality, reducing poverty, and improving education (Table 12.2).

Some health-related targets outside of the SDG-3, which are equally essential to *ensure healthy lives and promote well*-being for all, are mentioned below:

- ***Target 2.2***: ending of all forms of malnutrition and ensure the nutrition needs of adolescent girls, pregnant and lactating women, and older persons.
- ***Target 6.1***: ensure universal and equitable access to safe and affordable drinking water for everyone
- ***Target 6.2***: ending of open defecation and ensure access to adequate and equitable sanitation and hygiene for everyone
- ***Target 11.6***: reduction of the adverse per capita environmental impact of cities.

Table 12.2 Sustainable Development Goal 3 and associated targets

Goal 3. Ensure healthy lives and promote well-being for all at all ages by 2030
• **3.1** Reduction of the global maternal mortality ratio to less than 70 per 100,000 live births
• **3.2** End preventable deaths of newborns and children under 5 years of age. it aims to reduce neonatal mortality to at least as low as 12 per 1,000 live births and under-5 mortality to at least as low as 25 per 1,000 live births
• **3.3** End the epidemics of AIDS, tuberculosis, malaria, and neglected tropical diseases and combat hepatitis, water-borne diseases and other communicable diseases
• **3.4** Reduce by one-third premature mortality from non-communicable diseases through prevention and treatment and promote mental health and well-being
• **3.5** Strengthen the prevention and treatment of substance abuse, including narcotic drug abuse and harmful use of alcohol
• **3.6** Halve the number of global deaths and injuries from road traffic accidents
• **3.7** Ensure universal access to sexual and reproductive healthcare services, including for family planning, information and education, and the integration of reproductive health into national strategies and programs
• **3.8** Achieve universal health coverage, including financial risk protection, access to quality essential healthcare services, and access to safe, effective, quality, and affordable essential medicines and vaccines for all
• **3.9** Substantially reduce the number of deaths and illnesses from hazardous chemicals and air, water, and soil pollution and contamination
• **3.a** Strengthen the implementation of the World Health Organization Framework Convention on Tobacco Control in all countries
• **3.b** support the research and development of vaccines and medicines for the communicable and non-communicable diseases that primarily affect developing countries, provide access to affordable essential medicines and vaccines, in accordance with the Doha Declaration on the TRIPS Agreement and Public Health
• **3.c** Substantially increase health financing and the recruitment, development, training, and retention of the health workforce in developing countries
• **3.d** Strengthen the early warning and risk reduction capacity of all countries, particularly in developing nations for management of national and global health risks

Source UNDP's sustainable development report (2019)

- ***Target 13.1***: strengthen resilience and adaptive capacity to increasing climate-related hazards and natural disasters in all countries
- ***Target 16.1***: reduction in violence and related death rates across the globe.

12.5 Community Health Workers in Achievements of Sustainable Development Goals

The task force of community health workers has played a significant role in reducing maternal and child mortality rates in different developing countries, including India. A successful universal vaccination program, implemented by the community health workers, helped India to be a polio-free nation (How India Managed to Defeat Polio, BBC NEWS, 13 January 2014). As they are part of the community, they have an easy reach to the community, which can be useful for universal health coverage. Basic rights of health for the remotest people can be ensured by community health workers' engagement as well as they can assist trained nurses and doctors at the time of health emergencies.

In India, more than 0.9 million ASHA workers have been designated to assist with the healthcare system as frontline warriors (Jain D. 2020). They reach up to the grassroot level of society with healthcare services. Several states have tasked ASHA workers to visit door to door, collect information about recent travel history of people and support home quarantine as per requirement. ASHAs are also involved in promoting safe practices among the community, e.g., frequent handwashing, maintaining social distancing, practicing good respiratory hygiene, etc. (Jain 2020).

This pandemic is a reminder that the hospital-centered health structure is fragile before communicable diseases. Policymakers have to rethink in different ways for the next pandemic preparedness. Institution-based treatments are not adequate due to limited beds and must be replaced by community-centered care (Nacoti et al. 2020). Treatment at the time of pandemic is required for the entire population. So those patients who could not access or did not get a chance to admit in the bed will be deprived of their right to health. It is against the concept of universal health coverage.

On the other hand, hospitals become the hotspot of communicable diseases as infected patients rapidly populate them. It causes transmission of disease to uninfected persons. This kind of disaster could be averted only by a massive deployment of outreach services of health care, which is only possible by implementing community-centered health care services. Home care and mobile clinics also avoid unnecessary movements and relax pressure from hospitals (Nacoti et al. 2020).

For nurses and doctors, it is hardly possible to provide services outside of institutions. And most of the developing nations are suffering from a shortage of doctors and nurses. So, in this context, community health workers can be the most useful human resources as they can visit each household in their area. They can provide home-based clinical care by providing early medicines according to symptoms of patients. A broad surveillance system can be created with the community health workers and local people (Nacoti et al. 2020). This approach limits hospitalization, thereby decreasing contagion, and protecting patients and health care workers (Nacoti et al. 2020).

WHO has declared deep concern about the spread and severity of the Corona Pandemic at the present time and about the alarming infection level. Strong actions are taken by countries to slow down the infection rate. However, it is difficult to believe that people will follow the health instructions and cooperate with officials to break the disease chain in developing countries like India. So, despite 68 days of lockdown and 3 months unlock period with regulations, total cases and death toll raised more than 3.5 million and more than 62 thousand, respectively, at the end of August in India. The health system has just collapsed. Now leaders are also suggesting home-based care for mildly ill Covid-19 patients.

12.6 Health-Related SDGs Achievement and Pandemic

Universal access to good quality health care services for good health and well-being without facing financial hardship got priority in Sustainable Development Goals, 2015. Before the pandemic started, uneven progress was there. However, the pandemic has disrupted the implementation of many SDGs, including health as well. Vaccines are considered as a powerful tool of public health but the pandemic has turned back remarkable progress on vaccination. The crisis has affected all segments of the population; it has engulfed all sectors of economy; it has reached all areas of the world. Not surprisingly, it is affecting the world's poorest and most vulnerable people the most. More than half (53%) of the 129 countries where data were available reported either moderate-to-severe disruptions or a total suspension of childhood vaccination services for arising pandemic situation during March–April 2020 (see Footnote 10).

Sustainable development goal report 2020 says that essential health services covered about one-third to one half of the global population in 2017, estimated to be between 2.5 billion and 3.7 billion of the global population. In contrast, only 12–27% of the population in low-income countries were fully covered by the essential health services that year. If current trends continue, only 39–63% of the global population will be covered by such services by 2030, which is far from the target, and disparity in health care will remain. The pandemic has a higher risk of exclusion for the vulnerable population. Now it is a challenge for global leaders to continue essential health services coverage at the rate of the pre-pandemic era.

As COVID-19 continues to spread, many health facilities are closed or provide only limited services on an emergency basis due to limited health infrastructure. What is more, many women and girls choose to skip important medical check-ups for fear of contracting the virus. Also, the global supply chain disruptions may lead to shortages of contraceptives. As a result, tens of millions of women may not be able to access contraceptive services, resulting in millions of unintended pregnancies.

In developing countries, the need for public health is neglected very much. Governments of the countries are reluctant to fund adequately in the health care system. This inadequate infrastructure and inadequate investment cause high morbidity and the fast-spreading of communicable diseases. Communicable diseases can be prevented from forming pandemic by breaking the chain at the proper time. Millions of the world are dying from preventable disease conditions each year without getting the treatment and care due to poor health infrastructure. Now this pandemic has created an extraordinary situation where extra burden has disrupted the essential health services. Keeping this in mind, our governments have to plan for the long term and revise their allocation in the health budget.

Until the end of 2019, advances in many health areas continued, but the progress rate was not sufficient to meet most Goal 3 targets (sustainable development goal report 2020). The COVID-19 pandemic is throwing progress even further off track. Now people are afraid to visit the healthcare institutions for check-ups, vaccinations, and even for urgent medical care, which could have potentially fatal consequences and threatens to reverse decades of improvements in health outcomes.

Childhood immunization efforts have been interrupted due to the COVID-19 crisis globally. Since March 2020, routine childhood immunization services have been disrupted on a scale not seen since the inception of the Expanded Programme on Immunization in the 1970s (sustainable development goal report 2020). Prevention and treatment services for non-communicable diseases have been severely disrupted since the COVID-19 pandemic began. The health care system of low-income countries is mostly affected by this pandemic. Many people are deprived of the basic health services and medicines they need. COVID-related interruptions in health care could cause a spike in illness and deaths from other communicable diseases like malaria, tuberculosis, etc.

According to the Sustainable Development Goals report 2020, due to COVID-19 pandemic, progress in the malaria eradication program has stalled after many years of impressive reductions in malaria's global burden (Fig. 12.2). The world is not on a trajectory to achieve the SDG target of ending malaria. Worse yet, malaria prevention services could be disrupted by the spread of COVID-19 in malaria-endemic countries, mostly developing countries. Recent modeling shows that the cancelation of prevention campaigns and severe disruptions in treatment in sub-Saharan Africa could lead to a 23% increase in cases and a cent percent increase in deaths by the end of 2020 (compared with a 2018 baseline) (sustainable development goal report 2020).

The sustainable development goals report, 2020 also apprehends more. COVID-19 pandemic if leads to a 25% global reduction in expected tuberculosis detection for three months, there would be a realistic possibility of a 13% rise in tuberculosis deaths (given, the levels of disruption being observed in multiple countries remain constant). This would bring the world back to the tuberculosis mortality levels of five years ago. Due to COVID-19, many regions face temporary suspension of community-based activities in neglected tropical diseases management, which may erode gains won by years of hard work and investment.

As Covid-19 has interrupted the basic essential health services, marginalized people have been mostly deprived of essential public healthcare services. To achieve universal health coverage by 2030, governments have to accelerate investment in the

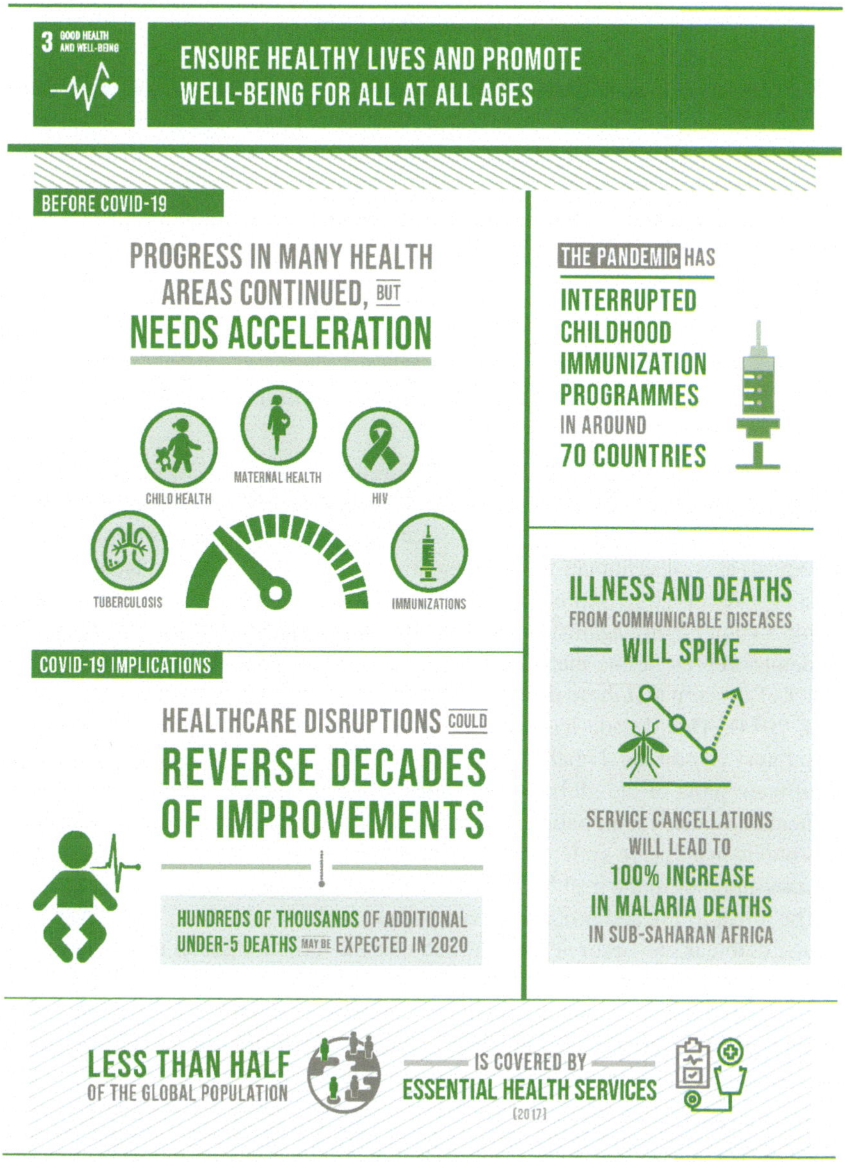

Fig. 12.2 Sustainable Development Goals Report 2020 (*Source* UNO 2020 [sustainable development goals report, 2020. UNO. Available under the URL: https://unstats.un.org/sdgs/report/2020/The-Sustainable-Development-Goals-Report-2020.pdf. Accessed on 16 September 2020])

health sector and make the healthcare system affordable and accessible for all. Income cut due to Covid-19 prevention purpose lockdown has forced many families to curb their expenditure on health, as well as this pandemic has highlighted the shortage of medical personnel worldwide. GDP of developing countries and the reluctance of government do not allow large investment in human resource development. So, for the countries where the health system is unable to cope with the surge of demand, the community health workers may be an effective alternative solution for this purpose. They can provide basic health care needs to a wider population based on their short-term training.

12.7 Preparedness for Health-Related Disaster

Disasters are unpredictable and inevitable events on the earth that cause serious disruption to human lives, resource management, and the environment over a short or long period. Disaster can have a natural origin (geological, hydrometeorological, and biological) as well as human-induced (environmental degradation and technological hazards). We can't stop the disasters from occurring, but our preparedness can reduce people's vulnerability against disasters. Large-scale infectious disease outbreaks or pandemics represent one manifestation of such events. Importantly, it should be a matter of concern that these diseases' outbreaks are increasing in frequency (Smith et al. 2014). This decade has witnessed the outbreaks of swine flu, Middle East Respiratory Syndrome, Nipah, Ebola, Cholera, Zika, yellow fever, rift valley fever in different parts of the globe as an epidemic. To handle the situation and ensure medical assistance for all patients, a well-built public health care system is required. This public health care system acts as a firewall to reduce the community spread of diseases and relieve the healthcare system's unsustainable pressure.

The threat to life and property is everywhere in the world in any form. Any new disease outbreak can occur at any time and take the form of disaster if we can't build up the capacity to resist it. Healthcare infrastructure in most of the countries is not sufficiently strong to combat the situation. This will increase the vulnerability of the population if they are not well prepared. *Vulnerability* is the state of being exposed to the possibility of being attacked or affected (Fig. 12.3). Vulnerability is defined in

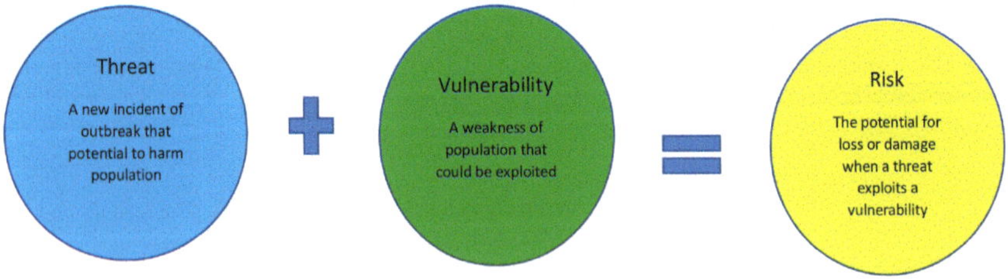

Fig. 12.3 Relation of threat, vulnerability, and risk for any disaster (*Source* Prepared by the author)

the Hyogo Framework for Action (2005–2015) as: 'The conditions determined by physical, social, economic and environmental factors or processes, which increase the susceptibility of a community to the impact of hazards'. Risk is the result of threats and vulnerability. A more vulnerable population to any disease can be exploited easily. So, to cope with the situation, *capacity building* is important, which means empowering the local people with information, training, and equipment to build up resilience in the community. For that, prior investment and proper planning in the health sector from the community level is required. The involvement of grassroots level people in capacity building is more useful to combat the situation.

The Hyogo Framework (2005–2015)[12] and Sendai Framework for Disaster Risk Reduction (SFDRR 2015) gave importance to build up robust disaster preparedness for effective response at all levels. SFDRR prioritized *resilience building* through supporting and training community health groups for disaster risk reduction. *Resilience* is considered as the ability to accommodate, resist, and recover from the effects of disasters. SFDRR had recommended the implementation of International Health Regulations (IHR 2005)[13] of WHO. It spoke for strengthening national preparedness, surveillance, and response capacities that were suggested for public health management and preventing the international spread of diseases. Timely surveillance and preparedness enable communities to build up resilience against the pandemic. Community health workers are an important asset in the *capacity building* and preparedness (Chatterjee 2020) as they have a wide reach within the community.

Since adopting the Sendai Framework, the WHO has encouraged the member countries to strengthen its approach to Health Emergencies and Disaster Risk Management (Table 12.3). The General Programme of Work 13 for 2019–2023 recognizes that 'the world faces threats from high-impact health emergencies (epidemics, pandemics, conflicts, natural and technological disasters) and the emergence of antimicrobial resistance' (WHO 2018).

In the context of health emergencies, the WHO promise to:

> work with the Member States and partners to increase all-hazards health emergency detection and risk management capacities across all phases of risk prevention and detection, emergency preparedness, response and recovery through the implementation of the IHR (2005) and the Sendai Framework for Disaster Risk Reduction. (WHO 2018)

The Health EDRM Framework aims to strengthen resistance capacity within and beyond the health sector. It confronts the health impacts of all types of emergencies and disasters and efforts collaboratively to reduce future events' health risks (Wright et al. 2020).

---

[12]Hyogo Framework for Action 2005–2015, was the global blueprint for disaster reduction effort which was held on 18–22 January 2005, Kobe, Hyogo, Japan. Available under the URL: https://www.preventionweb.net/files/1037_hyogoframeworkforactionenglish.pdf. Accessed on 31 August 2020.

[13]International Health Regulations, 2005 are not limited to specific diseases but apply to new and ever changing public health risks.

Table 12.3 Functions and Components of the World Health Organization's Health Emergency and Disaster Risk Management Framework

• Policies, strategies, and legislation define the structures, roles, and responsibilities of governments and other actors for Health EDRM; Includes strategies for strengthening Health EDRM capacities
• Planning and coordination emphasize effective coordination mechanisms for planning and operations for Health EDRM
• Human resources include planning for staffing, education, and training across the spectrum of Health EDRM capacities at all levels and personnel's occupational health and safety
• Financial resources support the implementation of Health EDRM activities, capacity development, and contingency funding for emergency response and recovery
• Information and knowledge management include risk assessment, surveillance, early warning, information management, technical guidance, and research
• Risk communications recognize that communicating effectively is critical for health and other sectors, government authorities, the media, and the general public
• Health infrastructure and logistics focuses on safe, sustainable, secure, and prepared health facilities, critical infrastructure (e.g., water, power), and logistics and supply systems to support Health EDRM
• Health and related services recognize the wide range of healthcare services and related measures for Health EDRM
• Community capacities for Health EDRM focus on strengthening local health workforce capacities and inclusive community-centered planning and action
• Monitoring and evaluation include processes to monitor progress toward meeting Health EDRM objectives, including monitoring risks and capacities and evaluating the implementation of strategies, related programs, and activities

Source (WHO 2019: x–xi)[a]

[a]World Health Organization. Disease Outbreak News Website. Available under the URL: http://www.who.int/csr/don/archive/year/2018/en/. Accessed on 12 July 2020

12.8 Conclusion

Health is growing as a core dimension of sustainable development. So, resilience building in the health system and ensuring health security on a global level is a necessity of the time. The shortage of trained nurses and doctors is a fundamental problem of the developing countries' health structure, which is a constraint of universal health coverage. It cannot be solved overnight. Even in this pandemic situation, many high-income countries have been suffering from a shortage of trained health personnel. So, these countries can try to minimize the health care gap by decentralizing the healthcare system. Investment in community health workers could be a significant step in this perspective. Strong and formalized Community Health Workers (CHW) can provide primary care that is effective for life-saving, increase easy access to health care, and keep health care affordable. The CHW systems deliver a positive economic return and help in women empowerment. Overall they are cost-effective solutions for the governments.

Covid-19 pandemic is a whistleblower to humankind regarding the necessity of a robust health care system that can provide essential services protecting against emerging epidemics. Policymakers have to prioritize the preventive aspects, e.g., awareness creation by CHWs of health care could reduce diseases' burden. However, they are not a remedy for all health-related problems. There are many flaws in these workgroups. Still, their effectiveness in the health system cannot be ignored. They took an important role in the 2015 Ebola outbreak management in west Africa. Their involvement should be applauded globally for their incredible role in improving public health security and pandemic preparedness.

References

Boniol M, McIsaac M, Xu L, Wuliji T, Diallo K, Campbell J (2019) Gender equity in the health workforce: analysis of 104 countries; WHO. https://apps.who.int/iris/bitstream/handle/10665/311314/WHO-HIS-HWF-Gender-WP1-2019.1-eng.pdf?ua=1

Boyce MR, Katz R (2019) Community health workers and pandemic preparedness: current and prospective roles

Bueno de Mesquita J, Thomas R, Gauter C, Havkwist A, Hoddy R, Larasati A, et al (2018) Monitoring the sustainable development goals through human right accountability reviews. Bull World Health Organ 96(9):627–633 (Sept 1)

Chatterjee PK (2020) Community Preparedness for COVID-19 and Frontline Health Workers in Chhattisgarh. Downloaded free from http://www.ijph.in on Wednesday, July 15, 2020, IP: 157.43.223.241

Dahn B, Perry H, Maeda A, Palazuelos D (2015) Strengthening primary health care through community health workers: investment case and financing recommendations, WHO, July 2015. https://doi.org/10.13140/RG.2.1.4865.0086

Declaration of Alma-Ata: International Conference on Primary Health Care, Alma-Ata, USSR, 6–12 September 1978. Available from: https://www.paho.org/English/DD/PIN/alma-ata_declaration.htm

Gowshall M, Taylor-Robinson SD (2018) The increasing prevalence of non-communicable diseases in low-middle income countries: the view from Malawi

Hall J, Taylor R (2003) Health for all beyond 2000: the demise of the Alma-Ata Declaration and primary health care in developing countries. Med J 178(1):17–20. https://doi.org/10.5694/j.1326-5377.2003.tb05033.x

https://www.jhsph.edu/research/centers-and-institutes/johns-hopkins-primary-care-policy-center/Publications_PDFs/E91.pdf

Jain D (2020) Time to bring some hope to ASHA workers fighting corona virus at frontline, 2 April 2020. outlookindia.com

Kane S, Kok M, Ormel H, Otiso L, Sidat M, Namakhoma I, Nasir S, Gemechu D, Rashid S, Taegtmeyer M, Theobald S, Koning K (2016) Limits and opportunities to community health worker empowerment: a multi-country comparative study. https://doi.org/10.1016/j.socscimed.2016.07.019

Mackey TK, Vianb A, Kohler J (2018) The sustainable development goals as a framework to combat health-sector corruption. Bull World Health Organ 96:634–643. http://dx.doi.org/10.2471/BLT.18.209502

Nacoti M, Ciocca A, Guipponi A, Barambillasca P, Lussana F (2020) At the epicentre of the Covid-19 pandemic and Humanitarian Crisis in Italy: changing Perspectives on Preparation and Mitigation. https://doi.org/10.1056/cat.20.0080

Non-communicable diseases progress monitor 2020. Available from: https://apps.who.int/iris/bitstream/handle/10665/330805/9789240000490eng.pdf?sequence=1&isAllowed=y
Opemo D (2020) Integrating disaster risk reduction and adaptation to Covid-19 in Kenya. Int J Med Sci Health Res 4(03). ISSN: 2581-3366
Smith KF, Goldberg M, Rosenthal S, Carlson L, Chen J, Chen C, et al (2014) Global rise in human infectious disease outbreaks. J R Soc Interface
WHO Director-General addresses the Executive Board. Geneva: World Health Organization; 2018. Available from: http://www.who.int/dg/speeches/2018/world-health-day/en/
WHO Regional Office for Africa (2020) Available under the https://www.afro.who.int/health-topics/communicable-diseases. Accessed on 24 September 2020
Wright N, Fagan L, Lapitan JM, Kayano R, Abrahams J, Huda Q, Murray V (2020) Health emergency and disaster risk management: five years into implementation of the Sendai. Framework. https://doi.org/10.1007/s13753-020-00274-x

Moumita Mondal is an Assistant Professor, Department of Geography at Rammohan College, Kolkata, since 2017. She completed her Post-Graduation in Geography from the University of Calcutta in 2014. Her area of interest is climatology, social geography, and medical geography.

Part IV
Preparedness and Policies

Chapter 13
The Public Health System Resilience Addendum: A Tool to Help Governments Manage Biological Hazards Better and Prepare for an Uncertain Future

Sanjaya Bhatia

Abstract This article explains how the Disaster Resilience Scorecard, and its Public Health Addendum is an easy to use tool for governments in managing the COVID-19 crisis and ensuring effective recovery and build back better. The underpinning concept is that pandemic management will be enhanced if cities are resilient. In this model, cities need to complete the Disaster Resilience Scorecard for cities and the public health addendum, and then develop and implement action plans. This multisectoral approach could be short-cut in the COVID-19 pandemic by rapidly completing the public health system Scorecard to develop, guide, and implement an action plan for sustainably managing response and recovery.

Keywords Multisectoral approach · Pandemic · Preparedness · COVID-19 · Management

13.1 Background

The "*Public Health System Resilience Addendum*" of the *Disaster Resilience Scorecard for Cities* aims to help local governments identify areas that require strengthening in relation to public health issues that are not adequately emphasized in the original version of the *Disaster Resilience Scorecard for Cities* (Fig. 13.1). This Addendum was designed by UNDRR, with the support of the World Health Organization (WHO) and other partners. The Addendum is better used in conjunction with the WHO's Health Emergency and Disaster Risk Management (Health EDRM) Framework.[1]

[1] View the official website of WHO for getting the detailed information on Health EDRM framework: https://www.who.int/hac/techguidance/preparedness/factsheets/en/.

S. Bhatia (✉)
Global Education and Training Institute (GETI), Incheon, South Korea
e-mail: bhatia1@un.org

M. Mishra and R. B. Singh, *COVID-19 Pandemic Trajectory in the Developing World*, Advances in Geographical and Environmental Sciences,
https://doi.org/10.1007/978-981-33-6440-0_13

Fig. 13.1 Disaster Resilience Scorecard for Cities: Public Health System Resilience Addendum: (Reproduced from UNDRR Making Cities Resilient Campaign Scorecard[2])

13.1.1 Making Cities Resilient Campaign 2010–2020

The Making Cities Resilient Campaign (MCR), launched in 2010, was very successful in raising awareness of disaster risk reduction (DRR) and resilience among cities and local governments globally. The work of the Campaign began in 2010, midway through the implementation of the Hyogo Framework for Action (2005–2015). Since 2015, it has been key for delivering Target E of the Sendai Framework for Disaster Risk Reduction 2015–2030 (Sendai Framework),[3] guided by the Ten Essentials for Making Cities Resilient (the Ten Essentials).[4] More than 4350 cities signed up to the Campaign, declaring a determination to build resilience.

Review research conducted during the Campaign indicates that cities that have joined the MCR campaign have made significant progress in reducing disaster risk than cities that have not gone through a similar process of disaster risk awareness.

[2]Visit UNDRR website: https://www.unisdr.org/campaign/resilientcities/toolkit/article/public-health-system-resilience-scorecard.

[3]Visit: https://www.undrr.org/publication/sendai-framework-disaster-risk-reduction-2015-2030.

[4]Visit the UNDRR website: https://www.unisdr.org/campaign/resilientcities/toolkit/article/the-ten-essentials-for-making-cities-resilient.

13.1.2 Changing Risk in a Changing World

Over time the nature of risk has changed. With the increasing complexity of the human, economic and political systems (including in the international financial system, communications, trade, and supply chains, and urbanization) and natural systems (marine, land, and air), risk has become increasingly systemic, which means risk is reflected in all sectors, in all walks of life and governance. It is impossible not to adopt a holistic approach to risk reduction.

This means while risk may start out one way, because of its dynamic nature, it may often combine with or cascade into another risk. An example is how Think of climate change is now contributing to environmental degradation and biodiversity loss, causing losses in crop yields and food production. Another example is where an extreme event creates an unanticipated technological risk, causing the partial or full disabling of a national power grid with cascading impacts on business, critical infrastructure, and civil security, and disruption of basic services. The nuclear disaster at Fukushima is an example where the tsunami caused a power outage, which caused the nuclear meltdown, which in turn caused evacuations and contamination.

We can no longer afford a hazard-by-hazard risk reduction approach. This is particularly important at the city level. Risk cannot be departmentalized or made the responsibility of just one public service provider or responder. Local planning bodies with representation must be involved that assume an all-of-service provider approach. Moreover, cities must plan not just to reduce risk but to invest in resilience-building: allowing systems, services, and people to respond to the crisis, cope with shocks and stresses and rebound.

13.2 The Tools

13.2.1 Disaster Resilience Scorecard for Cities

The Making Cities Resilient Campaign introduced a number of tools for cities and local governments. The key tool is the Disaster Resilience Scorecard for Cities ("the Scorecard") which is freely downloadable at the UNDRR website.[5]

The Sendai Framework for DRR 2015–2030 identifies the role of the local governments in reducing disaster risks, including the risks from biological hazards and health emergencies, such as the current pandemic. The Making Cities Resilient Campaign has simplified the message from the Sendai Framework by creating a Ten Essentials, breaking down the message from the Sendai framework into ten checklist points, which, when followed by local governments, can help them achieve the objectives of the Sendai Framework. These ten points are divided into issues

[5] Visit: https://www.unisdr.org/campaign/resilientcities/toolkit/article/disaster-resilience-scorecard-for-cities.

of governance, planning, and response. The ten essentials are relatively holistic and treat the local government as a "system of systems" (technological, social, economic), encouraging interconnectedness among the different sectors. However, public health system issues do not emerge clearly from this structure.

13.2.2 Public Health Addendum

A limitation of the Scorecard is that public health issues are not adequately emphasized. While the more obvious health factors such as hospital services, capacities, and structural and non-structural safety are covered in the Scorecard (under Essential 8), other disaster-related public health issues have not been well addressed. The Public Health Addendum was developed by partners of the Campaign, and UNDRR, with the support of the World Health Organization (WHO) to mitigate this gap. The Addendum should be used in conjunction with the WHO's Health Emergency and Disaster Risk Management (Health EDRM) Framework.

The Public Health Addendum is available free in English, Arabic, and Spanish. The term "public health issues" is used to cover the generalized impacts of disasters on the health systems of local governments. These may include:

- Events (including, disease outbreaks drought, earthquake, flood, tornado, cyclone, famine, wild-fires, and other disasters);
- Immediate consequences of a disaster (including mass casualties and psychosocial impacts);
- Secondary consequences of disasters (including malnutrition, disruption to livelihoods and vaccination programs, psychological impacts and more; all these issues have been witnessed in the COVID-19 emergency);
- Disruptions in health care services for persons with chronic health issues (for example, access to medications for chronic conditions such as HIV, diabetes, and cancer treatment);
- Consideration of the vulnerable populations (for example, the extremely poor, infants, people with disabilities, senior citizens, women).

A broader description of the health system includes people, institutions, and resources who are in the health and other sectors. The responsibility to handle a health emergency is not that of the health department alone, nor that of the civil defense by itself. It is the responsibility of all departments, as the impact is on all. The City Scorecard and the Addendum promote a system of the systems approach to resilience.

13.3 Structure of the Public Health System Resilience Assessment

13.3.1 The Structure

The Public Health Addendum uses the same "Ten Essentials"[6] structure as the Scorecard—in the context of health emergencies and focuses on the wider issues of management and recovery. It is not a medical or epidemiological tool and addresses the "non-medical" issues of a medical emergency including—hospitals, quarantine facilities; nursing homes, health clinics, doctors' offices; mental health facilities; laboratory and testing facilities; public sector health departments; water and sanitation; food distribution mechanisms; pharmaceutical and medical supply distribution; community information and alert systems; technical skills, staff, and equipment required; and other aspects of a fully functioning public health system. In this manner the tool covers both the hardware and the software (human resources) required to handle and recover from a health emergency. The Public Health Scorecard links hospitals, communities, schools, community centers, non-profit organizations, law enforcement, faith institutions, and elected officials. Some examples of what the Scorecard covers include:

1. Organization & Governance
   - Are public health and medical professionals involved in disaster planning and management?
   - Are other professionals (e.g., sanitation, water, energy, comms) involved in public health planning?
2. Risk understanding
   - Inclusion of a pandemic scenario in risk planning
   - Inclusion of pandemics (and pre-existing chronic health stresses—malaria, malnutrition) as a complexity factor, alongside "disasters as usual"—floods, earthquakes, fire, etc.
3. Financial architecture
   - Adequacy and protection of funding
   - Resilience "dividends"—other benefits that arise from resilience spending
4. Land use and building codes
   - Code and zoning compliance of key facilities

[6]"Ten Essentials for Making Cities Resilient" Scorecard integrates the public health with the governance (i.e., Essential 1), disaster scenarios (i.e., Essential 2), finances (i.e., Essential 3), land use/building codes (i.e., Essential 4), management of ecosystem services (i.e., Essential 5), institutional capacity (i.e., Essential 6), societal capacity (i.e., Essential 7), infrastructure resilience (i.e., Essential 8), disaster response (i.e., Essential 9), recovery/building back better (i.e., Essential 10).

5. Ecosystem services
   - Protection of ecosystem services with health benefits—natural water filtration, tree cover, recreation space
6. Capacity
   - Availability of public health skills—medical and other
   - Availability and sharing of required data with (that is, to and from) other stakeholders
7. Social capacity
   - Community engagement processes and effectiveness
   - Community trust of information provided
   - Community mental health and mental stress management
8. Infrastructure
   - Resilience of key health infrastructures
   - Resilience of other relevant infrastructures—water, power, communications, sanitation, trash collection
   - Surge capacity
   - Continuity of care facilities for those already sick
9. Disaster response
   - Early warning systems
   - Integration with emergency management
   - Education, rehearsals, drills, public health supplies
10. Recovery planning
   - Offsetting long run impacts on health
   - Learning and improving

In total, there are 23 questions/indicators, each with a score of 0–5, where 5 is best practice (see Fig. 13.2 as an example).

13.3.2 Required Data

Data required to complete the Addendum will include:

- Public health system capacity, stakeholders, planning and procedural documentation;
- Emergency management planning and procedural documentation;
- Public health infrastructure (see Essential 8);
- Data on healthcare outcomes of previous disasters, if available;
- Demographic data, including for vulnerable populations;

Essential 06: Strengthen Institutional Capacity for Resilience
Addendum – Integration of public health and institutional capacity

Ref	Subject / Issue	Question / Assessment Area	*Indicative* measurement scale	Comments
A6	Integration of public health and institutional capacity (Essential 6)			
A6.1	**Availability of public health workforce with relevant competencies and skills for disaster resilience**	To what extent are the workforce, competencies and skills required to plan and maintain public health systems and services for disaster resilience available to the city?	5 – All relevant workforce competencies and skills identified and assessed to be adequate for disaster planning, health services and post disaster recovery, both in terms of skill depth and numbers. 4 – All relevant skills identified, and some minor shortfalls known to exist in certain skillsets or numbers thereof. 3 – All relevant skills identified, and more significant shortfalls known to exist in depth and numbers. 2 – Incomplete skills identification and significant shortfalls in those that are known, in depth and numbers. 1 – Rudimentary attempt at skill identification – shortfalls in depth and numbers suspected to be universal. 0 – No consideration given to the issue.	Essential 8 in the main Scorecard deals with doctors', nurses' and first responders' numbers and skills – users may choose to include that data in the assessment here. As set out in the Health EDRM framework referenced earlier, key public health skills include, but are not restricted to: • Doctors, nurses and other health workers where not addressed under Essential 8; • First responders where not addressed under Essential 8; • Other hospital or health facility staff; • Psychiatric care – doctors, nurses; • Care home staff; • Pharmacists; • Environmental health specialists (includes water and sanitation experts, food inspectors and vector control) • Epidemiologists; • Testing and laboratory staff; • Supply chain workers.

Fig. 13.2 An example from Essential 6 question (*Source* Reproduced from UNDRR Making Cities Resilient Campaign Scorecard)

- Community and professional feedback on system capacity and effectiveness.

13.3.3 Methodology to Use the Addendum

The user needs to understand that resilience is a long-term objective. It cannot be achieved in a short time. Depending on the objectives, the resilience functions may play out over a long time, e.g., strengthening of health infrastructure and establishing surge capacities of doctors and nurses may take years, while emergency planning and simulations for a pandemic may be conducted on an annual basis. We need to keep this perspective while using a tool such as the Health Scorecard—actions proposed in an action plan need to be considered on a timeline as short, medium, and long term for the actions to be realistic.

The user also needs to understand that the Health Scorecard Addendum is not about the health sector alone. As the current pandemic demonstrates, all aspects of socio-economic development are impacted by the pandemic. The consultations must include other relevant sectors and stakeholders to use the Scorecard effectively. Without a multi-stakeholder approach, the resulting action plan will be left lacking ownership and not be implemented. Apart from sectors and stakeholders within the local government, the national government will have a key role in providing data and support. In the current situation, national governments are providing guidance, financial support, surge capacity, and data, which helps the local governments in the response.

Another important consideration is that hazards do not act alone, and the impact of a hazard may become amplified due to underlying stresses, e.g., a community already impacted with poverty will likely be more impacted by a pandemic, as their livelihoods opportunities are reduced, as has been seen in the case of migrant labor.

Chronic economic, societal, or environmental stresses can worsen the impact of earthquakes, floods, and pandemics, and the hazards can worsen chronic stresses. This relationship between chronic stresses and hazards will continue in the recovery stage and delay recovery.

Economic stress (poverty) leads to a higher risk of contracting diseases such as COVID-19 through population density, poor sanitation, and underlying ill-health, and the pandemic can cause enforced closures on local businesses, so causing social and economic stress through lack of facilities and unemployment.

13.3.4 The Workshop

The Scorecard is to be used in a workshop-based approach intended to be used to:

- Build consensus and working relationships between the different sectors and departments in the government, as well as with other stakeholders who may be external to the government;
- Share results from a detailed study with the sectors and stakeholders;
- Explore specific essentials for further detailed assessment if required, and to develop a draft action plan of activities the government should undertake to reduce the gaps as identified by the use of the Scorecard.

The workshop methodology includes a number of steps, including:

1. Determine the objectives: All should be clear what the objectives of the exercise are—to develop an action plan and strategy to reinforce and strengthen the public health systems in the government. This can also be an opportunity to provide an orientation, and resources, on terminology, the methodology to be adopted, etc. to ensure all participants in the exercise and the workshops are on the same page and level of understanding and competency.
2. Identify the stakeholders: The list of stakeholders is not the usual health department personnel alone, as economic and planning, finance, public transport, utilities, etc. all play a role and are impacted by the biological hazards, including pandemic. External stakeholders such as civil society organizations and their networks, private sector service providers, business councils and chambers of commerce, etc. need to be involved to ensure the plan and strategy developed is relevant, practical, and effective. The COVID-19 situation demonstrates the role of the transport sector, supply chains, commerce, and economic and financial planning.
3. Carry out initial research: To determine and answer questions such as—what are pre-existing resources that could be used, what data is required, what are the sources of data, etc. This step would require a study of the Scorecard, to determine who should be invited and contacted, what data is required, and what are the potential source of the required data.

4. Conduct the workshop: The Scorecard exercise includes going through each question, discussing the options for the answer, and arriving at the most suitable score via consensus. Each session can be designed around this methodology. The workshop can be conducted in a number of short sessions virtually, in place of face to face. Each session could be 2–3 hours long, perhaps covering one or two essentials at a time.
5. The first step is the scoring. Once the scores are agreed upon, the participants can decide on whether they want to develop an action plan for all the questions and gaps or only for the questions which scored below a certain criterion, e.g., the score of 3 or below. Such a prioritization ensures the weakest areas are strengthened first.
6. The action plan must have actions to address the gap identified through the scoring, the time frame, and the roles and responsibilities—who will lead, who will support. Normally the plan should be valid for a period of 3–5 years.
7. Implement: Once the action plan is developed and finalized, it should be endorsed by an appropriate level of government to ensure its legal status. This could be the Mayor's office or the city council. Such endorsement guarantees the strategy and action plan will become part of the documents considered when the development plan of the government is budgeted and funded. Endorsement ensures continuity even with a change in political leadership.

13.4 Conclusion

The benefits of completing the Scorecard extend far beyond the development of a plan to manage biological hazards and to report. The conversations and dialogue the process enables, and the silos it can break, are more important than the scores. Often, informal relationships are developed between the departments, which can greatly improve coordination during emergencies. The Scorecard encourages inter-sectoral discussions, which means departments that may not regularly exchange information are empowered to discuss planning and implementation issues with each other. This adds to the resilience of the city, especially if we remember that resilience is an issue of governance. When completed in a collaborative and consultative mechanism, local governments can:

- Understand how resilient their public health systems are
- Create and increase awareness and identify the remaining challenges
- Enable dialogue, interactions, and breaking of silos between key city stakeholders, bringing together those who may otherwise not collaborate
- Enable discussions to decide the priorities for investment and action promote and enable the development of a resilience strategy or an action plan that integrates public health issues with other issues of resilience
- Promote the development of projects that enhance the resilience of the public health systems of the local government and contribute to the overall resilience.

This tool can be used by local governments, cities as well as national governments, as the indicators are sufficiently generic for all levels of administration. The use of this tool can help governments prepare for biological hazards and health emergencies, including pandemics in the future. Being prepared is better than getting caught unaware, as happened during this pandemic.

Key Resources

A Handbook for Local Government Leaders [2017 Edition], UNDRR. https://www.unisdr.org/campaign/resilientcities/toolkit/article/a-handbook-for-local-government-leaders-2017-edition

Disaster Risk Reduction Terminology, UNDRR. http://www.preventionweb.net/english/professional/terminology/

Health Emergency and Disaster Risk Management Framework, WHO. https://www.who.int/hac/techguidance/preparedness/health-emergency-and-disaster-risk-management-framework-eng.pdf?ua=1

Making Cities Resilient Campaign Website, UNDRR. https://www.unisdr.org/campaign/resilientcities/

Sendai Framework for Disaster Risk Reduction 2015–2030, UN. http://www.preventionweb.net/drr-framework/sendai-framework

The Global Assessment Report & Global Risk Atlas, UNDRR. http://www.preventionweb.net/gar/

The Ten Essentials for Making Cities Resilience, UNDRR. https://www.unisdr.org/campaign/resilientcities/toolkit/article/the-ten-essentials-for-making-cities-resilient

Understanding Disaster Risk—based upon GARs 2009, 2011, 2013, 2015, UNISDR—including risk models, viewers, and data, UNDRR. http://www.preventionweb.net/risk

Sanjaya Bhatia is the Head of the UNDRR (United Nations Office for Disaster Risk Reduction) Global Education and Training Institute (GETI) and the Office for Northeast Asia (ONEA) in Incheon, Republic of Korea. He is responsible for the training of over 8000 government officials globally. His areas of focus are disaster risk reduction, disaster recovery, safe schools and hospitals, and disaster risk management for local governments. He holds a degree in law and a master's degree in Public Administration from New York University. He has authored a number of publications. He is a certified trainer for the Incident Command System (Emergency Management) and functioned as a trainer and resource person for the Government of India and the United Nations. He has worked with the Government of India (Indian Administrative Service), the World Bank, and the United Nations in the field of disaster risk reduction and climate change adaptation for over 30 years.

Chapter 14
Instilling Self-Sustainment: The Key Survival Strategy Amid the Pandemic

Suresh Kumar and Raj Kumar

Abstract After a hundred years, another epidemic has touched the global community and challenges the platinum grade science and technology development. The COVID-19 did not spare the people belonging to the developed world or developing countries. The corona virus has cautioned us about our unhealthy lifestyle, breathing polluted air, consuming pesticides filled vegetables and fruits, drinking untreated water and questions our pathetic immunity status. The pandemic effect has made us realize our bad food habits, non-eco-friendly working environments, and our actions leading to the devastation of nature. This pandemic has given a lesson not to leave the ancient knowledge of texts scientifically, using traditional medicine for a healthy life, and live a pro-environment lifestyle. This chapter explores the traditional sources of rich knowledge and critically examines its uses for a sustainable environment, a sustainable lifestyle, and will develop a sustainable green earth.

Keywords *Atmanirbhar Bharat* · Pro-environment · Sustainable development · AYUSH · Green earth

14.1 Introduction

COVID-19, or commonly known as coronavirus disease, has spread in most parts of the world, including India. It results in infecting more than one crore (28 million) people across the globe, with meshing more than 917,417 to fatality. The world community faces a war-like situation (Or it may be Third World War in economic terms) due to the virus's hostile spread. It has already invaded 53 countries in the Africa continent, 44 in Europe, 8 in East Asia, 20 in Latin America, 9 in South Asia,

[1] See BBC News dated 15 September 2020). Visit: https://www.bbc.com/news/england/london.

S. Kumar (✉)
Department of African Studies, University of Delhi, Delhi, India
e-mail: skafrica2003@gmail.com

R. Kumar
Department of Political Science, SPM College for Women, Delhi, India

M. Mishra and R. B. Singh, *COVID-19 Pandemic Trajectory in the Developing World*, Advances in Geographical and Environmental Sciences,
https://doi.org/10.1007/978-981-33-6440-0_14

18 in West Asia, 14 in Oceania world, the United States of America, and Canada.[1] 'Originating' from China, the virus runs amuck around the world and challenges the peaceful global community and the economic order of the world on the one hand and hegemony the financial trade & commerce in the global market as an end product of this pandemic on the other hand. The coronavirus is exacerbating the silent death in the society that leads to mass devastation globally.

Patanjali defined ways to moral conduct in the Sanskrit language is worth mentioning in this context. This verse[2] says that 'humanity should respect all living creatures' feelings and behave with love and grace'. Nonviolence is the biggest weapon to solve all the disputes in the world. Even the Brundtland Commission mentions:

> Humanity has the ability to make development sustainable to ensure that it meets the needs of the present without compromising the ability of future generations to meet their own needs. (WBCSD Global Scenario 2009: 4)

The COVID-19 pandemic and the lockdown of the whole global activities have given us a lesson to recall the efficiencies of the natural process of healing the environment, leading to environmental sustainability in real terms. In a Sanskrit sloka,[3] it is rightly said, emphasizing the strength of natural healing power, that 'one should chew the nature's food & eatable things and get good health once the food is digested.' Similarly, our nature heals the air, water, soil, and habitat naturally, considering the best for nature and world society and ultimately leads toward yogic life.

The post-pandemic period could perceive the real meaning of 'climate change.' Long global lockdown shows the signs of progressive improvement in the environment. The global community has come forward to save the lives of human beings by distributing food, medicine, and other essential items. For example, the Sikh religion's worship places, commonly known as the Gurudwaras, worldwide (as we find them in Europe, USA, Canada, Australia, India, East Asia, and many other places) have been continuously distributing foods to the needy people for the crisis months (March to June 2020). So, there is neither shortage of food nor the good-hearted people donating cash and goods for the needy people in the world. The question arises here that why we are not caring about our nature and protect them from misuse. Nature has the capability to heal itself, and the post-pandemic period

---

[2]See *the Mahabharata*, Chapter *Shanti Parva*, Para *265*,

तस्माद् प्रमाणतः कार्यो, धर्मः सूक्ष्मो विजानता ।
अहिंसा सर्वभूतेभ्यो,धर्मेभ्यो ज्यायसी मता ।।

[3]See *the Mahabharata*, Chapter *Shanti Parva*, Para *321*,

यच्छक्यं ग्रसितुं ग्रस्यं ग्रस्तं परिणमेच्च यत् ।
हितं च परिणामे यत् तदाद्यं भूतिमिच्छता ।।

emphasizes adopting the natural healing for the physical ailments of human beings and pausing a stop racing blindly for market goods. Humankind needs to inculcate within their attitudes to adopt a natural way of life.

Moreover, the post-lockdown period has witnessed the heavy rains in the different parts of the world, blue sky, and fresh air, and it is the direct result of nature healing itself during the lockdown period. It is further observed that the winter season of 2020–2021 has brought more happiness and cheers in the lives of the human being, flora, and fauna. Along with it, the post-pandemic lockdown period has created millions of unemployment among youth due to retrenchment in different companies. As per their skill, the global youth should inculcate them in developing their own entrepreneurs and generate employment rather than depending on the companies, which in India is known as *Atmanirbhar* (i.e., Self-sustainment).

This chapter critically examines the post-pandemic period by analyzing the natural healing process, which is a step toward a sustainable environment. It tries to explore the avenues for survival through practicing *Atmanirbhar* in a real sense. Also, it advocates to minimize the dependency on the blind market economy products and promotes creating a delicate balance in nature; a balance between the need and the supply.

14.2 Socio-Political Dialogues in Post-Pandemic India

The pandemic affects the socio-political fresco of India in varied ways. The labor migration from industrial hubs to their hometowns, closure of industrial production and loss of agricultural income, and stalemate in day-to-day activities led to a severe social and economic crisis. 'The lockdown bound virtualization of life caused by the lockdown is already generating important social, economic, fiscal, and urban impacts, which are partially irreversible, need to formulate an assertive policy. The success of India's fight against COVID-19 would be an upshot of multiple sector coordination among different agencies of health, technology, and diplomacy. They play an extremely coherent role in supplementing sector efforts' (Chaturvedi 2020: 27). As a result, due to lockdown, most of the sectors are adversely affected. The investment banks and international agencies have projected India's revised downward economic growth. Rashmi (2020) pointed out that 'the latest projections of GDP growth for the Indian economy in FY21 could range from –0.5 percent (Nomura's) to 2.5 percent (Moody's). The Economic Intelligence Unit projects a *quarter-on-quarter* growth of –9.3 percent in Q2, 2020. The GDP growth in FY21 might well be toward the lower end of the range if the social distancing measures remain in place for a relatively extended period. It would lead to the business and consumer confidence taking time to return, resuming the normal business operations' (Rashmi 2020: 30). The pandemic effect needs a long-term strategy to revive the lives of people. India comprises a dominant economy in South Asia, and it needs effective government strategies for the post-pandemic plan of action and the necessary funding. For example, Indian pharmaceutical companies are producing and exporting vaccines throughout the world. Indian pharmaceutical companies have grabbed big requisitions from Russia to manufacture the COVID-19 vaccine bottles billion in number. Such time-oriented decisions will bring economic benefit to the people of India.[4] The nation has been running along the right track in the Science, Technology & Innovation (STI) policies,[5] and it looks forward to developing cooperation at the regional level, which could create a synergy among the countries in building the *Atmanirbhar* program for the youth.

The pandemic damage of human development done so far across the world needs an indigenous technology to become *Atmanirbhar* (Self-sustainable), and the continent is looking toward India being all-time supporters. The Government of India has announced several *Atmanirbhar* programs for the youth as per their needs, abilities, and requirement of their societies living in the rural, urban, and metropolitan areas. The COVID-19 post-pandemic period will be the time to drive forward with a shared understanding of what is meant by a sustainable and responsible consumption from the yoga perspective. The global community faces rising fuel prices, severe issues of hunger and poverty, scarcity of water, arable land, and other natural resources that question human consumptions, and how one may deal with it has never been more crucial. The progressively degrading environments and iniquitous social structures seriously threaten the future of mankind. Exacerbating these challenges today is the new danger posed by global warming and climate change. The global community bears the most severe brunt of these environmental and societal imbalances. Today, the warning to unstoppable competition for survival in the liberal market economy should be taken care of. The post-pandemic world community in 2020 should start thinking for the present and future generations, failing which nobody could stop the pandemic consequences and destruction of the Mammalian Class and our environment. Yoga attracts the worldly governments toward more sustainable choices and sharing this information with their citizens. This information will trigger innovation and ultimately set the trend toward a sustainable environment through Yoga among the global community in the post-pandemic period.

The economic relief packages, introduced by the government of India to stimulate the country's pandemic hit the economy, have been analyzed from multiple lenses since they were announced—economic, social, political, etc. However, environmental and sustainability lenses seem to be missing from these analyses. This chapter will highlight the environmental aspects of the current economic packages, discuss how these packages could be further leveraged for green growth, and deliberate upon a climate-conscious economic rebuilding strategy going forward.

---

[4]See the Deccan herald dated August 23, 2020, available under the URL: https://www.deccanherald.com/national/coronavirus-news-live-updates-unlock-30-rules-india-maharashtra-karnataka-delhi-tamil-nadu-mumbai-bengaluru-chennai-ahmedabad-new-delhi-total-cases-deaths-recoveries-today-covid-19-coronavirus-vaccine-covid-vaccine-updates-869265.html#3.

[5]See Mygovt dated 24 September 2020, URL: https://www.mygov.in/campaigns/stip-2020/.

14.3 Meaning of Naturalization

India and the whole world went under the process of lockdown, and as a result, the ingredients of the nature such as Air, Water, and Light pollutant levels have come down drastically. Compared with the same period last year, 'levels of nitrogen dioxide have fallen by 40 percent as PM2.5 is down by 10 percent, which means that people without COVID-19 can breathe easier. These two forms of pollution, which weaken the heart and respiratory system, are together chiefly responsible for about 470,000 deaths in Europe each year'.[6] Health experts confirm the findings echoed their experience during the pandemic. 'We have seen many fewer patients admitted with asthma and Chronic Obstructive Pulmonary Disease (COPD) over the last month, and there is no doubt that a fall in air pollution is part of the reason,' said Dr. L.J. Smith, a consultant in respiratory medicine at King's College hospital in London.[7] The improvement in air quality over the last month of the corona virus lockdown has led to 11,000 fewer deaths from pollution in the UK and elsewhere in Europe.[8]

People have changed their diehard habits that helped in purifying the environment in India. For example, Hindu people, after death, put ashes in the holy river Ganga. However, the lockdown during the COVID-19 outbreak has changed their habits that save the largest river of India to have unwanted elements. Similarly, in the absence of industrial waste, most of the rivers in India (such as the Yamuna in Delhi) are cleaned through the process of naturalization. Nature heals the environment, and India is the leader across the world and ready to supply the indigenous natural methods of Ayurveda, Yoga, and Naturopathy, Unani, Siddha, and Homoeopathy in healing the human being, society, and global environment. Yoga and Ayurvedic science is the end product of the history of ancient Indian spirituality. This paper explains how traditional Yoga and Ayurveda science is helpful for human development by strengthening their body immunity and in healing the environment in the post-pandemic world. Today, the world community practices human development. Sustainable production and consumption policy learning from Yoga may become an important medium for the development of global society. Yoga concerns around the world should address global challenges related to body immunity, shortage of resources, water scarcity, climate change, and loss of biodiversity and planning, promotion, coordination, and overseeing the implementation of environment and forestry program by connecting it with *yogic sutras* and Ayurvedic sciences.

---

[6]See the Guardian dated 30 April 2020, report entitled 'UK - Clean air in Europe during lockdown leads to 11,000 fewer deaths' which is available under the URL: https://www.theguardian.com/environment/2020/apr/30/clean-air-in-europe-during-lockdown-leads-to-11000-fewer-deaths.

[7]See the New York Times dated 17 June 2020, URL: https://www.nytimes.com/issue/todayspaper/2020/06/17/todays-new-york-times.

[8]See the Guardian dated 30 April 2020, URL: https://www.theguardian.com/environment/2020/apr/30/clean-air-in-europe-during-lockdown-leads-to-11000-fewer-deaths.

14.4 Naturalization Healing in India

The nature cure treatment through the process of naturalization cannot be ignored during this global health crisis, and it is observed scientifically in the world. Ganga and Yamuna are the two most prominently polluted rivers of the country; however, visuals of the rivers that were taken during the lockdown show them relatively cleaner. The media report (NDTV) observed that 'many industries and offices were closed due to the lockdown, the Yamuna is looking cleaner. The stoppage of industrial pollutants, industrial waste, and household garbage has definitely had a positive effect on water quality. Along with it, there is increased water flow due to unseasonal rainfall and snowfall in some parts'.[9] The epidemic naturalization process has given a lesson to shape our future approach, and the industrial pollutants should be minimized or restricted to dispose of in the water bodies. This naturalization process resurfaces the missing birds, animals, insects, and aquatic species in the rivers, ponds, and sea. For example, lions, leopards, Sambar-deer, endangered civet in Kerala, Neelgai, number of fishes in Byas River, Navi Mumbai saw thousands of flamingos and migratory birds.

The restrictions in the vehicular movement didn't only help the air pollution to decrease but also had a very noticeable impact on noise pollution. For at least two months, there was a drastic drop in the honking, no sounds of vehicular engines, no machinery running in the factories, and no loudspeakers. The noise of vehicles' honking and the airplanes has been replaced by the sounds of birds chirping. The West Bengal Pollution Control Board has released noise pollution specific data for Kolkata. It reports that there has been a 50–75 percent decrease in noise pollution during the lockdown. With the improvement in nearly all aspects like air, water, noise, and wildlife, trees and plants are roping the benefits of the clean environment and the lack of human activities too. As the world unlocks in phases, experts urge everyone to consider these environment gains and treat the 'Mother Earth' with compassion and gratefulness.[10]

Ayurveda meant life (Ayur) and knowledge or science (Veda) explains about scientific life. It is further explained in the Encyclopedia Britannica that says 'any system of knowledge that is concerned with the physical world and its phenomena and that entails unbiased observations and systematic experimentation' (Britannica: 2020).[11] Today, all the Ayurvedic medicines are provided under the ISM drugs and are supported under 'The Drugs & Cosmetics Act and Rules, 1940' (23 of 1940, as amended up to December 31, 2016), Ministry of Health and Family Welfare, Government of India in India. The Government of India has supported the systematic research in this field known as AYUSH (Ayurveda, Yoga, Unani, Siddha, and Homeopathy). It further encouraged the Indian Systems of Medicine and Homeopathy (ISM&H) and Central Council for Research in Indian Medicine and Homeopathy (CCRIMH)

---

[9]See the NDTV Report dated 17 August 2020, URL: https://www.ndtv.com/topic/news-reporting.

[10]See the NDTV Reports, 10 July 2020. https://swachhindia.ndtv.com/world-environment-day-2020-this-is-how-the-environment-is-healing-during-the-covid-19-lockdown-45613/.

[11]See Britannica, URL: https://www.britannica.com/science/science.

to work extensively in fighting against the modern-day diseases, including COVID-19. The Ministry of Ayush[12] provides the bottles and pouches (a mixture of natural herbal ingredients for the soup) in fighting against the COVID-19 as it stimulates the immunity in the human body. The Government of India also prefers advertising to go Natural Way for All Treatments, including COVID-19 and heal the body, mind, and nature. This is further true that avoiding all chemicals in the food and in daily use, which will minimize the pollutants in our daily lives and save the environment. T. C. James and Apurva Bhatnagar further explain that:

> although there is no deniability that there is a vast pool of traditional medicinal knowledge and traditional medicinal practices of immeasurable value in India, one cannot overlook the advancement in technology and modern medical sciences and how it has changed our lives for the better. It is imperative that traditional systems of medicine should not be looked as an alternative but rather as a complementary science to modern medicine. With increasing levels of education and knowledge expansion in other fields, it is necessary that the ISMs keep abreast of them. The practitioners of these systems should be able to converse and dialogue with them in the jargon of science with ease and felicity. This will be possible only if the systems infuse into them science and technology in a big way. Wide-spread application of science and technology will naturally lead to a great number of innovations. (James and Bhatnagar 2020: 24)

AYUSH science is the end product of the history of ancient India spirituality and this traditional science is helpful for human development by strengthening their body immunity and in healing the environment in the post-pandemic world. This is the major reason that the Parliament of India has passed two bills: the National Commission Homeopathy Bill 2020 and the National Commission for Indian System of Medicine Bill 2020 in the post-pandemic session of the Parliament.[13] It will strengthen the idea of naturalization in the lives & livelihood of human beings in India.

Yoga concerns around the world should address global challenges related to body immunity, shortage of resources, water scarcity, climate change, and loss of biodiversity and planning, promotion, coordination, and overseeing the implementation of environment and forestry program by connecting it with AYUSH. This is the process of naturalization healing the environment in brief, and the young geographers must advertise this method for future survival.

---

[12]See the Official Website of the Ministry of AYUSH, Govt. of India: https://health.ncog.gov.in/ayush-covid-dashbaord/.

[13]See the Economic Times dated 15 September 2020 on 'MSP will stay, farmers will get investment, technology: Govt on farm sector bills'.

Table 14.1 *Atmanirbhar Bharat* Program by the Government of India

Pillars of *Atmanirbhar Bharat*	Different Phases of *Atmanirbhar Bharat*	Different Channels of *Atmanirbhar Bharat*
• Economy • Infrastructure • System • Vibrant Demography • Demand	• Businesses including MSMEs • Poor, including migrants & farmers • Agriculture • New Horizons of Growth • Government Reform and Enablers	• PM-CARES Fund • Health Advisory to Elders • All India Helpline Numbers • Volunteers

(*Source* Govt. of India)

14.5 Self-Sustainment as the Lifesaver Mechanism

The Government of India has started the *Atmanirbhar*, i.e., self-reliance program of '*Building Atmanirbhar Bharat & Overcoming Covid-19*' (Table 14.1).[14] India indeed had zero production of Personal Protection Equipment (PPE) before March 2020, and under this *Atmanirbhar* scheme, India produces more than 2 lakh PPE kits everyday in fighting against the COVID-19. Along with it, many automobile production centers were motivated to prepare life-saving ventilators and strengthened the self-reliance program. The Start-up program, keeping the social demography of India in mind, has become a vibrant hub for the Indian youth.

India is considered the third-largest start-up hub in the world after China & the USA. The report states that:

> the number of start-ups in India was expected to cross the 9300 mark in 2019, and has the potential to grow up to four times this number by 2025. India's large consumers base with access to the cheapest internet in the world provides an ideal platform for the start-up businesses to take off. In the health sector, projects like Net-Meds and 1 mg, etc. have penetrated the Indian markets, ensuring availability of affordable medicine in the remotest part of the country. Most of these online pharmacies also employ doctors to cross check the prescription and provide online consultations to customers. Furthermore, there are also projects like Practo that provides easy information about nearest consulting and specialized doctors, rated and reviewed by verified patients. The app also allows for prior booking at clinics and hospitals, along with offering attractive discounts on consulting fees. (Banerjee 2020: 22)

[14]See the *Atmanirbhar* Program's Campaign by the Government of India which is available under the URL: https://www.india.gov.in/spotlight/building-atmanirbhar-bharat-overcoming-covid-19#:~:text=Government%20of%20India%20is%20taking,threats%20posed%20by%20COVID%2D19.&text=The%20Five%20pillars%20of%20Atmanirbhar,Infrastructure.

14.5.1 Vibrant Demography

India being an agricultural country, has taken essential steps in the post-pandemic conditions to procure the interests of the farmers. 'The plans and programs are drawn up for the rural sector and Rs. 17,500 Crore (175 million rupees) will be transferred online under the Direct Benefit Transfer in the accounts of the farmers. Organic farming will promote among the farmers to get better prices of their products and facilitate exports.'[15] The rural infrastructure is on the priority of the government, and the farmers will be connected directly to buyers that will strengthen buyer–seller relations in the long term in post-pandemic India.

Along with it, the government has passed two legislations in the interests of the farmers entitled 'The Farmers Produce Trade and Commerce (Promotion and Facilitation) Bill, 2020' and 'The Farmers (Empowerment and Protection) Agreement on Price Assurance and Farm Services Bill, 2020' in the post-pandemic session of the Parliament on September 17, 2020. Mr. Narender Singh Tomer, the Minister of Agriculture, and Farmers' Welfare; Rural Development and Panchayatiraj, has explained that:

> the bills are aimed at making farming profitable. These bills will increase competition and promote private investment, which will help in development of farm infrastructure and generate employment. The farmers will be able to connect directly with big business and exporters and will be able to make farming profitable. It will strengthen farmers and help India to become *Atmanirbhar*.[16]

Similarly, the human development of vibrant demography urgently needs interconnectivity in the Post-COVID period. A household looks at the options of a healthy family, schools and colleges look for their classes, small businesses to move on, offices to run, the police needs to check crime, the army needs to secure the border; and all need to be done under complete lockdown, partially lockdown or under any restricted conditions. Here, the Indian geographers' mission is to organize the world's information and communicate to the desired mass by working under this *Atmanirbhar* vision to connect India, strengthen their self-sustenance, and visualize for the connectivity. Rahul Bhasin, the Managing Director of Baring Private Equity Partners, has outlined:

> IT actually has a big broad sort of horizon at this moment and everything which feeds into it – the enabling ecosystem of telecom and fibers etc. are big beneficiaries. It is really about everyone who facilitates gathering large amounts of data whether it is the Internet of Things (IoT) which gathers data from objects or whether it is the industry, the cell phone industry,

---

[15] See the Economic Times dated 8 August 2020. Delhi, available under the URL: https://economictimes.indiatimes.com/news/economy/agriculture/minimum-support-price-mechanism-for-farmers-will-continue-agriculture-minister-narendra-singh-tomar/articleshow/78170493.cms?utm_source=contentofinterest&utm_medium=text&utm_campaign=cppst.

[16] See the Economic Times dated 17 September 2020, available under the URL: https://economictimes.indiatimes.com/news/economy/agriculture/minimum-support-price-mechanism-for-farmers-will-continue-agriculture-minister-narendra-singh-tomar/articleshow/78170493.cms?utm_source=contentofinterest&utm_medium=text&utm_campaign=cppst.

> the variable industry which gathers data about people, the cloud which helps store data securely. Then the layer of artificial intelligence comes, which actually exploits that data, gets insights and takes us forward. Covid-19 is on everybody's top of mind and there is a lot of data floating around. However brilliant people are, no one could actually figure out what it was all about, till a gentlemen called Dr. Jacobson in Tennessee who runs a Oak Ridge National Laboratory, which has one of the fastest supercomputers in the world, took 17,000 samples and fed it into the computer putting it in a self-learning algorithm and came out with the hypothesis that this is something like a Bradykinin storm which is completely different from what everybody felt it was. They have isolated drugs which can be used to treat and which are straight away being taken into clinic. They have also figured out that this is the effect of Covid-19 and the IT have harnessed a lot of data, used the new computation tools and been able to come out with insights which are going to differentiate how that industry works and how we as a human society evolve. (The Economic Times, 18 September 2020)[17]

The young generation of Indian geographers should utilize their technological skills (such as Remote Sensing & GIS, Simulations, Computer technology of Spatial and Pattern Analyses like geophysical, sociological, demographic, hydrological, meteorological, or political, Data analysis with SPSS, SYMAP, GIMMS, and others). It will bring opportunities to earn and stabilize their lives under *Atmanirbhar* vision and employ others by sitting at home. The use of millions of smartphones in the rural–urban profiles needs to connect through these geographical technologies in their daily lives in serving multi-purpose utilities. The digitalization of small businesses under *Atmanirbhar* vision scheme establishes the online business using maps, driving connections, and become a part of the formal economy.

The Goggle has supported this Atmanirbhar vision and the post-pandemic period is the beginning of the geographers to work in his discipline and get connected digitally. Sunder Pichai, CEO of Google India has announced the Google initiative for India Digitization Fund, confirming that his company would invest INR 75,000 crore, or approximately $10 billion, into India over the next 5–7 years. The company would do this through a mix of equity investments, partnerships, and operational, infrastructure, and ecosystem investments.[18] This will add confidence to the future of India' digital economy. Investments will focus on four areas critical to India's digitization:

- First, enabling affordable access and information for every Indian in their own language, whether it is Hindi, Tamil, Punjabi, or any other
- Second, building new products and services that are deeply relevant to India's unique needs

[17]See the Economic Times dated 18 September 2020, available under the URL: https://economictimes.indiatimes.com/markets/expert-view/rahul-bhasin-on-how-covid-could-prove-to-be-a-game-changer-for-it/articleshow/78186611.cms?utm_source=contentofinterest&utm_medium=text&utm_campaign=cppst.

[18]See the India Today dated 13 July 2020, available under the URL: https://www.indiatoday.in/technology/news/story/google-announces-india-digitization-fund-will-invest-rs-75-000-crore-in-india-in-5-7-years-1700056-2020-07-13#:~:text=Google%20CEO%20Sundar%20Pichai%20on%20Monday%20announced%20that.%20as%20part,of%20Google's%20digital%20India%20mission.

- Third, empowering businesses as they continue or embark on their digital transformation
- Fourth, leveraging technology and Artificial Intelligence for social well-being, in areas like health, education, and agriculture.

Similarly, the EU–India Summit has finished on July 15, 2020 and discussed the global cooperation and solidarity to protect lives, to mitigate the socio-economic consequences, and to strengthen preparedness and response capacities. This EU-India Strategic Partnership has discussed climate, environment, digital economy, and connectivity and developed a symbiotic relationship on geo-economic issues such as:

- 'Public health initiatives;
- climate change;
- connectivity;
- sustainable green economies in energy and resource efficiency;
- artificial intelligence and data protection and;
- The market economy, firmly based on the principles of multilateralism' (EU-India Summit July 15 2020).

14.5.2 Climate Change and Green Economies

The international discussion and consensus on climate agenda had come out in the form of the Paris Agreement and EU-India Clean Energy and Climate Partnership. Our government has enthusiastically embraced European technology to develop clean and smart cities, clean rivers under the India transformation project. The government in the post-pandemic COVID-19 situation should include promoting and de-risking investment in renewable energy, green infrastructure, joint research and development, and work on Green Growth Equity Fund created by the UK and India or private financial initiatives like Tata CleanTech Capital Limited. India and EU partnerships will develop long-term strategies on emission reduction, adaptation, and building resilience in dealing with climate diplomacy. The stronger climate action requires the young Geographers' active participation in building the different responsibilities and development needs of the country as per their expertise areas. This participation will encourage the young Geographers at the global level to build consensus for the meaningful Climate Change action program.

The post-pandemic COVID-19 did not show the sign of encouragement in terms of production and made the world dependent on the USA or China's supply chains, willingly or unwillingly. This Geopolitical shift of dependency compulsorily designs the landscape of Geo-economics as far as industrial and innovation policies, trade agreements, and the market requirement are concerned. The interests of the EU in the current Geo-economic situation offer vast possibilities for the futures market, including the academia of Geographers on the one hand, and it will work in the post-pandemic period that may extend to another three to five crucial years on the other hand. It will strengthen the India–EU connectivity in real terms and help in coming out of the crisis period.

The self-sustaining mechanism needs a strong desire or ways to moral conduct with the universe and practicing nonviolence, truth & honesty, non-stealing, meditation, and non-possessiveness in their lives. It persuades everybody to get rid of painful life by adopting non-violent practices. The *30.15 Sutras* of Kautilya has pointed out that nonviolence is the best solution to save and strengthen lives (Chaturvedi 2020: 13). The donation toward life (*Pran-daan*) is the wisest technique to strengthen the green environment-friendly world. Gandhian philosophy is to avoidance of harm to any living creature in thought or deed. This is the only way to solve the established problems and work for the naturalized healing to the environment. The self-sustaining policy during the post-pandemic period today needs truth and honesty in our daily lives.

The industrial sector technology innovations have driven the over-growth that stands as a challenge before the *Atmanirbhar* (Self-sustainment). The developed world needs to redefine the value and corporate value in line with its *Atmanirbhar* (Self-sustainable) vision of a sustainable and prosperous society on the one hand and decide an optimal balance between self-regulating market mechanisms and legislative initiatives on the other hand. It is the time to unite sustainable production with sustainable consumption that needs respect to Atmanirbhar (Self-sustaining) and active participation in implementing it in the activities of their industrial houses. This participation linkage will understand current and future consumption patterns, harnessing innovation to develop more sustainable products, services, and behavioral change initiatives through *Atmanirbhar* (Self-sustainment). The whole world community should adopt the healthy lifestyles in the post-pandemic today and critically analyze the industrial producers, need of consumers, and society at large and adopt the sustainable life pattern such as:

- Human social systems and well-being do not necessarily rely on high consumption levels and should practice the self-sustainable mechanism.
- More and more consumers are concerned about environmental, social, and economic issues and increasingly willing to act on those concerns. Consumer willingness needs awareness program and motivates them to act in line with sustainable consumer behavior. The young generation should understand self-sustainment and keep away these elements such as availability, affordability, convenience, product performance, conflicting priorities, skepticism, and force of habit.
- The self-sustainable idea of sustainable lifestyles should be based on informed purchasing decisions and changes in the behavior of consumers that should get the support of active participation of businesses, governments, and civil society in implementing Atmanirbhar (Self-sustain) *Sutras*.
- The self-sustainable mechanism should formulate a consensus on the legal, fiscal, and cultural environment for sustainable businesses to flourish and implement globally.

14.6 Conclusion

The post-pandemic period will pose significant challenges before the world community to sustain biodiversity by recognizing the biodiversity-related risk and the value of ecosystem services. The industrial track record in managing biodiversity should be accessed by retaining land and other natural resources, capital, markets, and skilled staff. The responsible management of biodiversity concerns can result in benefiting directly to meet consumer demand for environmentally sustainable products. This applies not only to companies that depend on primary production, such as forestry, fisheries, agriculture, oil and gas development, and mining but to others too. All governments should promote.

- *Firstly*, the leading businesses seeking to incorporate biodiversity concerns through self-sustainable equation into their investment decisions, notably into the design and implementation of their business practices and supply chain management systems. No one can replicate the natural ecosystem, but businesses need to maintain characteristic biodiversity so that the right conditions can be created through complex interactions in a sustainable way.
- *Secondly*, all the governments should promote the scope of the biodiversity business case extending beyond the marketing departments of companies promoting the environment and Atmanirbhar (Self-sustainment) to improve the images of corporate brands. The areas of opportunities in India relate to Atmanirbhar (Self-sustainment) is rich in biodiversity.
  The industries of India can engage with biodiversity issues. It can contribute to the objectives of the CBD. It will be helpful in raising employee awareness of biodiversity issues and managing environmental impacts responsibly. It could address in enhancing the conservation value of habitats on landholdings, helping to fund conservation initiatives by adhering to principles of sustainable use of biodiversity and promoting the thesis of self-sustainment.
- *Thirdly*, there is a need to form a uniform biodiversity policy through Yoga and naturalization for leading companies in the world that will strengthen the approach toward biodiversity issues. All the major industrial houses of the world need a high-quality corporate policy on board level, and implement biodiversity action plans connecting with the environmental management system will work under the guidance of AYUSH. 'India has developed a number of policies and laws to promote the conservation and sustainable use of biodiversity and protect the environment'. The challenge now is to assess and plug the gaps in these policies and laws, and in their implementation, so that economic development processes are reoriented to protect the natural resource base and associated livelihoods. A real solution to the biodiversity challenge must include contributions by all stakeholders, from government agencies to businesses.

Summing up, the post-pandemic period will be the time to drive forward with a shared understanding of what is meant by sustainable, responsible consumption from the AYUSH perspective. The global community faces rising fuel prices, serious issues of hunger and poverty, scarcity of water, arable land, and other natural resources; questions human consumption and how one may deal with it has never been more crucial. The progressively degrading environments and iniquitous social structures seriously threaten the future of humankind. Exacerbating these challenges today is the new danger posed by global warming and climate change. The global community bears the most severe brunt of these environmental and societal imbalances. It warns the post-1990 unstoppable competition for survival in the liberal market economy to either stop, considering the future of the coming generations, or face the natural destruction of the Mammalian Class and natural environment. The adoption of naturalization attracts the worldly governments toward more sustainable choices and sharing this information with the common man. This information will act as a trigger for innovation and ultimately set the trend toward a sustainable environment through Yoga among the global community in the post-pandemic period.

References

Biswajit Banerjee (2020) Impact of Covid-19 on the World Economy. RIS Diary 3 29 rd Special Issue on COVID-19. RIS Publications, Delhi

Chaturvedi BK (2020) Chanakya Neeti with Sutras of Chanakya. Cloudtail India

Chaturvedi S (2020) RIS Diary 619rd. Special Issue on COVID-19. RIS Publications, Delhi

EU-India Summit (15 July 2020). https://www.consilium.europa.eu/en/press/press-releases/2020/07/15/joint-statement-15th-eu-india-summit-15-july-2020/#:~:text=The%2015th%20Summit%20between,EU%20was%20represented%20by%20Mr. Accessed on 15 July 2020

James TC, Bhatnagar A (2020, April 15) Science, technology, and innovation in Indian systems of medicine: an exploration in the context of COVID-19 Pandemic. RIS Diary, 3rd Special Issue on COVID-19

Rajani Ranjan Rashmi (2020, June 20) Global climate change: challenges for India. G 20 DIGEST. GIS Publication, Delhi

WBCSD Global Scenario 2000–2050 (2009) Exploring sustainable development. World Business Council for Sustainable Development

Suresh Kumar is a Professor in the Department of African Studies, Faculty of Social Sciences at the University of Delhi. He shouldered the responsibilities as the Head of the Department and Coordinator of Centre for African Studies, UGC Area Study Program for 2015–2019. He is the Chief Editor of Peer-Reviewed & UGC indexed journal—Indian Journal of African Studies and Africaindia.org (http://www.africaindia.org). Professor Kumar is a passionate traveler who has visited forty countries for research and teaching assignments. He is the Chief Collaborator of the Indian Council of Social Science Research-National Institute for the Humanities and Social Sciences (ICSSR-NIHSS), International Joint Research Project on Religion, Yoga, and Education in India and South Africa. He articulated his opinion on different aspects of Africa before the International Geography Union (2004 & 2008), Ronald H. Brown Institute (2008), United Nations Organisation (2012), World Policy Council (WPC, 2014, 2016 & February 2021), European Commission of African Studies (ECAS), Council for the Development of Social Science Research in Africa (CODESRIA, 2015), IBSA, BRICS-IKS (2016, 2018 and 2019) and many others. About 146 articles have been published in different books, journals, and newspapers. He has actively contributed his opinions on India and Africa in the different international media channels of BBC, Aljazeera (English), ABC (USA), Yomiuri Shinbun (Japan), and All India Radio.

Raj Kumar is an Assistant Professor in the Department of Political Science, SPM College, University of Delhi.

Chapter 15
Community Resilience: A Potential Answer to the Emerging Pandemic

Somenath Halder

Abstract There is a growing concern, on a global scale, about the 'resilience' and especially 'community resilience' amidst the emergency of new-pandemic (COVID-19) until the discovery of lifesaving drug(s). Though the term 'resilience' is widely used in various disciplinary platforms but metaphorically adopted in this piece to address peoples' struggle for surviving and coping capacity under the vulnerable contagious environment, and restarting post-pandemic life. Thereafter, the encoded term 'community' coherently included for focusing the world's community and also inherently scoping for addressing the issue of resilience hierarchically from large to small communities or like urban to rural communities. There is no doubt that the intruded COVID-19 brings a new threat to human civilization. Collectively various communities are adopting new adaptation strategies. While the methodology adopted is a review of literature from renounced contributors after rational examining, modification, and appropriate reconstructions of dimensions, as well as components or indicators. From a pandemic perspective, the four major dimensions of community resilience and their underside indicators are considered—*Society and Economy, Environment and Climate Change, Infrastructure*, and *Administration and Governance*. After a rigorous review of available focussed literature, it would not be inappropriate to state that the conceptual framework presented in this chapter could help measure the resilient power during the onset of a pandemic and reframe the policy proposal post-COVID survival.

Keywords Health hazard · Risk factor · Composite score assessment CSA · Standardized score · Global phenomena

15.1 Introduction

The global scene of mass-scalar epidemic and loss of human lives, thousand of millions, from a historical perspective, is not new. Alongside this, there is a 'vague'

S. Halder (✉)
Department of Geography, Kaliachak College, Malda, West Bengal, India
e-mail: somnatgeo@gmail.com

M. Mishra and R. B. Singh, *COVID-19 Pandemic Trajectory in the Developing World*, Advances in Geographical and Environmental Sciences,
https://doi.org/10.1007/978-981-33-6440-0_15

idea or grey statement about the occurrence of the pandemic in every hundred years around (Morens and Taubenberger 2018; Kertscher 2020; Frost 2020) becomes widespread. A well established universal nature of pandemic is its worldwide outbreak. In addition, a 'new-pandemic' happens when a new pathogen emerges to infect humans and between communities with having contagious nature (Porta 2008; Merriam-Webster Dictionary n.d.). Again, the similar phonetic term 'endemic' is never so like a pandemic. The concurrence of a new-pandemic, *i.e.*, COVID-19, in the twenty-first century, is reportedly started from Wuhan, China (Dryhurst et al. 2020). It gradually reaches each nook and corner of the globe. The current fatal impact of this pandemic is around more than 691,013 confirmed deaths and more than 18 million confirmed infected cases (WHO 2020). The practical situation of human fatalities across affected nations or geographical areas (small or big) might be more than the available official records. Although there are controversies regarding the gaps between recorded data and ground realities, however, there might not be any debate on the challenging stressors by the new-pandemic to the global community. Not only these, but the current health emergency has also pushed the global community toward several crises *like* international relations and trade, economic, socio-political, socio-cultural, food and nutrition, health and wellbeing, and many more. However, this present situation will remain beyond the reach of effective tackling, instead of having improved science and technology of human care, unless the discovery of reliable and clinically tested lifesaving drug(s). Thus until the research and innovation of nectar-like drugs in medical science, the 'community resilience' would be a subordinate and helpful instrument for the survival of humans.

This study purposefully confers to relate out between three concepts like 'pandemic', 'community', and 'resilience'. Moreover, it attempts to highlight the conceptual and diagnostic outlay of this very discourse (*community resilience*) not only for academic contribution but also to reframe policy in view of the sustainability of civil societies in the post-pandemic era. Instead of not being a case study, this piece is closely examining the existed and developed concepts around the arena of pre-mentioned terminologies and also seeks to judge the *focussed term* under the pandemic scenario. Besides this, the present work is attempting to throw light on *community resilience*, specifically the dimensions and scale of investigation. Other than these, after coherent and systematic cross-analysis of appropriate dimensions or major components with subcomponents, some more suitable subcomponents are developed further. Finally, a modified kind of conceptual model of community resilience, under the shadow of new-pandemic, has been built.

The remainder of this volume has acquainted with the relevant qualitative and quantitative discussion of said theme. The next sections are broadly comprised of the cross-examinations of the terms from numerous scholarly literature. Moreover, it also essentially cites the inherent qualities of both terms. The enhanced subsection is focused on exploring the direction of the study specifying community, resilience, and spatial differentiation. *Section* 15.3 is simply devoted to researching the validity as well as the significance of community resilience (CR) under the severe pandemic phenomenon. The later part, *Sect.* 15.4, is functionally representing the calibration of robust and sound composite score assessment (CSA) and its different connected techniques. Finally, *Sect.* 15.5 is interpreting the roundup importance and future usage of *this* customized model for reframing existing policies which usually addressing human risk segments.

15.2 'Community' and 'Resilience': Partner or Opponent

In an academic platform, during the description of the *man-nature interface,* most commonly, the term 'resilience' is applied as metaphoric sense. However, the genesis of this terminology or concept has been traced from the sciences of physics and mathematics. Along with the core of natural sciences, the concept is defined as the capacity of a substance or system to bounce back to its original state after the relocation of stressors. In another way, the definition may confer as a material with resilient power may bend or get squeezed, caused by stressors, and later on, return to its earlier status, rather than broken, after releasing the stressors (Gordon 1978; Bodin and Wiman 2004). Since its evolution and more comprehensive applications and due to ever-continuous development, there is no concrete definition of the said concept suitably applied to social sciences and human risk assessment. Whenever 'resilience' as a concept with growing concern, has been adopted to describe the capabilities or power of adaptations of individuals (*e.g.*, Orencio et al. 2015; Bonanno 2004; Butler et al. 2007; Rutter 1993; Werner and Smith 1982), groups or communities (*e.g.*, Brown and Kulig 1996/1997; Sonn and Fisher 1998; Smith et al. 2012; DasGupta and Shaw 2015; Dinh et al. 2016; Sharifuzzaman et al. 2018), and wider societies (*e.g.*, Pfefferbaum et al. 2015; Godschalk 2003; Adger 2000). On the other hand, the emphasizing subject areas (other than communities) are covered by the very *concept* like pandemic (*e.g.*, Naja and Hamadeh 2020; Keenan 2020), disaster and hazard management (*e.g.*, Paton et al. 2001; Ainuddin and Routray 2012; Bergstrand et al. 2014; Danar and Pushpalal 2014; Alshehri et al. 2015; Kulig and Botey 2016; Li et al. 2016; Parson et al. 2016; Anwar et al. 2017; Qin et al. 2017; Mayer 2019; Song et al. 2019; Almutairi et al. 2020), crisis analysis (*e.g.*, Camfield et al. 2013; Simpson 2020) and so forth.

When any kind of external shocks in the form of disaster (natural, human-made, or health-related) comes to the habitable areas, the members of the residing communities are probably the first who face and react upon it. So, 'community' is the principal actor in the non-implicit recognition under the approach of community resilience (CR). Besides, for this very reason, the community members should not only be recognized as 'victims' but also as prime means to deal with the extreme stressors. Moreover, it is worthy of being mentioned, in this context, to those who work with this concept (CR), should spontaneously acknowledge the dynamic qualities of a community like *strength* and *power* rather than *weakness* and *defenselessness* (Adger 2000; Gallopin 2006; Norris et al. 2008; Ungar 2011). Within a due course of time, epistemologically, the term 'resilience' is originated from natural sciences but presently becomes familiar after adjoining with the community (*e.g.*, 'community resilience') and has been popularly adopted in the different academic and disciplinary milieu. In the broader section, like disaster or hazard management, the approach of CR is encountered with a broad definition. It is regarded that the word *resilience* possesses the qualitative and quantitative angles, and it incorporates the measurement of the degree or magnitude of vulnerability, power of struggle, and recovering capabilities of any target group or community (Norris et al. 2008; Engle 2011; Ungar 2011). Simultaneously, the COVID-19 is concurring onto present civilization, globally, and threatening the existence of humanity, which should be regarded as global-health-disaster.

Moreover, according to Visser et al. (2012), *resilience* at the individual level is apprehended with an individual's characteristics like optimistic attitude, self-discipline, personal control, better self-esteem, the problem facing and solving skills, and practical views amidst the problematic environment. Community resilience (CR) is somewhat different from the previous, having important, worthy components. In another way, it can be said that CR is a healthier form of agglomeration of various components or subcomponents of individuals under a focussed human-group, which can able to cope up with sudden shock(s). This collective concept is concerned with individual members of the community rather than the community as a whole if the absorption of the shocks is concerned. CR is broadly linked with some dimensions like the magnitude of social connectedness, social relations and its quality, helpful community norms, the problem-solving capability of a community, and features of social networking threads (Norris et al. 2008; Aldrich 2012). *Resilience* and *community resilience*—both are dynamic and ever-changing concepts connected to a complex set of capacities, processes, and results that can interactively support one another. Nevertheless, it would be appropriate to explain CR in a single sentence:

> community resilience measures the extent to which community capital can absorb immediate disturbances or chronic stressors as well as the community's adaptive capacity to self-organize into a stable, functional community system. (Acevedo 2014)

In the perspectives of social science and disaster management, it is easily understandable that both the terms 'resilience' and 'community' are not opponents rather than partners to each other. Furthermore, in a joint mode (community + resilience = CR) it would be a potential way-out to combat against the emerging pandemic (COVID-19) and adaptation of a community, to immediate and long-term environmental changes, *following* positive initiatives with collective manner, co-operative engagement of community members, affirmative response and recovery actions with proper coordination, non-individual level problem identification, and potable solution, future planning with priority setting.

Re-conferring the above agenda in systematize way forward is now became necessary in this part. Eventually, the scope of the appraisal of the concept of CR in the current global context, *i.e.*, under pandemic condition, is three-dimensional. From the illustration (Fig. 15.1), it can be readily observable that the X-axis shows *resilience*, Y-axis represents *community*, and Z-axis shows the *geographical* or *spatial* dimension. Thus incorporating the above three dimensions in a single frame would make

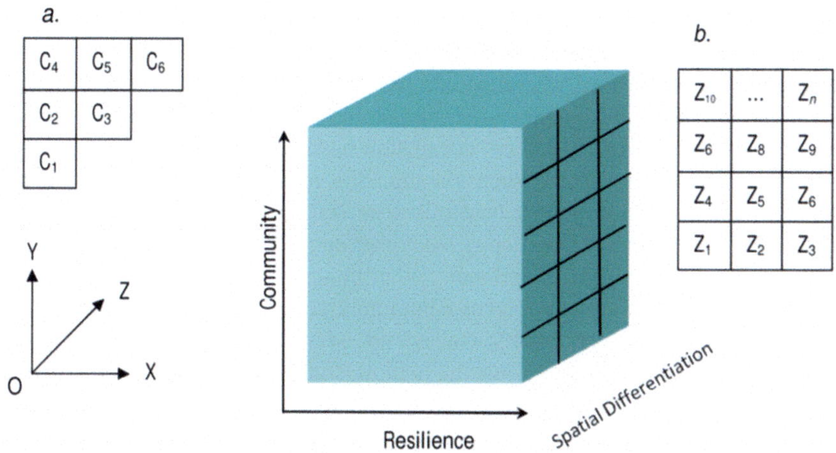

Fig. 15.1 Dimensions of study of CR (*a.* community differentiation, *b.* spatial differentiation) (*Source* Developed by author)

it more convenient to highlight the crisis corners of CR under the pandemic situation for any specific community or community in the broad sense and coherent policymaking as well.

In addition to this, the idea of the community under the purview of the present volume refers to the group of people, sorted by numerous categories. Figure 15.2 replicates the classification of the community on which the present concept may throw light. The reason behind is the COVID-19, the most contagious epidemic, impacts

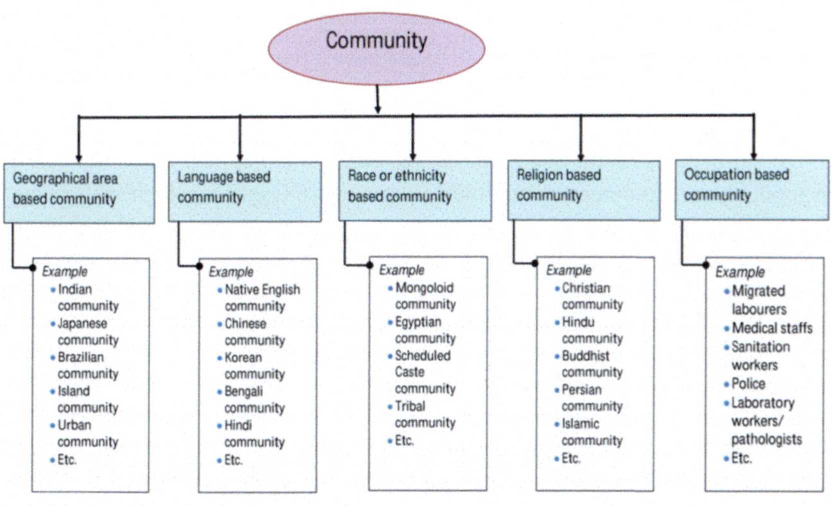

Fig. 15.2 Classified communities useful for studying CR under pandemic condition (*Source* Developed and conceptualized by author)

almost each and every human-group beyond their caste, creed, class, religion, and language and also each nook and corner of the globe. Therefore, the broad term 'community' is purposively classified into five major categories, *e.g.*, geographical area based community, language-based community, race or ethnicity-based community, religion-based community, and occupation-based community (Fig. 15.2). The underneath philosophy behind such classification is that among the categorized group of peoples, regarding their base of categorization, the power of resilience is supposed to vary from one community to another due to their varied lifestyle, livelihood, culture, knowledgebase, and many other things. Moreover, here lies another opportunity to study *this* with a cross-community comparative point of reference. The side-by-side cross-community analysis also may contribute some positive or more substantial points for policymaking to those communities who may have some lacuna(s) to fight against the pandemic. The upcoming paragraph focuses on the three dimensions, more intrinsically, and also the scale of research.

15.3 Re-Evaluating the Concept CR Under the Pandemic Lens

From the academic discussions in the preceding sections, it is now clear that the concept, as well as the approach of community resilience (CR) have not been applied to the new-pandemic scenario the world is presently facing. Thus during this crisis period (when due to lack of appropriate medication mortality rate is so high and turned into a global health emergency and throws the question of survival), it is ardent to keep the concept of CR under the scanner of the emerging vulnerable pandemic. In the surveyed literature, community resilience (CR) had been proven as a socio-political and strategic means for resolving hazardous or any human risk phenomenon, due to exterior stressors, from local to a global scale. Thus, in view of the above discussions, this section is attempting to re-evaluate the *concept* under the emerging pandemic condition. In order to boost physical, social, economic, cultural, and overall peoples' daily living, during the pandemic and post-pandemic period, this model-based commentary (multilevel) framework (Fig. 15.3) highlights human sustainability at the individual, community, regional, national, and global levels. This multilevel framework is broadly inspired by the work of Naja and Hamadeh (2020). They purposively focused on 'nutrition' amidst the COVID-19, considering the current global outreach of the pandemic, household (individual), and community resilience appeared as the mainstay while remaining the first-combating agents in public health emergency and preparedness (Reissman et al. 2006). At the household level, the basic reliant denominator that helps to cope up with any health-disaster, including COVID-19 or SARS-2, is the family with its structure, income potential, awareness of members, culture (in a broad sense), and standard safety and security measures followed by the members.

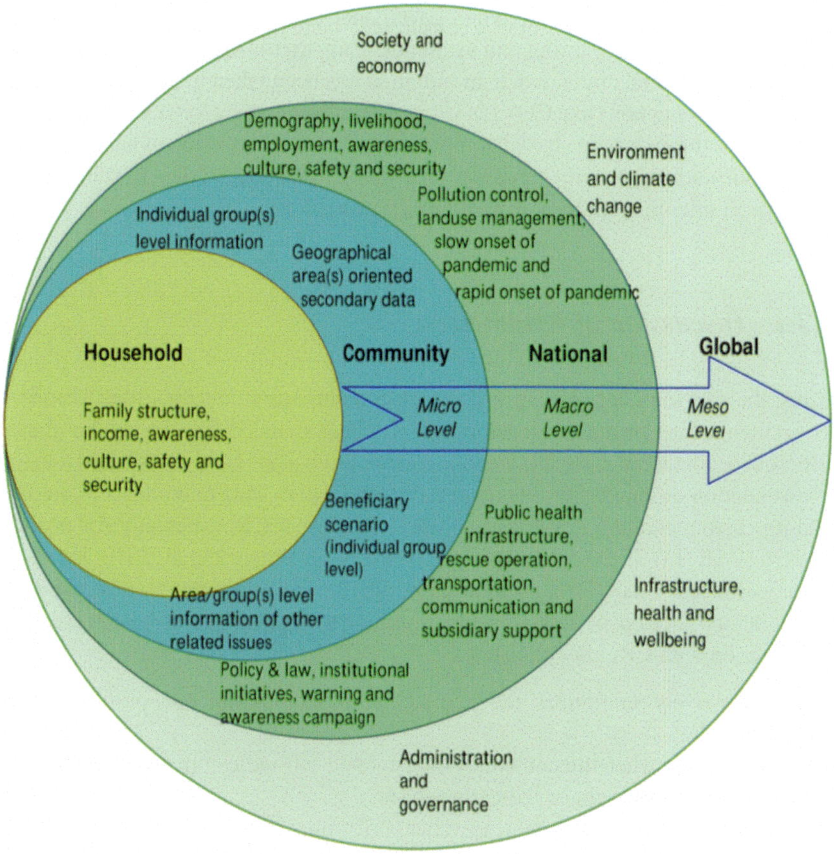

Fig. 15.3 A multilevel framework of research action on community resilience (*Source* Developed by author, based on Naja and Hamadeh 2020)

On the second layer, when it comes upon with *community-level* measures, the crucial bases are individual or group level information (primary or secondary), which helps to build the foundation of CR, then geographical area wise secondary information, especially for comparative areal reassessment. In addition to these, the beneficiary scenarios of the targeted community by any governmental or non-governmental institutions are also noteworthy for the same assessment, attaching with area or group level other related issues (helpful for the formation of robust subcomponents). Alongside the third layer (Fig. 15.3), contains the regional as well as national level analysis of significant indicators like society and economy (*indicators like*: demography, occupation, employment, awareness and education, safety and security), environment and climate change (*indicators like*: pollution control, landuse management, slow onset of pandemic, rapid onset of pandemic), infrastructure, health, and wellbeing (*indicators like*: public health infrastructure, rescue operation, transportation, communication, subsidiary support), administration and governance (*indicators like*: policy and law, institutional initiatives, warning and awareness campaign). Apart from these, in the fourth layer or the end circle, when an initiation has been taken to discuss the world level crisis phenomenon (like COVID-19) and determining the global level resilience measurement through score-based computation after customization of suitable indicator and sub-indicators, with necessary modification and country-level or vulnerable zone-wise, a scientific analysis would only be possible.

15.3.1 Discussion of Parameters

During the previous decades, in respect of contemporary concepts like 'community resilience', a continuous development has been evinced throughout the globe. Numerous academicians, social scientists, and policymakers have contributed toward the formulation of qualitative and quantitative indicators and components related to building customized and developed composite index of CR, especially for natural disasters (USIOTWSP 2007; Cutter et al. 2008, 2010; Peacock et al. 2010; Uy et al. 2011; Joerin and Shaw 2011; Joerin et al. 2012; Teo et al. 2013). The primary reasons behind such encouragement and aspiration to adopt indicator-based research, in connection with CR, may be listed as:

- It progressively minimizes the complexity of computing one particular key concept (CR);
- It sounds better when the number of inter-related sub-indicators is included under every single component or indicator; and
- It makes it easy to indulge any researcher performing comparative analysis among the correlated parameters, which indirectly contributes to further policymaking (Cutter et al. 2008).

In comparison with the conclusive recommendation of community resilience approach as disaster recovery (Shea 2018) or "the ability of the community to bounce back, respond to, recover from and absorb the impacts and cope with" (Ainuddin and Routray 2012: 911) the present context seemed to be little bit dissimilar. Thus the proceeding discussions about the referred *components and subcomponents* may not be exclusively related to the current pandemic.

15.3.2 Modified Structure of CR Under Pandemic

In this present section, based on the previous studies and also with the best of literature review experiences, an ensemble suitable 'community resilience index' (CRI) is developed to perform composite score assessment (CSA) of these, mentioned earlier. This focussed index is comprised of four major dimensions, including socioeconomic resilience (society and economy), environmental resilience (environment and

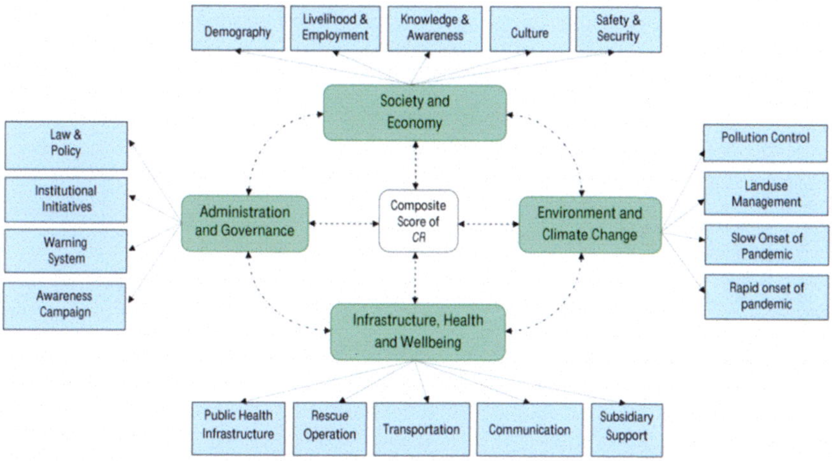

Fig. 15.4 Structure of community resilience and customized index (*Source* Developed by the author *after*, Norris et al. 2008; Qin et al. 2017)

climate change), infrastructure and health resilience (infrastructure, health, and well-being), institutional resilience (administration and governance) (Fig. 15.4). Afterward, these dimensions are divided into 22 components (demography, occupation, employment, pollution control, landuse management, slow onset of pandemic, rapid onset of pandemic, public health infrastructure, warning and awareness campaign, etc.). Moreover, 80 quantitative subcomponents or variables are incorporated for analyzing CR, under any spatial scale (Tables 15.1, 15.2, 15.3, and 15.4). These incorporated subcomponents or variables are altered into well-recognized forms (percentage, per person, binary value, density, etc.). Correspondingly, the effects or relation to each variable to CRI, either *positive* or *negative*, are identified and referred based on pioneering studies and review observations. Thus in the tables, the given '+' means positive effect, and '–' means the negative effect. Despite all of these, the other descriptions and references are given in the respective tables.

15.4 Socioeconomic Resilience

It is an established fact that the CR is a multidimensional concept, counting sociological, economic, ecological, and institutional elements (Cutter et al. 2008; Norris et al. 2008; Sherrieb et al. 2010; Qin et al. 2017). However, very few of the academic contributors have considered the clubbing of the 'social' and 'economic' parameters into a single dimension, especially in the case of present pandemic (COVID-19). To say elaborately, in case of this new-pandemic, humans are regarded as violent agents of contiguity indirectly. It threatens the foundation of *social resilience*, *i.e.*, social relation, social bondage, social networks, and belongings. Thus it is perceived that

Table 15.1 Variables composing socioeconomic resilience

Dimension	Component	Subcomponent	Unit	References	Relation to CR
Society and Economy	Demography	Average population growth rate	Point value	Cutter et al. 2003	(−) value > growth rate < resilience
		Population density	Population per thousand	Hou et al. 2016	(−) value > density < resilience
		Percentage of elderly population (> 60 years)	Percentage	Su et al. 2015	(−) value > proportion < resilience
		Percentage of disable population to total population	Percentage	Cutter et al. 2014; Sharifi 2016	(-) value > proportion < resilience
	Livelihood	Percentage of workers dependent solely on natural resource	Percentage	–	(+) value > proportion > resilience
		Percentage of workers dependent on single earning source	Percentage	–	(−) value > proportion < resilience
		Percentage of traditional workers become jobless during pandemic	Percentage	–	(−) value > proportion < resilience
		Percentage of traditional workers trapped under debt during pandemic	Percentage	–	(−) value > proportion < resilience
	Employment	Percentage of workers fall under unorganized sector	Percentage	–	(−) value > proportion < resilience
		Percentage of workers drawing half/partial salary	Percentage	–	(-) value > proportion < resilience
		Percentage of unemployed youths (> 25 years)	Percentage	–	(−) value > proportion < resilience

(continued)

Table 15.1 (continued)

Dimension	Component	Subcomponent	Unit	References	Relation to CR
		Percentage of workers adopted new means of earning during pandemic	Percentage	–	(+) value > proportion > resilience
	Culture	Frequency of cases not avoiding religious gathering	Point value	–	(−) value > magnitude < resilience
		Frequency of cases not avoiding marriage/death-ceremony/party like mass gathering	Point value	–	(−) value > magnitude < resilience
		Regional/local level growth of selling of herbal product	Point value	–	(+) value > magnitude > resilience
	Awareness	(inverse) Percentage of cases disobey local lockdown and roaming unnecessarily	Point value	–	(+) value > magnitude > resilience
		(inverse) Percentage of cases of hectic fines for not covering faces by 'face-mask/cloth'	Point value	Sing and Avikal 2020	(+) value > magnitude > resilience
		(inverse) Percentage of cases not avoiding heavy crowd	Point value	Sing and Avikal 2020	(+) value > magnitude > resilience
		Increments of sales at (regional/local level) sanitation items	Point value	–	(+) value > magnitude > resilience
	Safety and Security	Number of households below minimum living standard per 10,000 households	Point value	Lixin et al. 2014	(−) value > magnitude < resilience

(continued)

Table 15.1 (continued)

Dimension	Component	Subcomponent	Unit	References	Relation to CR
		Percentage of population above minimum living standard	Percentage	Lixin et al. 2014	(+) value > proportion > resilience
		Percentage of homeless family or household	Percentage	Despotaki et al. 2018, Chung et al. 2020	(−) value > proportion < resilience
		Expenditure for social welfare per person	$ per person	Zhang and Huang 2013	(+) value > magnitude > resilience
		Social welfare budget per person	$ per person	Zhang and Huang 2013	(+) value > magnitude > resilience

Source Compiled by author

Table 15.2 Variables composing environmental resilience

Dimension	Component	Subcomponent	Unit	References	Relation to CR
Environment and Climate Change	Pollution Control	Expenditure for air pollution control/forestation or green management per 10,000 persons	Point value	Markard and Rosenbloom 2020	(+) value > magnitude > resilience
		Expenditure for water pollution control/improving water quality per 10,000 persons	Point value	–	(+) value > magnitude > resilience
		Expenditure for household, market and waste management per 1000 households	Point value	–	(+) value > magnitude > resilience
		Sewage management centralized treatment rate	Percentage	Qin et al. 2017	(+) value > proportion > resilience
		Expenditure for green space/parks management per 10,000 persons	Point value	Qin et al. 2017	(+) value > magnitude > resilience
		Ratio of industrial solid waste treatment comprehensively utilized	Percentage	Qin et al. 2017	(+) value > proportion > resilience
	Landuse Management	Percentage of buildup area	Percentage	–	(+) value > proportion > resilience
		Percentage of area under commercial use	Percentage	Sherrieb et al. 2010; Cutter et al. 2016	(+) value > proportion > resilience
		Percentage of area under green and open space	Percentage	Qin et al. 2017	(+) value > proportion > resilience

(continued)

Table 15.2 (continued)

Dimension	Component	Subcomponent	Unit	References	Relation to CR
		Percentage of area under massive dense settlement	Percentage	–	(−) value > proportion < resilience
		Percentage of area under vulnerable old buildings	Percentage	–	(−) value > proportion < resilience
	Slow Onset of Pandemic	Percentage of community asset (school or other buildings) utilized for quarantine	Percentage	–	(+) value > proportion > resilience
		Frequency (per month) of area sanitization	Unit value	–	(+) value > magnitude > resilience
		Percentage of household (among total) advised for home quarantine	Percentage	–	(+) value > proportion > resilience
	Rapid Onset of Pandemic	Frequency (per month) of complete lockdown	Unit value	Qian et al. 2020	(+) value > magnitude > resilience
		Percentage of additional community asset (club or public community hall) utilized for quarantine	Percentage	–	(+) value > proportion > resilience
		Percentage of testing (per day) for Corona in local area	Percentage	Qian et al. 2020	(+) value > proportion > resilience
		Frequency (per month) of door-to-door health survey by local administration	Unit value	–	(+) value > magnitude > resilience

Source Compiled by author

Table 15.3 Variables, composing the infrastructure and health resilience

Dimension	Component	Subcomponent	Unit	References	Relation to CR
Infrastructure and Health	Public Health and Health Infrastructure	COVID recovery rate (total recovered patients/total COVID patients)	Point value	WHO 2020	(+) value > magnitude > resilience
		Number of days of active opening of OPD section in hospital/healthcare institutions (for non-COVID patients)	Point value	–	(+) value > magnitude > resilience
		Number of COVID dedicated hospitals/healthcare institutions per 10,000 people	Point value	Zhou et al. 2014 (modified)	(+) value > magnitude > resilience
		Number of COVID dedicated hospitals/healthcare beds per 1000 people	Point value	Ge et al. 2013 (modified)	(+) value > magnitude > resilience
		Average doctor-patient ratio, during pandemic (per 1000 people)	Ratio	Zhou et al. 2014 (modified)	(+) value > magnitude > resilience
		Average nurse-patient ratio, during pandemic (per 1000 people)	Ratio	Zhang and Huang 2013 (modified)	(+) value > magnitude > resilience

(continued)

Table 15.3 (continued)

Dimension	Component	Subcomponent	Unit	References	Relation to CR
		Percentage of health workers (doctor, nurse, health staffs, etc.) attacked by COVID-19	Percentage	–	(+) value > proportion > resilience
	Rescue Operation	Rescue of emigrants from abroad by air/water ways per 100 people	Percentage	–	(+) value > proportion > resilience
		Rescue of migrated workers by train/bus per 100 people	Percentage	–	(+) value > proportion > resilience
		Rescue of quarantined (suspected) COVID patients escaped from quarantine center (comparing with total COVID patient)	Percentage	Qian et al. 2020	(+) value > proportion > resilience
	Transportation	Principal road density (principal road in km./total buildup area km^2)	Point value	Frazier et al. 2013; Sharifi 2016	(+) value > magnitude > resilience
		Number of emergency vehicle available per 10,000 people	Point value	DasGupta and Shaw 2015	(+) value > magnitude > resilience
		Number of civil vehicles owned per 10,000 people	Point value	Liu and Li 2016	(+) value > magnitude > resilience

(continued)

Table 15.3 (continued)

Dimension	Component	Subcomponent	Unit	References	Relation to CR
		Number of public transportation vehicles per 10,000 people	Point value	Zhou et al. 2014	(−) value > magnitude < resilience
	Communication	Number of mobile users per 10,000 people	Point value	DasGupta and Shaw 2015	(+) value > magnitude > resilience
		Number of internet subscribers per 10,000 people	Point value	Lixin et al. 2014	(+) value > magnitude > resilience
	Electricity	Percentage of households using electricity for lighting	Percentage	Cimellaro et al. 2014; Mishra et al. 2017	(+) value > proportion > resilience
	Cooking gas	(Inverse) timely supply of cooking gas and home delivery	Point value	–	(−) value > proportion < resilience
	Water source	Percentage of households using tap/pipe water source	Percentage	Cimellaro et al. 2014; Mishra et al. 2017	(+) value > proportion > resilience
	Subsidiary Support	Provision of distribution of face-mask and sanitizer among poor (Yes = 1, No = 0)	Point value	Zou et al. 2020	(+) value > magnitude > resilience
		Percentage of poor benefited by subsidized rationing system (basic foodstuff) by NGO/Govt.	Percentage	Qian et al. 2020	(+) value > proportion > resilience

(continued)

Table 15.3 (continued)

Dimension	Component	Subcomponent	Unit	References	Relation to CR
		Provision of any monetary help for the poor (Yes = 1, No = 0)	Point value	Zou et al. 2020	(+) value > magnitude > resilience

Source Compiled by author

Table 15.4 Variable composing institutional resilience

Dimension	Component	Subcomponent	Unit	References	Relation to CR
Administration and Governance	Policy and Law	Implementation of strict price regulation rule(s) in regular market (Yes = 1, No = 0)	Numeric value	–	(+) value > proportion > resilience
		Implementation of law & order during violence against health workers (Yes = 1, No = 0)	Numeric value	–	(+) value > proportion > resilience
		Amendment of penalty rule for not using face-mask/cloth in public place (Yes = 1, No = 0)	Numeric value	Zou et al. 2020	(+) value > proportion > resilience
		Provision of any alternative job for reemployment of poor worker (Yes = 1, No = 0)	Numeric value	–	(+) value > proportion > resilience
	Institutional Initiatives	Deployment ratio of police force (number) during pandemic (per 10,000 people)	Point value	Zou et al. 2020	(+) value > magnitude > resilience
		Deployment ratio of volunteers (number) during pandemic (per 10,000 people)	Point value	–	(+) value > magnitude > resilience
		Deployment ratio of temporary health workers (number) during pandemic (per 10,000 people)	Point value	–	(+) value > magnitude > resilience
		Increment of 'bed-patient ratio' (pre-pandemic ratio—post-pandemic ratio)	Point value	Qian et al. 2020	(+) value > magnitude > resilience
		Provision of 'work from home' initiative, (Yes = 1, No = 0)	Numeric value	Qian et al. 2020	(+) value > proportion > resilience

(continued)

Table 15.4 (continued)

Dimension	Component	Subcomponent	Unit	References	Relation to CR
		Frequency of lockdown within six months (two weeks) for breaking the chain of contagion	Point value	–	(+) value > magnitude > resilience
	Warning Initiatives	Provision of high-tech warning system about COVID-19 in mobile ringtone (incoming & outgoing), (Yes = 1, No = 0)	Numeric value	Dako-Gyeke et al. 2020	(+) value > proportion > resilience
		Provision of warning system about COVID-19 in virtual social network, (Yes = 1, No = 0)	Numeric value	Dryhurst et al. 2020	(+) value > proportion > resilience
		Provision of traditional warning system about COVID-19 in public place, (Yes = 1, No = 0)	Numeric value	–	(+) value > proportion > resilience
	Awareness Campaign	Provision of up-to-date awareness campaign about COVID-19 in newspapers/television, (Yes = 1, No = 0)	Numeric value	–	(+) value > proportion > resilience
		Provision of small-scale awareness campaign about COVID-19 in marginal/remote areas (Yes = 1, No = 0)	Numeric value	–	(+) value > proportion > resilience
		Provision of praising and thanksgiving events for COVID-front-level-warriors (Yes = 1, No = 0)	Numeric value	Dako-Gyeke et al. 2020	(+) value > proportion > resilience

Source Compiled by author

the combination of 'social' and 'economic' factors may construct a more significant dimension of resilience as these two dimensions boost each other. Socioeconomic resilience is formed by six major components, including demography, livelihood, employment, culture, awareness, safety, and security (Table 15.1). Demographic characteristics refer to average population growth, population density, the proportion of the elderly, and the disabled population (Cutter et al. 2003, 2014; Hou et al. 2016; Su et al. 2015; Sharifi 2016). Whereas, under the broad economic angle, livelihood and employment cooperatively help to build a more suitable parameter, seemed to be reasonable for the current pandemic. Walking along the different way, under the major component of livelihood, four subcomponents are taken into account, *i.e.*, percentage of workers solely dependent on natural resource; the percentage of workers dependent on the single earning source; the percentage of traditional workers become jobless due to pandemic cause; and, percentage of workers severely trapped under debts due to pandemic.

Similarly, under the component of employment also four subcomponents are included, *i.e.*, percentage of workers under unorganized sector; the percentage of workers drawing partial salary due to pandemic; and percentage of unemployed youths (over 25 years age of both genders); and, percentage of workers adopted new kind of earning during the pandemic period. From the previous experiences, it has been observed that the above-mentioned phenomena (pointing out as subcomponent) are more or less true in the developing and underdeveloped countries.

So it is thought to be ardent for replicating one of the important corners of economic resilience. Unlike conventional resilience studies, this work purposively put culture and awareness as important components rather than direct inclusion of formal education or other allied factors. Three subcomponents are included under the major component like culture (Table 15.1), these are frequency of cases (recorded) violating mass-gathering rule for religious cause; frequency of cases (recorded) violating mass-gathering rule for social functions causes (e.g., marriage ceremony; death ritual, or rejoicing party); and, regional or local level growth of selling herbal product. It also indirectly helps to measure the variable of consciousness of any community for immunity boosting at the individual level. In case of another component like awareness among included subcomponents (four), the first three subcomponents are considered in the inverse scale, and only the last one is involved with the increment of sales of sanitation items during the pandemic, which replicates the growing consciousness of any community. The rest three subcomponents are the percentage of cases of disobeying local lockdown, percentage of cases of penalty for not using any face-mask or cloth, and percentage of cases of disobeying social-distancing norm (Singh and Avikal 2020). Under the broad dimension of socioeconomic resilience, the last important component is safety and security, and this component is formulated with five subcomponents. The first three are the proportion of households below minimum living standard, percentage of population above minimum living standard, and the percentage of homeless family or household. These above-discussed subcomponents are vital in case of resilience because the proportion of 'below-poverty-line' population and homeless family counts directly indicate the less resilience capability (Lixin et al. 2014; Despotaki et al. 2018; Chung et al.

2020). On the other hand, the rest two subcomponents like expenditure for social welfare per person and social welfare budget per person directly confirm the security measures (Zhang and Huang 2013).

15.5 Ecological Resilience

Environmental and climate change resilience comes under the holistic categorization of resilience, like ecological resilience. This kind of resilience dimension is comprised of important components like environmental pollution control; landuse management; slow onset of pandemic; and, rapid onset of pandemic. As this emerging pandemic is thought to be linked with environmental degradation and global climate change (Markard and Rosenbloom 2020), there are four major components incorporated to form the ecological or say, environmental, and climate change resilience. This is especially focussed on assessing the resilience under a public health emergency. Any kind of disease detrimentally carries and spreads through the basic earth's elements. Though in case of COVID-19 it is carried through water droplets in the air. Hence, the inclusion of the major component of 'environmental pollution control' is coherent. This component is constructed with six subcomponents like expenditure patterns (per 10,000 populations in a given area) for air pollution control, water pollution control, and waste management (Table 15.2).

Previously recognized three subcomponents i.e., sewage management centralized treatment rate; expenditure for green or open space, or parks management (per 10,000 persons); and, the ratio of industrial solid waste treatment which are comprehensively utilized and they are found instrumental (Qin et al. 2017). In order to build the second component of landuse management, five subcomponents are taken under consideration. The first three are the percentage of built-up area (as a general consensus); the percentage of area under commercial use; and, the percentage of area under green space having a positive effect on CR (Sherrieb et al. 2010; Cutter et al. 2016; Qin et al. 2017). The rest two variables are crucial. The percentage of area under the huge dense settlement is the common characteristic feature of human habitation in developing and underdevelope countries and also considered responsible for the furious spreading of endemics and epidemics. Again the percentage of area under old and vulnerable buildings also weakens the resilience quality as they extend the threat of unhygienic living and other related issues. Remaining two major components like the slow onset of pandemic and rapid onset of the pandemic are very reasonable for community resilience measures during the pandemic period, existing under the environmental management dimension. There are three subcomponents under the component of slow onset of pandemic, *i.e.*, percentage of community asset utilized for quarantine; frequency of area sanitization; and, percentage of households advised to be home quarantine. On the contrary, four different types of subcomponents are included under the component of rapid onset of pandemic, *i.e.*, frequency of complete lockdown (for breaking the chain of contiguity of COVID-19); the percentage of additional community assets utilized for quarantine (important for an enhanced number of cases in a heavily populated area); the percentage of Corona testing (a vital issue for combating coronavirus diseases) (Qian et al. 2020); and, frequency of door-to-door survey for monitoring health of citizens in the local area (Table 15.2).

15.6 Health and Infrastructural Resilience

Health and infrastructural resilience is the fundamental dimension of CR under pandemic conditions. It helps to gauge public health and health infrastructure, lifeline options (transportation, communication, electricity, cooking gas, water supply, etc.), rescue operation, and subsidiary support (specifically valid for underdeveloped and developing worlds). Furthermore, this very dimension critically serves the CR by providing support for better resilience and sustainability. Within this dimension, there are eight vital components—public health and health infrastructure; rescue operation; transportation; communication; electricity; cooking gas; water supply; and, subsidiary support (Table 15.3). The first component (public health and health infrastructure) is very much crucial for the pandemic-related composition of CR. This component has seven significant variables, i.e., area or region-specific COVID recovery rate (WHO 2020); the number of active opening of OPD for non-COVID patients; the number of COVID dedicated health institutions per 10,000 people (Zhou et al. 2014); COVID bed-patient ratio (Ge et al. 2013); doctor-patient ratio per thousand (Zhou et al. 2014); nurse-patient ratio per thousand (Zhang and Huang 2013); and, the percentage of health workers attacked by COVID-19. Additionally, the component like 'rescue operation' is seemed valuable for developing countries. It is composed of three subcomponents, *i.e.*, rescue of emigrants from outside country per 100 people; the rescue of migrated workers to their native villages per 100 people; and, the most vital one—the percentage of the rescue of quarantined COVID patients escaped from quarantine centers (ardently responsible for the mass-scale spreading of pandemic) (Qian et al. 2020). Among the components which provide lifeline options for better tolerance during a pandemic are transportation, communication, electricity, cooking gas, and water supply. Transportation is considered as an important lifeline because it not only maintains the supply of necessary daily-goods but also provides support in emergency communication and movements of passengers and patients in acute morbid condition for availing better health services. This component has been formulated with four subcomponents, i.e., principal road density (Frazier et al. 2013; Sharifi 2016); the number of emergency vehicle available per 10,000 people (DasGupta and Shaw 2015); the number of civil vehicles owned per 10,000 people (Liu and Li 2016); and, the number of public transportation vehicles per 10,000 people (Zhou et al. 2014). The subcomponents like the number of mobile users per 10,000 people (DasGupta and Shaw 2015) and the number of internet subscribers per 10,000 people (Lixin et al. 2014) build the valued component of *communication*, during a pandemic when changed virtually. Social network is the time-demand. The next two components are built with singular subcomponents, e.g., timely home delivery of cooking gas (inverse scale) and percentage of household using pipe or tap water (Cimellaro et al. 2014; Mishra et al. 2017). Under the dimension of health and infrastructural resilience, the last component, *i.e.*, subsidiary support hypothesized to be crucial for enhancing peoples' resilience, especially in poverty dominated areas (Qian et al. 2020). This component is composed with three respective variables, like percentage of poor (households) benefited by subsidized rationing system (basic foodstuff) by NGO or Govt., provision of distribution of *free* face-mask and sanitizer among poor (binary scale), and provision of any monetary help for the poor (binary scale) (Zou et al. 2020).

15.7 Institutional Resilience

Institutional resilience, here modified as administration and governance resilience, has been recognized as the ability of functioning of an organization or administrative body to foresee, organize for and adjust with new 'code' of dynamic change or strategic alteration due to emergence of any new kind of external shocks (Qin et al. 2017). However, the concurring situation, the worldwide public health emergency, is quite different from the previously held natural or human-made disasters. The COVID-19 outbreak demands the full attention of the almost entire machinery of any administration, local to national levels. Targetting the best and comprehensive output, four major components have been incorporated under institutional resilience, i.e., policy and law; institutional initiatives; warning initiatives; and, awareness campaigns (Table 15.4). The first component—policy and law, is composed of four subcomponents that would be measured by binary value analysis. The subcomponents are the implementation of price control in regular markets, implementation of law and order during violence against health workers, implementation of penalty advisory for not using face-mask, or piece of cloth in public places (Zou et al. 2020) and provision of any alternative job for the redeployment of vulnerable workers. Under the second important component, i.e., institutional initiatives, six subcomponents are assumed to be vital. These are the deployment ratio of local police force during the pandemic; deployment ratio of additional volunteers for pandemic management (Zou et al. 2020); deployment ratio of additional (temporary) health workers for pandemic management; increment of bed-patient ratio during pandemic; provision of 'work from home' initiative (binary scale) (Qian et al. 2020); and lastly, the frequency of lockdown within six months for breaking the chain of contagion (Table 15.4). In the case of the components of warning initiatives, three coherent subcomponents are taken into account.

These are the provision of high-tech warning system about COVID-19 in mobile ringtone (incoming and outgoing) (Dako-Gyeke et al. 2020); provision of warning system in virtual social platforms (Dryhurst et al. 2020); and, the provision of traditional warning system in public places (like street drama and loudspeaker announcements). The last component, under this major resilience dimension, i.e., the awareness campaign has incorporated three subcomponents—provision of an up-to-date awareness campaign about COVID-19 in newspaper and television programs; provision of a small-scale awareness campaign about coronavirus in marginal and remote areas and provision of praising; and, thanksgiving events for COVID-front-level-warriors (Dako-Gyeke et al. 2020).

Although the vast array of incorporated variables and adjacent components, along with their variety of units, seemingly create complexity to form a unique index, it is essential because of all-round coverage of newly intruded challenge and coping capacity in real sense. In the next section, the study approach, as well as the computational breakthrough, has been interpreted.

15.8 Study Approach of CR: Method and Analysis

Now the discussion comes to the point where it is necessary to underline, after knowing what should be the lines and orders and also the scale of database needed for community resilience (CR) study under pandemic, the corners and direction of study method of CR as well as calibration steps for the end result of CSA. After following the guidance of Fig. 15.1, it becomes an easy job for understanding. If the 'X' axis represents the degree of *resilience* and 'Y' axis represents *community*, and when a researcher is focused to a single geographical area ('Z_1') then binding with that particular singular region (micro or macro), he/she should go for inter-community (formally categorized) analysis of CR to the emerging pandemic. Apart from the above, when a researcher is motivated to study community resilience but with a geographical perspective, he/she must have to choose more than single 'area' or 'region' (*e.g.*, 'Z_1', 'Z_2', 'Z_3' … 'Z_n'). However, rest of the items like *community* ('Y_1') should be included as a whole (without any formal categorization), and the base item like *resilience* should be included as 'X'. But a researcher should be cautious about more than one 'area' or 'region' inclusion under a particular study. He/she either take into consideration the pre-classified formal regions under a micro (e.g., number of districts under a state)/macro (e.g., number of states under a nation or country) region or purposively do the systematic classification of the region (big or small) accordingly. The third option would be more complicated than the former two. Here, a researcher must choose the base axis 'X' for representing the targeted item *resilience* and its several resilience dimensions. Then he/she may choose the number of formally or systematically classified communities (*e.g.*, 'Y_1', 'Y_2', 'Y_3' … 'Y_n'). Also further, he/she may include the number of 'area' or 'region' (*e.g.*, 'Z_1', 'Z_2', 'Z_3' … 'Z_n') for mapping the differential community-wise magnitude of CR with the areal diffusion. It would not be wrong to say that the third option is proved to be more suitable for geographers who are much concerned with human risk assessment.

According to Qin et al. (2017) the development of composite score assessment *like* CR can be shown in a simplified format as follows:

$$CR = f\,(SocEcon\ Res,\ Ecol\ Res,\ Infras\ and\ Heal\ Res,\ Inst\ Res) \quad \text{(Eq. 15.1)}$$

Where, *CR* stands for community resilience; *SocEcon Res* stands for socioeconomic resilience; *Ecol Res* stands for ecological resilience; *Infras and Heal Res* stands for infrastructural and health resilience; and *Inst Res* stands for institutional resilience.

In the case of building the whole *Community Resilience Index* (CRI) each subcomponent or variable contributes uniformly for the formation of each major component as an overall index value, using a balanced weighted approach. In short, after a clubbing of specific subcomponents, the formation of major components is possible (Hahn et al. 2009; Pandey and Jha 2012; Alam 2017). Thus, assessment of the dimensions of CR should be determined to be 0 to 1 with equal weight to all associated subcomponents (Pandey and Jha 2012). The appropriateness behind the adaptation of the 'balanced weighted approach' for the composite score assessment (CSA) is, *this, the* approach gives equity to each and every subcomponents when any study contended with varied geographical regions (from micro to macro). On the contrary, few incorporated subcomponents (modified) may react differently and somewhere extraordinarily on varying geophysical environments. From the above point of view, it is justified to adopt the above-mentioned approach. Thus it would be authentic to ensemble those subcomponents with equal weightage. Because all associated subcomponents are measured on a varied scale so it is reasonable to standardize each subcomponent as an index (Hahn et al. 2009), and index building Eq. (15.2) as follows:

$$Indexsuc = \frac{SuC_V - SuC_{\min}}{SuC_{\max} - SuC_{\min}} \quad \text{(Eq. 15.2)}$$

Where, *Indexsuc* is the index value for each subcomponent,

SuC_v is the original subcomponent or indicator value for an area or v^{th} village,

SuC_{max} and SuC_{min} is the maximum and minimum value of each subcomponent.

These two maximum and minimum values of each subcomponent are used to convert the indicator value into a standardized index. For instance, variables that measure frequencies—such as 'percentage of unemployed youths' and 'percentage of the disabled population to total population'—the minimum value set at 0 and the maximum at 100.

The subcomponents are aggregated after being standardized using Eq. 15.3 (Pandey and Jha 2012; Alam 2017) as follows:

$$Cv = \frac{\sum_{i=0}^{n} Index_{suc}}{n} \quad \text{(Eq. 15.3)}$$

Where, C_v is one of the major components for *CRI*,

$Index_{suc}$ is the ith subcomponent, belonging to major component C_v for v^{th} village,

n is the number of subcomponents under the major component.

Once the computation of the values of each of the *four major components* and after average the CRI (Community Resilience Index), following Livelihood Vulnerability

Index model, would be derived as the following:

$$CRIa = \frac{\sum_{z=1}^{4} Wc_z C_{az}}{Wc_z} \quad \text{(Eq. 15.4)}$$

Equation 15.4 (Alam 2017; Alam et al. 2017) may also be articulated in the following way:

$$CRIa = \frac{W_{SE}SE_a + W_{ECC}ECC_a + W_{IH}IH_a + W_{AG}AG_a}{W_{SE} + W_{ECC} + W_{IH} + W_{AG}} \quad \text{(Eq. 15.5)}$$

Where, *CRIa* represents the CRI for a village or area a, which equals the weighted average of the four major dimensions. Side by side, weighted are included to assure that all dimensions contribute equally to the overall *CRI*. Along with this, the weight of each of the major components (*WCz*) is built by using the number of subcomponents that build each major component (Alam 2017; Alam et al. 2017). Sequentially based on the above-discussed pathways one can solve the research problem(s). The aftermath section is devoted toward the summary, caveats, and policy layout of the very concept.

15.9 Summary, Caveats, and Suggestions

Starting with the common epidemic symptom, or influenza-like symptom(s), among a small number of people in Wuhan, China (Chen et al. 2020; Li et al. 2020), this pathogenic attack has turned into pandemic as a result of rapid and huge infected cases with high human fatality and with ever-growing numbers world over. Subsequently, COVID-19 is the vital issue that has to be examined at all platforms (Singh and Avikal 2020). In the advent of this crucial time, it is needful to prescribe a 'policy-medicine' as community resilience, which will not only help to battle against this new kind of challenge but also overall preparedness of any community for the upcoming unknown threats. It is true that after the global experiences about the ongoing poignant impact of COVID-19, this post-modern society may survive if it can beat the human health issues, food & nutrition, economic, financial, and socio-political pressure in a highly competitive world (Sharfuddin 2020). Under this subjection, the real ray of hope would be *community resilience* (CR). After a systematic and rigorous review of literature, the present customized index has been composed which may be capable of addressing the every possible aspect of human life for combating against the utmost challenges and also for re-boosting coping capabilities for future, through the engagement of appropriate policies. Apart from this, the detailed discussions of each and every dimension, as well as major components (with correlated sub-indicators), may help to develop further suitable co-assisting variable(s) for the required field of study. Side by side, the entire volume also may be supportive of the new action of research running smoothly.

Alongside this, it has some caveats too. *First*, for composing the above said overall index, a vast array of datasets is needed, and from data-acquaintance and calibration point of view, it is complicated and troublesome. *Second*, in studying any geographical area, especially related to the underdeveloped or developing world or any middle-income (MICs) and low-income countries (LICs), all the required parameters may not be suitable, or datasets are easily accessible. *Third*, in some cases, the dataset on the temporal scale may not be available, which is proved to be valuable for making any ratio scale (*e.g.*, the increment of bed-patient ratio = pre-COVID bed-patient ratio—post-COVID bed-patient ratio). *Fourth*, the most important is the policy agenda; it is the way by which 'policy-medicine' would be applied for the best resilience. However, it is a well-known phenomenon that ill-management and poor implementation of the proposed policy and plans or programs in developing and underdeveloped nations are still going on. But the present compulsion of humanity, as a whole, is to minimize stressors as well as to boost up community resilience.

Amidst the din and bustle, a question may arise that whether the human-world is prepared for this vulnerable scenario. And it is the basic instinct of humanity innately carrying weakness and defenselessness with its journey of struggle (Sharfuddin 2020). The possible answer is affirmative. Yes, human society as a legislative form, like any nation, can reach its goal. In order to reach to its best part of policy suggestion, it can be said that based on (any) case study outcomes and its derived demanding corners or gaps a more suitable and appropriate reframed policy design may fruitfully contribute toward resilience power of community as a part or as a whole. Here, in Fig. 15.5 a sample policy model (in short and simple format) has been produced, which exclusively focuses on humanistic prioritization during the pandemic, and may be in post-pandemic period. Additionally, the proposed beneficiary policy suggestion might be customized, and the mode of prioritization might be reset in accordance with community-specific and geographical region-specific study findings. A researcher or a policymaker should always deal with the value of human lives and humanity in the true sense of the term to make policies working for the benefit of the humankind.

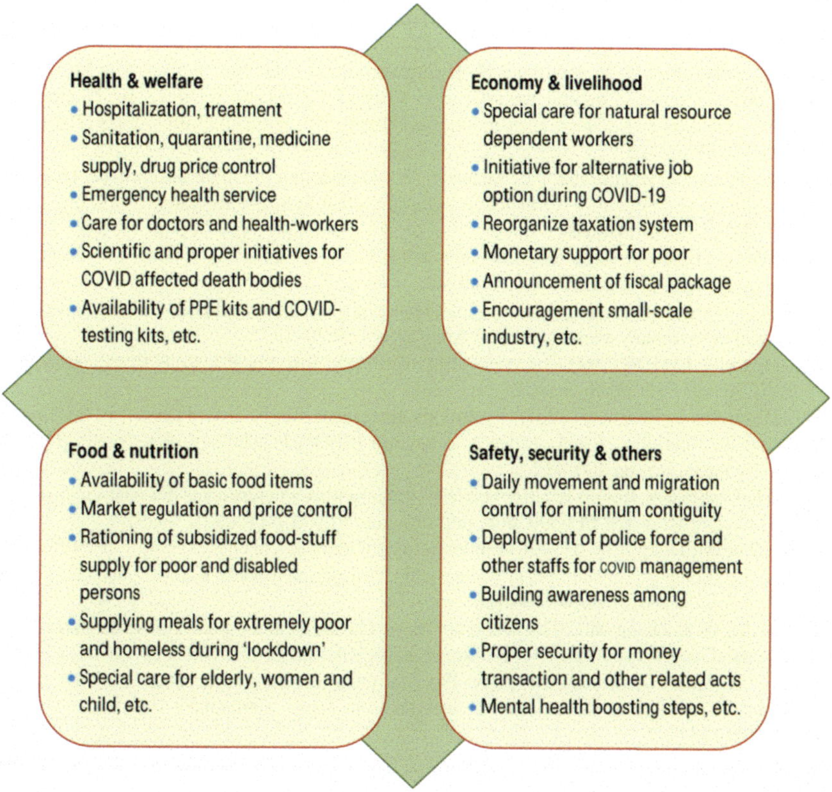

Fig. 15.5 Policy suggestion sector-wise prioritization after CSA (*Source* Developed by author)

References

Acevedo S (2014) Community resilience. In: Michalos AC (ed) Encyclopedia of quality of life and wellbeing research. Springer, Dordrecht. http://doi.org/10.1007/978-94-007-0753-5

Adger WN (2000) Social and ecological resilience: are they related? Prog Hum Geogr 24(3):347–364

Ainuddin S, Routray JK (2012) Earthquake hazards and community resilience in Baluchistan. Nat Hazards 63:909–937. https://doi.org/10.1007/s11069-012-0201-x

Alam GMM (2017) Livelihood cycle and vulnerability of rural households to climate change and hazards in Bangladesh. Environ Manag 59(5):777–791. https://doi.org/10.1007/s00267-017-0826-3

Alam GMM, Alam K, Shahbaz M, Clarke ML (2017) Drivers of vulnerability to climatic change in riparian char and river-bank households in Bangladesh: implications for policy, livelihoods and social development. Ecol 72:23–32

Aldrich DP (2012) Building resilience: social capital in post-disaster recovery. The University of Chicago Press, Chicago

Almutairi A, Mourshed M, Ameen RFM (2020) Coastal community resilience frameworks for disaster risk management. Nat Hazards 101:595–630. http://doi.org/10.1007/s11069-020-038 75-3

Alshehri SA, Rezgui Y, Li H (2015) Disaster community resilience assessment method: a concensus-based Delphi and AHP approach. Nat Hazards: 1–22. http://doi.org/10.1007/s11069-015-1719-5
Anwar HZ, Yustiningrum E, Andriana N, Kusumawardhani DTP, Sari AM (2017) Measuring community resilience to natural hazards: case study of Yogyakarta province. In: Djalante R, et al (eds) Disaster risk reduction in Indonesia, disaster risk reduction. http://doi.org/10.1007/978-3-319-54466-3_25
Bergstrand K, Mayer B, Brumback B, Zhang Y (2014) Assesing the relationship between social vulnerability and community resilience to hazards. Social Indicator Research: 1–19. http://doi.org/10.1007/s11205-014-0698-3
Bodin P, Wiman B (2004) Resilience and other stability concepts in ecology: notes on their origin, validity, and usefulness. ESS Bulletin 2:33–43
Bonanno G (2004) Loss, trauma, and human resilience: have we underestimated the human capacity to thrive after extremely aversive events? Am Psychol 59:20–28
Brown D, Kulig J (1996/1997) The concept of resiliency: theoretical lessons from community research. Health and Canadian Society 4:29–52
Butler L, Morland L, Leskin G (2007) Psychological resilience in the face of terrorism. In: Bongar B, Brown L, Beutler L, Breckenridge J, Zimbardo P (eds) Psychology of terrorism. Oxford University Press, NY, pp 400–417
Camfield L, Crabtree A, Roelen K (2013) Poverty, vulnerability and resilience in a post-2015 world. Social Indicator Research 113:599–608. http://doi.org/10.1007/s11205-013-0291-1
Chen N, Zhou M, Dong X et al (2020) Epidemiological and clinical characteristics of 99 cases of 2019 novel coronavirus pneumonia in Wuhan, China: a descriptive study. Lancet 395(10223):507–513. https://doi.org/10.1016/S0140-6736
Chung CKL, Xu J, Zhang M (2020) Geographies of Covid-19: how space and virus shape each other. Asian Geographer 37(2):99–116. https://doi.org/10.1080/10225706.2020.1767423
Cimellaro GP, Solari D, Bruneau M (2014) Physical infrastructure interdependency and regional resilience index after the 2011 Tohoku earthquake in Japan. Earthquake Eng Struct Dynam 4312:1763–1784
Cutter SL, Ash KD, Emrich CT (2014) The geographies of community disaster resilience. Glob Environ Change 29:65–77
Cutter SL, Ash KD, Emrich CT (2016) Urban–rural differences in disaster resilience. Annals of the American Association of Geographers 106(6):1236–1252
Cutter SL, Barnes L, Berry M, Burton C, et al (2008) A place-based model for understanding community resilience to natural disasters. Glob Environ Chang 18(4):598–606
Cutter SL, Boruff BJ, Shirley WL (2003) Social vulnerability to environmental hazards. Soc Sci 84:242–261
Cutter SL, Burton CG, Emrich CT (2010). Disaster resilience indicators for benchmarking baseline conditions. J Homel Secur Emerg Manag 7(1):1–22. https://doi.org/10.2202/1547-7355.1732
Dako-Gyeke P, Amazigo UV, Halpaap B, Manderson L (2020) Social innovation for health: engaging communities to address infectious diseases. Infectious Diseases of Poverty 9(98):1–4. https://doi.org/10.1186/s40249-020-00721-3
Danar OR, Pushpalal D (2014) Building community resilience: conceptual framework and its application in post tsunami resettlement. Procedia Economics and Finance 18:489–496
DasGupta R, Shaw R (2015) An indicator based approach to assess coastal communities' resilience against climate related disasters in Indian Sundarbans. Journal of Coastal Conservation 19:85–101. https://doi.org/10.1007/s11852-014-0369-1
Despotaki V, Sousa L, Burton CG (2018) Using resilience indicators in the prediction of earthquake recovery. Earthquake Spectra 34(1):265–282
Dinh H, Freyens B, Daly A, Vidyattama Y (2016) Measuring community economic resilience in Australia: Estimates of recent levels and trends. Soc Ind Res: 1–20. http://doi.org/10.1007/s11205-016-1337-y
Dryhurst S, Schneider CR, Kerr J et al (2020) Risk perceptions of COVID-19 around the world. J Risk Res. https://doi.org/10.1080/13669877.2020.175

Engle NL (2011). Adaptive capacity and its assessment. Global Environ Change 21:647–656
Frazier TG, Thompson CM, Dezzani RJ, Butsick D (2013) Spatial and temporal quantification of resilience at the community scale. Applied Geography 42:95–107
Frost P (2020) Hundred years of pandemics. Morning Star Online News, UK, pp 1–10. Accessed 1 July 2020 from. https://morningstaronline.co.uk/article/f/hundred-years-pandemics
Gallopin GC (2006) Linkages between vulnerability, resilience, and adaptive capacity. Global Environ Change 16:293–303
Ge Y, Dou W, Gu Z, Qian X et al (2013) Assessment of social vulnerability to natural hazards in the Yangtze River Delta, China. Stoch Environ Res Risk Assess 27:1899–1908
Godschalk D (2003) Urban hazard mitigation: creating resilient cities. Natural Hazards Review 4:136–143
Gordon J (1978) Structures. Penguin Books, Harmondsworth, UK
Hahn MB, Riederer AM, Foster SO (2009) The livelihood vulnerability index: a pragmatic approach to assessing risk from climate variability and change—a case study in Mozambique. Global Environ Chang 19:74–88
Hou J, Lv J, Chen X, Yu S (2016) China's regional social vulnerability to geological disasters: evaluation and spatial characteristics analysis. Nat Hazards 84(S1):97–111
Joerin J, Shaw R (2011) Mapping climate and disaster resilience in cities in community, environment and disaster risk management. Emerald Publications, UK 6:47–61
Joerin J, Shaw R, Takeuchi Y, Krishnamurthy R (2012) Action-oriented resilience assessment of communities in Chennai. India Environ Hazards 11(3):226–241
Keenan JM (2020) COVID, resilience, and the built environment. Environment Systems and Decisions: 1–6. https://doi.org/10.1007/s10669-020-09773-0
Kertscher T (2020, April 10) Fact watch: has a pandemic occurred every 100 years? News: Austin American-Statesman, Austin, TX, pp 1–4. Accessed 11 July 2020 from. https://www.statesman.com/news/20200410/fact-check-has-pandemic-occurred-every-100-years
Kulig J, Botey AP (2016) Facing a wildfire: what did we learn about individual and community resilience? Nat Hazards: 1–11. http://doi.org/10.1007/s11069-016-2277-1
Li Q, Guan X, Wu P, et al (2020) Early transmission dynamics in Wuhan, China, of novel coronavirus-infected pneumonia. N Engl J Med 382:1199–1207. https://doi.org/10.1056/NEJMoa2001316
Li X, Wang L, Liu S (2016) Geographical analysis of community resilience to seismic hazard in Southwest China. International J of Disaster Risk Science 7:257–276. https://doi.org/10.1007/s13753-016-0091-8
Liu D, Li Y (2016) Social vulnerability of rural households to flood hazards in western mountainous regions of Henan province, China. Nat Hazards Earth SystSci 16(5):1123–1134
Lixin Y, Xi Z, Lingling G, Dong Z (2014) Analysis of social vulnerability to hazards in China. Environ Earth Sci 71(7):3109–3117
Markard J, Rosenbloom D (2020) A tale of two crises: COVID-19 and climate. Sustainability: Science, Practice and Policy 16(1):53–60. http://doi.org/10.1080/15487733.2020.1765679
Mayer B (2019) A review of the literature on community resilience and disaster recovery. Current Environmental Health Reports: 1–7. https://doi.org/10.1007/s40572-019-00239-3
Merriam-Webster (n.d.) Pandemic. In: Merriam-Webster.com dictionary. https://www.merriam-webster.com/dictionary/pandemic. Accessed 14 July 2020
Mishra A, Ghate R, Maharjan A, Gurung J et al (2017) Building ex ante resilience of disaster-exposed mountain communities: drawing insights from the Nepal earthquake recovery. International Journal of Disaster Risk Reduction 22:167–178
Morens DM, Taubenberger JK (2018) The mother of all pandemics is 100 years old (and going strong). AJPH (Sept. 25, 2018). http://doi.org/10.2105/AJPH.2018.304631
Naja F, Hamadeh R (2020) Nutrition amid the COVID-19 pandemic: a multilevel framework for action. European Journal of Clinical Nutrition: 1–5. https://doi.org/10.1038/s41430-020-0634-3

Norris FH, Stevens SP, Pfefferbaum B, Wyche KF, Pfefferbaum RL (2008) Community resilience as a metaphor, theory, set of capacities, and strategy for disaster readiness. Am J Community Psychol 41:127–150

Orencio PM, Endo A, Taniguchi M, Fujii M (2015) Using thresholds of severity to threats to and the resilience of human systems in measuring human security. Social Indicator Research: 1–21. http://doi.org/10.1007/s11205-015-1152-x

Pandey R, Jha SK (2012) Climate vulnerability index-measure of climate change vulnerability to communities: a case of rural lower Himalaya, India. Mitig Adapt StratGl 17:487–506. https://doi.org/10.1007/s11027-001-9338-2

Parson M, Glavac S, Hastings P, Marshall G, et al (2016) Top-down assessment of disaster resilience: a conceptual framework using coping and adaptive capacities. International Journal of Disaster Risk Reduction 19:1–11. http://dx.doi.org/10.1016/j.ijdrr.2016.07.005

Paton D, Millar M, Johnston D (2001) Community resilience to volcanic hazard consequences. Nat Hazards 24:157–169

Peacock WG, Brody SD, Seitz WA, Merrell WJ, et al (2010) Advancing resilience of coastal localities: developing, implementing, and sustaining the use of coastal resilience indicators: a final report. Hazard reduction and recovery center final report for NOAA CSC grant no. A07NOS4730147

Pfefferbaum B, Van Horn RL, Pfefferbaum RL (2015) A conceptual framework to enhance community resilience using social capital. Clin Soc Work Jour: 1–9. http://doi.org/10.1007/s10615-015-0556-z

Porta M (ed) (2008) Dictionary of Epidemiology. Oxford University Press, p 179. https://books.google.com/books?id=3Dr8dyuzvTkCandpg+PA179. Accessed 14 July 2020

Qian X, Ren R, Wang Y, et al (2020) Fighting against the common enemy of COVID-19: a practice of building community with a shared future for mankind. Infectious Diseases of Poverty 9(34):1–6. https://doi.org/10.1186/s40249-020-00650-1

Qin W, Lin A, Fang J, Wang L, Li M (2017) Spatial and temporal evolution of community resilience to natural hazards in the coastal areas of China. Nat Hazards: 1–19. http://doi.org/10.1007/s11069-017-2967-3

Reissman DB, Watson PJ, Klomp RW, Tanielian TL, Prior SD (2006) Pandemic influenza preparedness: adaptive responses to an evolving challenge. J Homel Sec Emerg Manag 3:1–24

Rutter M (1993) Resilience: some conceptual considerations. J Adolesc Health 14:626–631

Sharfuddin S (2020) The world after Covid-19. The Round Table 109(3):247–257. https://doi.org/10.1080/00358533.2020.1760498

Sharifi A (2016) A critical review of selected tools for assessing community resilience. Ecological Indicators 69:629–664

Sharifuzzaman SM, Shahadat Hossain M, Rahman Chowdhury S, Sarker S, et al (2018) Elements of fishing community resilience to climate change in the coastal zone of Bangladesh. Journal of Coastal Conservation: 1–10. https://doi.org/10.1007/s11852-018-0626-9

Shea J (2018) Community resilience approach to disaster recovery: strategies communities can use. In: Cnaan RA, Milofsky C (eds) Handbook of community movements and local organizations in the 21st century, Handbooks of Sociology and Social Research. AG: Springer International Publishing, pp 371–389. https://doi.org/10.1007/978-3-319-77416-9_23

Sherrieb K, Norris FH, Galea S (2010) Measuring capacities for community resilience. Social Indicators Research 992:227–247

Simpson V (2020) Community resilience: one community's view of recovery from opioid epidemic. International Journal of Community Well-Being: 1–13. https://doi.org/10.1007/s42413-020-00064-9

Singh R, Avikal S (2020) COVID-19: a decision-making approach for prioritization of preventive activities. International Journal of Healthcare Management. https://doi.org/10.1080/20479700.2020.1782661

Smith JW, Moore RL, Anderson DH, Siderelis C (2012) Community resilience in Southern Appalachia: a theoretical framework and three case studies. Human Ecology 40:341–353. https://doi.org/10.1007/s10745-012-9470-y
Song J, Huang B, Li R, Pandey R (2019) Construction of the scale-specific resilience index to facilitate multiscale decision making in disaster management: A case study of the 2015 Nepal earthquake. Social Indicator Research: 1–35. https://doi.org/10.1007/s11205-019-02188-8
Sonn C, Fisher A (1998) Sense of community: community resilient responses to oppression and change. Journal of Community Psychology 26:457–472
Su S, Pi J, Wan C, Li H, Xiao R, Li B (2015) Categorizing social vulnerability patterns in Chinese coastal cities. Ocean Coast Manag 116:1–8
Teo M, Goonetilleke A, Ziyath AM (2013) An integrated framework for assessing community resilience in disaster management. In proceedings of the 9th Annual International Conference of the International Institute for Infrastructure Renewal and Reconstruction, Risk-informed Disaster Management: Planning for Response, Recovery and Resilience.https://www.researchgate.net/publication/258225443_An_integrated_framework_for_assessing_community_resilience_in_disaster_management. Accessed 10 July 2020
Ungar M (2011) The social ecology of resilience: addressing contextual and cultural ambiguity of a nascent construct. Am J Orthopsychiatry 81(1):1–17
US Indian Ocean Tsunami Warning System Program (USIOTWSP) (2007) How resilient is your coastal community? A Guide for evaluating coastal community resilience to tsunamis and other coastal hazards. US Indian Ocean Tsunami Warning System Program supported by the United States Agency for International Development and partners, Bangkok, Thailand. https://www.crc.uri.edu/download/CCRGuide_lowres.pdf. Accessed 30 June 2020
Uy N, Takeuchi Y, Shaw R (2011) Local adaptation for livelihood resilience in Albay, Philippines. Environ Hazards 10(2):139–153
Visser M, Finestone M, Sikkema K, Boeving-Allen A, Ferreira R, Eloff I et al (2012) Development and piloting of a mother and child intervention to promote resilience in young children of HIV-infected mothers in South Africa. Evaluation and Program Planning 35:491–500
Werner EE, Smith RS (1982) Vulnerable but invincible: a longitudinal study of resilient children and youth. McGraw-Hill, New York
WHO (2020, August 4) Corona virus disease (COVID-19), Situation Report-197. World Health Organization: Geneva, pp 1–18. https://www.who.int/docs/default-source/coronaviruse/situation-reports/20200804-covid-19-sitrep-197.pdf?sfvrsn=94f7a01d_2
Zhang N, Huang H (2013) Social vulnerability for public safety: a case study of Beijing, China. Chin Sci Bull 58:2387–2394
Zhou Y, Li N, Wu W, Wu J (2014) Assessment of provincial social vulnerability to natural disasters in China. Nat Hazards 71(3):2165–2186
Zou H, Shu Y, Feng T (2020) How Shenzhen, China avoided widespread community transmission: a potential model for successful prevention and control of COVID-19. Infectious Diseases of Poverty 9(89):1–4. https://doi.org/10.1186/s40249-020-00714-2

Somenath Halder is an Assistant Professor in the Department of Geography at Kaliachak College, West Bengal, India. He has been working especially on marginal and deprived communities since 2008, in the home state and other major states across India. He obtained his Ph.D. from the Visva-Bharati, India. He has several publications in journals of international repute.

The manufacturer's authorised representative in the EU is Springer Nature Customer Service Centre GmbH, Europaplatz 3, 69115 Heidelberg, Germany. If you have any concerns regarding our products, please contact ProductSafety@springernature.com

Printed and bound by CPI Group (UK) Ltd, Croydon, CR0 4YY
30/09/2026
02225241-0010